Perspectives on Ethical Hacking and Penetration Testing

Keshav Kaushik
University of Petroleum and Energy Studies, India

Akashdeep Bhardwaj
University of Petroleum and Energy Studies, India

A volume in the Advances in
Information Security, Privacy, and
Ethics (AISPE) Book Series

Published in the United States of America by
IGI Global
Information Science Reference (an imprint of IGI Global)
701 E. Chocolate Avenue
Hershey PA, USA 17033
Tel: 717-533-8845
Fax: 717-533-8661
E-mail: cust@igi-global.com
Web site: http://www.igi-global.com

Library of Congress Cataloging-in-Publication Data

Names: Kaushik, Keshav, editor. | Bhardwaj, Akashdeep, 1971- editor.
Title: Perspectives on ethical hacking and penetration testing / edited by
   Keshav Kaushik, Akashdeep Bhardwaj.
Description: Hershey, PA : Information Science Reference, 2023. | Includes
   bibliographical references and index. | Summary: "Perspectives on
   Ethical Hacking and Penetration Testing familiarizes readers with
   in-depth and professional hacking and vulnerability scanning subjects.
   The book discusses each of the processes and tools systematically and
   logically, so that the reader can see how the data from each tool may be
   fully exploited in the penetration test's succeeding stages. This
   procedure enables readers to observe how the research instruments and
   phases interact. This book provides a high level of understanding of the
   emerging technologies in penetration testing, cyber-attacks, and ethical
   hacking and offers the potential of acquiring and processing a
   tremendous amount of data from the physical world. Covering topics such
   as cybercrimes, digital forensics, and wireless hacking, this premier
   reference source is an excellent resource for cybersecurity
   professionals, IT managers, students and educators of higher education,
   librarians, researchers, and academicians"-- Provided by publisher.
Identifiers: LCCN 2022062311 (print) | LCCN 2022062312 (ebook) | ISBN
   9781668482186 (h/c) | ISBN 9781668482193 (s/c) | ISBN 9781668482209
   (eISBN)
Subjects: LCSH: Penetration testing (Computer security) | Computer
   crimes--Prevention.
Classification: LCC QA76.9.A25 P44173 2023 (print) | LCC QA76.9.A25
   (ebook) | DDC 005.8--dc23/eng/20230125
LC record available at https://lccn.loc.gov/2022062311
LC ebook record available at https://lccn.loc.gov/2022062312

This book is published in the IGI Global book series Advances in Information Security, Privacy, and Ethics (AISPE) (ISSN: 1948-9730; eISSN: 1948-9749)

British Cataloguing in Publication Data
A Cataloguing in Publication record for this book is available from the British Library.

All work contributed to this book is new, previously-unpublished material.
The views expressed in this book are those of the authors, but not necessarily of the publisher.

For electronic access to this publication, please contact: eresources@igi-global.com.

Advances in Information Security, Privacy, and Ethics (AISPE) Book Series

ISSN:1948-9730
EISSN:1948-9749

Editor-in-Chief: Manish Gupta, State University of New York, USA

MISSION

As digital technologies become more pervasive in everyday life and the Internet is utilized in ever increasing ways by both private and public entities, concern over digital threats becomes more prevalent.

The **Advances in Information Security, Privacy, & Ethics (AISPE) Book Series** provides cutting-edge research on the protection and misuse of information and technology across various industries and settings. Comprised of scholarly research on topics such as identity management, cryptography, system security, authentication, and data protection, this book series is ideal for reference by IT professionals, academicians, and upper-level students.

COVERAGE

- Cookies
- Network Security Services
- Security Information Management
- Computer ethics
- IT Risk
- Security Classifications
- Privacy-Enhancing Technologies
- Telecommunications Regulations
- Risk management
- Tracking Cookies

IGI Global is currently accepting manuscripts for publication within this series. To submit a proposal for a volume in this series, please contact our Acquisition Editors at Acquisitions@igi-global.com or visit: http://www.igi-global.com/publish/.

Titles in this Series

For a list of additional titles in this series, please visit:
http://www.igi-global.com/book-series/advances-information-security-privacy-ethics/37157

AI Tools for Protecting and Preventing Sophisticated Cyber Attacks
Eduard Babulak (National Science Foundation, USA)
Information Science Reference • copyright 2023 • 233pp • H/C (ISBN: 9781668471104)
• US $250.00 (our price)

Cyber Trafficking, Threat Behavior, and Malicious Activity Monitoring for Healthcare Organizations
Dinesh C. Dobhal (Graphic Era University (Deemed), India) Sachin Sharma (Graphic Era University (Deemed), India) Kamlesh C. Purohit (Graphic Era University (Deemed), India) Lata Nautiyal (University of Bristol, UK) and Karan Singh (Jawaharlal Nehru University, India)
Medical Information Science Reference • copyright 2023 • 206pp • H/C (ISBN: 9781668466469) • US $315.00 (our price)

Emerging Perspectives in Systems Security Engineering, Data Science, and Artificial Intelligence
Maurice Dawson (Illinois Institute of Technology, USA)
Information Science Reference • copyright 2023 • 315pp • H/C (ISBN: 9781668463253)
• US $250.00 (our price)

Global Perspectives on the Applications of Computer Vision in Cybersecurity
Franklin Tchakounte (University of Ngaoundere, Cameroon) and Marcellin Atemkeng (University of Rhodes, South Africa)
Engineering Science Reference • copyright 2023 • 300pp • H/C (ISBN: 9781668481271)
• US $250.00 (our price)

Handbook of Research on Data Science and Cybersecurity Innovations in Industry 4.0 Technologies
Thangavel Murugan (United Arab Emirates University, Al Ain, UAE) and Nirmala E. (VIT Bhopal University, India)
Information Science Reference • copyright 2023 • 600pp • H/C (ISBN: 9781668481455)
• US $325.00 (our price)

For an entire list of titles in this series, please visit:
http://www.igi-global.com/book-series/advances-information-security-privacy-ethics/37157

701 East Chocolate Avenue, Hershey, PA 17033, USA
Tel: 717-533-8845 x100 • Fax: 717-533-8661
E-Mail: cust@igi-global.com • www.igi-global.com

Table of Contents

Detailed Table of Contents

Chapter 1

    Preeti Sharma, University of Petroleum and Energy Studies, India
    Manoj Kumar, University of Wollongong, Dubai, UAE
    Hitesh Kumar Sharma, University of Petroleum and Energy Studies,
        India

A branch of forensic science called "digital forensics" deals with the utilization of digital data received, maintained, and conveyed by electronic devices as evidence in inquiries and legal proceedings. It is a growing field in computing that frequently necessitates the intelligent analysis of large amounts of complex data. Rapid advancements in computer science and information technology enable the development of novel techniques and software for digital investigations. Initially, much of the analysis software was unique and proprietary, but over time, specialised analysis software for both the private and governmental sectors became available. The aim of this chapter is to deliver a comprehensive overview of digital forensics phases, applications, merits, and demerits and widely used software of the domain. The chapter also discusses legitimate and legal considerations, followed by the scope and role of artificial intelligence for solving complex problems of digital forensics.

Chapter 2

    Ayushi Malik, University of Petroleum and Energy Studies, India
    Shagun Gehlot, University of Petroleum and Energy Studies, India
    Ambika Aggarwal, University of Petroleum and Energy Studies, India

Online applications hold sensitive data that is valuable, and hackers strive to identify weaknesses and exploit them in order to steal data, pose as users, or disrupt the application. Web applications are increasingly exposed to dangers and attack vectors that are more advanced. Additionally, theft of private information—such as user credentials or billing information for credit cards—occurs frequently. Attackers initially concentrated on obtaining personal information that was accidentally exposed through poorly built or poorly protected web apps. Insecure design, security misconfiguration, vulnerable and outdated components, identification and authentication failures, etc. are some of the most prominent web application vulnerabilities that will be covered in this chapter. The authors are using countermeasures including fuzz testing, source code review, and encoding approaches to get around these vulnerabilities. As a result, this chapter offers information on the various attacks that website visitors who use web applications encounter.

Chapter 3

With the National Cybercrime Reporting portal witnessing an increase of 15.3% increase in the cyber cases in second quarter of 2022, the post pandemic world has seen a tremendous rise in the number of cybercrime cases. The cases have increased many folds, 125% from 2021, and still continues beyond. The targets of crime are not limited to any age group and innocent children have not been spared. The exposure to online classes, conferences, meetings, etc. has opened the door to criminal activities in a humongous way. Thus, there comes the need for each and every one of us to be well aware of the recent practices that these criminals use and not fall into the clutches of these nefarious cyber criminals.

Chapter 4

The previous method that has been suggested is to increase the automatic closing of security holes in networks that are vulnerable. This process is the amalgamation of various phases, which begin with collecting information and end with mitigating vulnerabilities. The network's internal domain name is considered input in the proposed method for internal audit purposes. The operational services collect live IPs. Exploits are typically created to gain access to a system, enable the acquisition of administrative privileges, or launch denial-of-service attacks. Each exploit is configured and executed after the list of exploits for each service has been obtained. Whether the endeavor can effectively approach the framework, the moderation step is conjured, checking the sort of access got. The suggestion evaluates the incoming data using the knowledge set stored in its memory. It maintains a table detailing the IP address incomings. The guilty detection is increased by 27.6%, and the security is increased by 36.8% compared to previous work.

Chapter 5

Qasem Abu Al-Haija, Princess Sumaya University for Technology, Jordan

Rahmeh Ibrahim, Princess Sumaya University for Technology, Jordan

Darknet is an overlay portion of the Internet network that can only be accessed using specific authorization using distinctively tailored communication protocols. Attackers usually exploit the darknet to threaten several world-wile users with different types of attack/intrusion vectors. In this chapter, we shed the light on the darknet network, concepts, elements, structure, and other aspects of darknet utilization. Specifically, this chapter will extend the elaboration on the darknet, the dark web components, the dark web access methods, the anonymity and confidentiality of the dark web, the dark web crimes, the cyber-attacks on the dark web, the malware on the dark web, the internet governance of dark web, the payment on the dark web, and the impacts of the dark web. This chapter will profound the knowledge of the dark web and provides more insights to readers about darknet attacks, malware, and their counter-measures.

Chapter 6

Qasem Abu Al-Haija, Princess Sumaya University for Technology, Jordan

Noor A. Jebril, Princess Sumaya University, Jordan

Ransomware can lock users' information or resources (such as screens); hence, authorized users are blocked from retrieving their private data/assets. Ransomware enciphers the victim's plaintext data into ciphertext data; subsequently, the victim host can no longer decipher the ciphertext data to original plaintext data. To get back the plaintext data, the user will need the proper decryption key; therefore, the user needs to pay the ransom. In this chapter, the authors shed light on ransomware malware, concepts, elements, structure, and other aspects of ransomware utilization. Specifically, this chapter will extend the elaboration on the ransomware, the state-of-art ransomware, the ransomware lifecycle, the ransomware activation and encryption processes, the ransom request process, the payment and recovery, the ransomware types, recommendation for ransomware detection and prevention, and strategies for ransomware mitigation.

Chapter 7

Aaeen Naushadahmad Alchi, Gujarat University, India

Kiranbhai R. Dodiya, Gujarat University, India

Malware, classified as ransomware, encrypts data on a computer, preventing individuals from accessing it. The intruder then demands a ransom from the user for the password that unlocks the files. Recent cyberattacks against prominent corporate targets have increased the extensive media attention on ransomware. The primary reason for computer intrusions is financial gain. Ransomware targets individual owners of information, keeping their file systems captive until a ransom is paid, compared to malware, which permits criminals to steal valuable data and then use it throughout the digital marketplace. Ransomware's terrifying complexity level heralds a paradigm shift in the cybercrime ecosystem. Ransomware has become more mysterious, with some latest forms working without ever connecting to the Internet. In this chapter, the authors will discuss the overview of ransomware, the history and development of ransomware, some of the famous cases, the anatomy of ransomware attacks, types of ransomware attack vectors, and the prevention of such kinds of attacks in cyberspace.

Chapter 8

Now, computers and smartphones have become more powerful than internet connectivity. Every 'smart' device in our environment now aspires to use digital interventions to solve real-world problems. The buzz around IoT is, of course, huge. This disturbing technology penetrates various industries, develops new IoT applications, and connects all internet-enabled devices around us. One survey shows that 61 billion connected devices are expected to be available by 2025. But some of them shine more than others, in the mad rush of "newer" and "better" IoT applications. The chapter aims to present a summary of the challenges in the applications that the internet of things must face in their research and development, which are to be explored in this book chapter.

Chapter 9

Research was conducted to increase the awareness of employees with regard to cyber security to fill the gap in the literature where few studies on how effective the measures implemented in organizations were reported. This research uses the outcome of the phishing drills that a public institution applied to its personnel, participation of said personnel in awareness training, and the reading statistics of regularly published information security bulletins. This has been beneficial in determining the methods to increase the cyber security awareness of personnel in organizations with 1,000 or more personnel; users were considered as a whole, and not individually evaluated. Findings report that organizations can increase users' cybersecurity awareness by systematically conducting phishing exercises, providing awareness training, and regularly publishing information security bulletins. The awareness of reading bulletins rapidly increased after phishing exercises and training and decreased in the following months; however, an increase was observed in the long term.

Threats on communication networks are proliferating. It is surprising that more than half of Indian organizations have experienced a breach in 2022. These breaches can occur due to various reasons, such as human error, software vulnerabilities, or malicious attacks. Tech executives frequently commit the sin of being unready for network data breaches despite the early indications. The consequences of a network security breach can be severe, including financial losses, damage to reputation, and loss of sensitive data. The punitive damages are severe, and it is easier to lose and much harder to restore the faith of customers. This chapter gives an overview on the evolution of network architecture, the grounds behind data leaks and violations. It also discusses the best approaches to counter the threats, governance structure and emerging developments in network security. This chapter also gives you the analysis on the methodologies adopted for network forensics and explores the viewpoint on where network forensics could be applied.

Penetration testing is an art, and the path of its mastery starts from having a good grasp on fundamental knowledge. Throughout this chapter, readers will be made familiar with the building blocks in an easy-to-interpret manner. During the course of this chapter, readers will learn about topics such as penetration testing types, phases, dos & don'ts, and types of OSINT (open source intelligence) methods including image OSINT, email OSINT, Google dorking and social media intelligence

(SOCMINT). Please note that this chapter will not be very comprehensive. Readers are recommended to do their own research on such topics in order to gain more in-depth knowledge. Links to additional reading materials will be added in reference section at the end of the chapter. Readers are advised to go through them at least once to learn more comprehensive information about some chapter topics.

Penetration testing is an ever-growing field that deals with a lot of products and services. This chapter will begin introduction of networking, Linux, and Bash. Furthermore, the common tools of the trade (Nmap, Nessus, Metasploit, Burp suite, etc.) for specific type of penetration testing assessments. Readers are recommended to do their own research on such topics in order to gain more in-depth knowledge. Links to additional reading materials will be added in the reference section at the end of the chapter. Readers are advised to go through them at least once to learn more comprehensive information about some chapter topics.

Chapter 13

This book chapter examines the increasing danger of social engineering attacks in cybersecurity. These attacks focus on exploiting human vulnerabilities instead of technical weaknesses and target the human element of organizations. The chapter outlines the various types of social engineering attacks such as phishing, pretexting, baiting, and quid pro quo, and explores the strategies employed by social engineers including the creation of urgency, trust, and fear. It also covers countermeasures that can be employed to guard against social engineering attacks, such as education and awareness programs for employees and technical solutions like spam filters and multi-factor authentication. By understanding the threat of social engineering attacks and taking proactive steps to mitigate this risk, individuals and organizations can protect themselves against this growing cybersecurity menace.

Chapter 14

The depth of the Internet extends well beyond the surface information that many people may quickly access in their routine searches. Some people may think of the web as only being made up of webpages that can be found using conventional search engines like Google. This information, referred to as the "Surface web," represents a very small percentage of the entire internet. The part of the internet that search engines and web crawlers do not index is known as the deep web. On the other hand, a subset of the deep web known as the "dark web" is only accessible using specialized software like Tor (The Onion Router). The surface web is primarily used for acceptable daily online activity, while the dark web is purely anonymous and is known for carrying out illicit transactions. The dark web is a small part of the deep web which can be accessed through the Tor browser. This chapter aims to examine current technology developments and some intriguing recent dark web statistics to evaluate the dark web's present state, technologies, usage, and current trends and data breaches.

Chapter 15

This chapter, per the authors, delves into the realm of server-side attacks within the context of ethical hacking and penetration testing. It aims to provide a comprehensive understanding of server-side vulnerabilities, focusing on web, database, and application servers. Additionally, the chapter explores prevalent server-side attack techniques, guiding readers through the process of identifying, exploiting, and mitigating server-side vulnerabilities in a responsible and ethical manner. Legal and ethical considerations surrounding server-side attacks are also discussed, emphasizing the importance of responsible disclosure and collaboration with vendors. Lastly, the chapter concludes by examining the role of server-side attacks in ethical hacking and highlighting future trends and challenges that ethical hackers may encounter in the ever-evolving digital landscape.

Chapter 16

The network has become portable as a result of digital modulation, adaptive modulation, information compression, wireless access, and multiplexing. Wireless devices connected to the internet can possess a serious risk to the information security. These devices communicate among themselves in a public domain which is very easily susceptible to attacks. These devices only depend upon the encryption and their shared keys to help them mitigate the risk when data is in transit. Also WEP/WPA (wired equivalent privacy/ wireless protected access) cracking tools are taken care to avoid break into attacks. Several wireless networks, their security features, threats, and countermeasures to keep the network secure are all covered in this chapter. It analyses various wireless encryption techniques, highlighting their advantages and disadvantages. The chapter also explores wireless network attack techniques and provides countermeasures to safeguard the information systems and also provide a wireless penetration testing framework for safeguarding the wireless network.

Preface

Welcome to *Perspectives on Ethical Hacking and Penetration Testing*, a comprehensive reference book that delves into the fascinating world of cybersecurity, hacking, and vulnerability assessment. As editors, it is our pleasure to introduce you to this collection of insightful chapters authored by experts and enthusiasts in the field.

In today's digitally interconnected landscape, where every facet of our lives is entwined with technology, the significance of cybersecurity cannot be overstated. Our reliance on interconnected devices, systems, and networks has opened unprecedented avenues for communication, convenience, and efficiency. However, this rapid expansion of connectivity has also brought forth a myriad of challenges, ranging from data breaches to sophisticated cyber-attacks.

The genesis of this book lies in our desire to explore the realm of ethical hacking and penetration testing, shedding light on the tools, techniques, and methodologies employed to secure and fortify our digital ecosystems. Through a meticulous curation of topics, we intend to provide a comprehensive resource for researchers, students, professionals, and anyone curious about the intricacies of cybersecurity.

Perspectives on Ethical Hacking and Penetration Testing is structured to guide you through a progressive journey, starting from fundamental concepts and gradually advancing to more complex themes. Whether you're a novice seeking an introduction to hacking and vulnerability assessment or a seasoned professional striving to enhance your skillset, this book is designed to accommodate learners at all levels.

We have meticulously organized the book's contents to ensure a coherent flow of knowledge. From the motivations behind hacking to the ethical considerations that drive penetration testing, each chapter offers unique insights into the multifaceted landscape of cybersecurity. The chapters delve into a wide array of subjects, including vulnerability assessment, penetration testing methodologies, exploits, attacks on web applications, IoT hacking, wireless security, and much more.

With an emphasis on practicality, each chapter elucidates the tools, techniques, and real-world scenarios that exemplify the concepts discussed. As you progress through the book, you'll gain an in-depth understanding of not only the methods employed by ethical hackers but also the mindset required to navigate the ever-evolving world of cybersecurity.

This edited volume brings together the expertise of researchers, practitioners, and educators who have dedicated themselves to unraveling the complexities of cybersecurity. The diverse perspectives and insights shared within these pages are a testament to the collaborative effort that fuels our field.

ORGANIZATION OF THE BOOK

Chapter 1

In this chapter, we delve into the fascinating realm of digital forensics, a burgeoning field that deals with utilizing digital data as evidence in legal proceedings. The chapter provides a comprehensive overview of digital forensics phases, applications, merits, and challenges, along with an exploration of widely used software. We also discuss the legal considerations, scope, and the role of artificial intelligence in addressing complex problems within digital forensics.

Chapter 2

With the ubiquity of online applications, safeguarding sensitive data has become paramount. This chapter focuses on the vulnerabilities inherent in web applications and the sophisticated techniques hackers use to exploit them. By addressing issues like Insecure Design, Security Misconfiguration, and more, the chapter illuminates various attack vectors. We also delve into countermeasures such as fuzz testing, source code review, and encoding approaches to mitigate these vulnerabilities.

Chapter 3

As cybercrime cases surge globally, this chapter underscores the imperative of understanding the practices used by cybercriminals. The post-pandemic world has witnessed an alarming increase in cybercrime cases, targeting individuals across all age groups. The chapter sheds light on recent practices employed by criminals, equipping readers to guard against nefarious cyber activities.

Chapter 4

This chapter introduces an innovative approach to closing security gaps in vulnerable networks. It presents a multi-phased process, beginning with information collection and culminating in vulnerability mitigation. By incorporating operational services and exploit configurations, the proposed method enhances guilty detection and overall network security compared to prior approaches.

Chapter 5

The Darknet's intriguing yet ominous world is explored in this chapter. Covering dark web components, access methods, anonymity, crimes, attacks, and more, we provide insights into the shadowy realm of the internet. The chapter delves into various aspects of darknet utilization, shedding light on its complexity and providing readers with a deeper understanding of its dynamics.

Chapter 6

This chapter dives into the ominous world of ransomware, exploring its lifecycle, activation processes, ransom requests, and payment dynamics. Readers gain an understanding of how ransomware functions, the types of attacks it encompasses, and the strategies to detect, prevent, and mitigate its impact. The chapter equips readers with knowledge to navigate the treacherous landscape of ransomware attacks.

Chapter 7

Building upon the previous chapter's insights, this chapter offers a comprehensive overview of ransomware, tracing its history, development, and infamous cases. It delves into the anatomy of ransomware attacks, different types of attack vectors, and strategies for prevention. With a focus on fostering cyber resilience, the chapter emphasizes the importance of safeguarding against these evolving threats.

Chapter 8

The rapid expansion of the Internet of Things (IoT) presents both opportunities and challenges. This chapter addresses the diverse challenges facing IoT applications and devices, exploring their impact on various industries. By discussing the evolution of IoT, the chapter provides a comprehensive understanding of the complexities and innovations shaping the IoT landscape.

Chapter 9

Recognizing the crucial role of human awareness in cybersecurity, this chapter presents research findings on enhancing employees' awareness of cyber threats. By systematically conducting phishing exercises, awareness training, and disseminating information security bulletins, organizations can bolster cybersecurity awareness. The chapter emphasizes the importance of proactive measures to defend against cyber threats.

Chapter 10

The alarming rise in network breaches calls for proactive strategies to secure organizational data. This chapter delves into the evolution of network architecture, analyzes the grounds behind data leaks, and presents effective approaches to counter threats. It discusses governance structures, emerging developments, and methodologies in network security, offering insights to navigate the ever-evolving threat landscape.

Chapter 11

Penetration testing is an art, and this chapter serves as a foundational guide. It introduces readers to the fundamental concepts, types, and dos and don'ts of penetration testing. The chapter extends beyond the basics, discussing various types of OSINT methods, guiding readers toward a deeper understanding of this vital practice in cybersecurity.

Chapter 12

Building on the previous chapter, this section offers insights into mastering penetration testing techniques. It provides an introduction to networking, Linux, and Bash, essential skills for successful ethical hackers. Additionally, the chapter covers common tools, methodologies, and offers readers opportunities for further research to enhance their penetration testing capabilities.

Chapter 13

Social engineering attacks exploit human vulnerabilities, posing significant risks to organizations. This chapter comprehensively examines social engineering attack types, strategies, and countermeasures. It emphasizes the importance of educating employees, implementing technical solutions, and ethical collaboration to defend against these manipulative tactics.

Chapter 14

In this chapter, we navigate the enigmatic realm of the dark web, providing insights into its structure, components, and activities. By examining the technologies, anonymity, crimes, and impacts of the dark web, readers gain a deeper understanding of this hidden facet of the internet. Current trends and statistics shed light on its evolving landscape.

Chapter 15

Server-side attacks pose significant threats to information security. This chapter delves into the realm of server-side vulnerabilities within ethical hacking and penetration testing. Exploring web, database, and application servers, the chapter discusses prevalent attack techniques and countermeasures. It also emphasizes the ethical and legal considerations of addressing these vulnerabilities.

Chapter 16

The ubiquity of wireless devices brings new challenges to network security. This chapter highlights the vulnerabilities and threats associated with wireless networks, exploring encryption techniques and attack vectors. It addresses the significance of wireless penetration testing and provides insights into countering evolving threats, offering guidance to safeguard information systems.

Perspectives on Ethical Hacking and Penetration Testing is not just a compilation of chapters; it's a journey that aims to empower readers with the knowledge, skills, and ethical considerations necessary to navigate the digital landscape responsibly. We invite you to explore the chapters within, engage with the content, and embark on your own quest to uncover the mysteries of ethical hacking and penetration testing.

Keshav Kaushik
University of Petroleum and Energy Studies, India

Akashdeep Bhardwaj
University of Petroleum and Energy Studies, India

Chapter 1
A Guide to Digital Forensic "Theoretical to Software–Based Investigations"

Preeti Sharma
iD https://orcid.org/0000-0001-7530-3821
University of Petroleum and Energy Studies, India

Manoj Kumar
iD https://orcid.org/0000-0001-5113-0639
University of Wollongong, Dubai, UAE

Hitesh Kumar Sharma
University of Petroleum and Energy Studies, India

ABSTRACT

A branch of forensic science called "digital forensics" deals with the utilization of digital data received, maintained, and conveyed by electronic devices as evidence in inquiries and legal proceedings. It is a growing field in computing that frequently necessitates the intelligent analysis of large amounts of complex data. Rapid advancements in computer science and information technology enable the development of novel techniques and software for digital investigations. Initially, much of the analysis software was unique and proprietary, but over time, specialised analysis software for both the private and governmental sectors became available. The aim of this chapter is to deliver a comprehensive overview of digital forensics phases, applications, merits, and demerits and widely used software of the domain. The chapter also discusses legitimate and legal considerations, followed by the scope and role of artificial intelligence for solving complex problems of digital forensics.

DOI: 10.4018/978-1-6684-8218-6.ch001

1. INTRODUCTION

Across the globe, people and associations are racing to implement new advances to improve and grow in an increasingly interconnected world. The convergence of the technological progressions in informative technology, for example, cloud computing, social networking, personal devices, example, smartphones, and so forth, and the pervasive utilization of it worldwide have resulted in numerous benefits for humanity, yet it additionally gives roads to misuse and has presented new challenges for policing cybercrimes. Cyber-crimes or digital crimes have increased in frequency with the advancement and more complex techniques being deployed by individuals and groups with intricate and advanced knowledge of the working of the internet, networks, and security architectures, and who use their specialist skills for negative and crimes, normally called the Dark Side of Internet. It presents significant implications and difficulties for national and economic security (Naick and Bachalla,2016).

Many associations are at huge risk. This statement has been proved by the number of complaints received and processed for instance by the Internet Crime Complaint Centre (IC3) of the Federal Bureau of Investigation (FBI). In 2017, the total quantities of complaints received are 301,580 with reported losses of $1,418.7 million. In this report, India is at second number in the list of top 20 victim nations with 2,819 complaints. This is significant additionally considering many personal and organizational data breaches and monetary losses go unreported in our nation and most complaints are by financial institutions like credit card organizations and banks. The list of top 20 countries by victim is depicted in figure 1 (sourced from https://www.bankinfosecurity.com/fbi-sees-internet-enabled-c rime-losses-hit-13-billion-a-10033)."

The number of incidents of cybercrime in India is rising pointedly. An IIT Kanpur study shows that the number grew from 71,780 in 2013 to 1.49 lac in 2014 to 3 lac in addition in 2015, in this way recording a yearly increment of more than 100% from 2014 to 2015. With the advent of various digital gadgets, the internet, and social media, the environment in which digital crimes are committed has fundamentally changed. It is currently insufficient to simply examine the victim's PC's hard drive, as additional evidence will be required for the successful prosecution of the perpetrator and determination of the root cause of the crime (Palmer,2001). The latter is fundamental for knowing about the new methods utilized by criminals and accordingly modified the investigation as additionally the investigation of future crimes. The development of the highly technical and sophisticated nature of digital crimes has made another part of science known as Digital Forensics. Because there were few specialized digital forensic tools available in the 1980s, investigators frequently performed live examinations on media, examining computers from within

Figure 1. List of top 20 victim countries of cyber crime
(Naick and Bachalla, 2016)

Top 20 Foreign Countries by Victim

Excluding the United States

1. Canada		6. Brazil		11. Germany		16. United Arab Emirates	
2. India	2,188	7. Mexico		12. South Africa		17. Malaysia	
3. United Kingdom	1,509	8. China		13. Turkey		18. Singapore	
4. Australia		9. Japan		14. Spain		19. Nigeria	
5. France		10. Philippines		15. Hong Kong		20. New Zealand	

the working framework, and extracting evidence using existing system admin tools. This practice conveyed the risk of inadvertently or intentionally modifying data on the plate, which prompted claims of evidence tampering. In the mid-1990s, many tools were created to solve the issue. The chapter involves an introduction, brief history, and objectives of digital forensics in section 1. Section 2 discusses its current issues followed by its various phases and categories in sections 3 and 4. Sections 5 and 6 introduce various tools and software like FTK, QRadar, Parrot securities, etc. used for analyses of different forensic cases. Section 7 discusses various advantages, disadvantages, and applications of Digital Forensics. Section 8 defines Legitimate and legal considerations of the forensic field followed by section 9 introduces about Role of Digital Forensics in Artificial Intelligence. Finally, section 10 concludes the chapter with a discussion about the future scopes of the forensic field.

1.1 Origin of Digital Forensics

In 1984, the FBI started creating tools to look at computer evidence, which is when the field of digital forensics was first born. To combat digital crime, digital forensic professionals acquire in-depth information, design specialized forensic software, and follow conventional methods from physical forensics. Computer crime is a significant criminal activity with rising incidence and frequency. Business organizations, law enforcement, and the government are all being put under pressure by this rise in illegal activities. Hence, a quick reformulation of standards and processes was required to move from document-based evidence to digital/electronic evidence. Many inquiry models have been put out over the years by various inventors.

The study found that certain models tended to only apply to extremely narrow scenarios, while others had a wider application. Some of the models are frequently fairly detailed, while others are overly broad. As a result, adopting the right or appropriate investigative methodology may prove challenging for forensic investigators. In order to cope with digital evidence analysis, a computer forensic investigating procedure was suggested in 1984. In 1984, a computer forensic investigative procedure was established for handling digital evidence inquiry such that the outcomes will be legally acceptable and scientifically credible. A computer forensic investigative procedure was established for handling digital evidence inquiry such that the outcomes will be legally acceptable and scientifically credible. The value of information has risen, as has the effect it has over digital evidence as a result of the growing growth in information technology applications in companies and the government. As information technology has developed, information security science has emerged as the primary driver and pillar of its use, as well as a tool for combating cybercrime. The key turning points in the development of digital forensics as a brief history is shown in Figure2.

1.2 Objectives of Digital Forensics

The fundamental objectives of utilizing Computer forensics includes the following points:

- "The recovery, dismantling, and safeguarding of computer and related materials in such a way that the examining organisation may submit them as evidence in an official courtroom is aided by digital forensic."
- "Hypothesising the motive for the crime and identifying the primary perpetrator is also supported by forensic."
- "It aids in ensuring that the digital evidence obtained is not compromised and also design measures at a suspected crime scene."

Figure 2. Brief history highlighting the important landmarks in the evolution of digital forensics

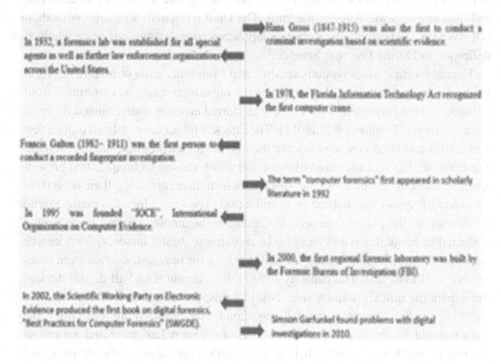

- "Recovering erased records and erased allotments from digital media in order to extract and approve them: Data acquisition and duplication is also aided by digital forensic."
- "Assessing the expected consequence of the harmful action on the individual in question and assists in quickly distinguishing the evidence as well."
- "A comprehensive report on the examination interaction provides by digital forensics."
- "The chain of custody to preserve the evidence is one of the main objectives of digital forensics."

2. DIGITAL FORENSICS AND ITS CURRENT ISSUES

"Digital forensics is primarily a practitioner-oriented field; due to a lack of standardization, a never-ending upgrade cycle, and a high degree of uncertainty, it is difficult to forecast and define how the field of digital forensics will advance. American Heritage Dictionary defines digital forensics as "Digital forensics is the utilization of science and innovation to investigation and build up facts in criminal and civil courts of law". The objective of any digital forensic investigation will be to distinguish harmful events and their impact, decide the main cause of an event, and discover who was responsible for it. The Digital Forensic Research Workshop (DFRWS) in 2001 provided the most widely accepted and commonly acknowledged definition of Digital Forensic Science.

Digital forensic investigations are different from other sorts of investigations in two ways. In the first place, they might be remote crimes might be committed from a distance. That implies that the crime was started at some undetermined distance from the target (Solomon et al.,2011). The attacker might have utilized quite a few procedures and strategies to obfuscate their actual area to confuse the investigator. Furthermore, the attacker could likewise use anti-forensics techniques that prevent forensic tools, investigations, and specialists from accomplishing their objectives or making leaps in the method of examination. The crime location could extend all throughout the planet. The second variable is the amount of data available and collected to break down and analyse. In an extreme digital incident, there can be terabytes of information to examine and that may (or may not) contain even bytes of shreds of evidence. The capacity limit of the computer and all digital devices are expanding quickly step by step. Not just this, presently the user can likewise store information somewhat on the web, cloud, and so on Along these lines, the enhancement in the capacity and remove information has presented enormous difficulties and it is extremely challenging to collect and examine the information."

The advances in the technologies like social media, smartphones, cloud computing, and so forth likewise forced difficulties in digital forensics. The recent concerns related to digital forensics as displayed in Figure 2 are:"

2.1 Prominent Issues of Digital Forensics

a.   "Social networking"

"Social Media forensics is another part of networking forensics. Its spotlights the checking and investigation of social media content. Social networking sites like Google+, Facebook, Twitter, WhatsApp, YouTube, and so forth have extended quickly in recent years. It is for the most part utilized for regular day-to-day existence updates, news, and significant data. Its availability, speed, and convenience of social media have made them significant sources of direct data so there is a requirement for forensic instruments that address such an important use. Its assortment, investigation, process, and assessment progressively could include various conventions and the amount of data might be exceptionally large and challenging."

b.   "The growing size of storage

Figure 3. Current issues of digital forensics

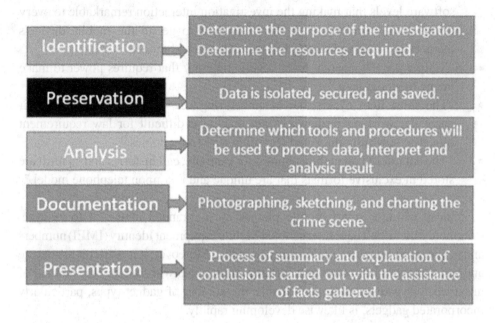

"Previously, digital forensics investigations were limited to the investigation of a single framework with small capacity discs; now, investigations are increasingly requiring the investigation of several frameworks with diverse big capacity discs, network storage, and cloud storage." Today a 2TB hard drive is very cheap anybody can easily get however it requires over 7 hours to picture. The information that must analyse additionally incorporate remote data and information found on network and internet. The use of the "cloud" for distant handling and capacity introduces additional obstacles because network data is difficult to come by, making acquisition challenging and complex. Because storage devices are becoming larger in size, there is frequently inadequate time and effort to create a forensic picture of a subject device, or to process and evaluate all of the data that is discovered."

c.   "Mobile and embedded devices"

"Digital Forensics incorporates the wide scope of digital gadgets that are regularly essential for an examination. Notwithstanding, the law requirement and forensics investigations have attempted to adequately oversee digital evidence got from gadgets like smart phones, including iPhone, Android and Blackberry, advanced media players and game control centre. A portion of the reasons are:"

- "The mobile phones have no standard interface, either at the hardware or software levels that making the investigation interaction remarkable to every device model. Indeed, even the links used to get to the mobile device's memory are diverse for different manufacture and model."
- "Mobile gadgets operate from volatile memory that requires power to make image of stored data."
- "The short item cycles that manufacturing new mobile phones and their particular working frameworks are making it difficult for law requirement offices to stay fully informed regarding new technologies."
- "The data stored on mobile phones, for example, call histories and so forth are stored in exclusive formats that are unique and rely upon telephone model."

"Cheap Chinese smart mobile phones are unbranded and hard to analyze. Commonly, they don't have International Mobile Equipment Identity (IMEI) numbers and hence, can't be followed (Hoelz et al.,2009). Moreover, it is difficult to analyze and handle new or less commonly utilized gadgets. Notwithstanding the number of unsatisfactory gadgets of a specific sort, the number of gadget types, particularly incorporated gadgets, is likewise developing rapidly."

d.   Encryption of course"

"Yahoo promises "encryption everywhere," Google switches to 2,048-bit certificates, and HTTP 2.0 will be natively encoded." We are, of course, approaching the era of encryption throughout the company. On the one hand, it will increase security, but on the other hand, it will make it impossible for law enforcement authorities to acquire data. The EnCaseTM Certified Examiner (EnCE) programme trains public and private sector professionals in the use of OpentextTM, EnCaseTM Forensic. Professionals who have mastered computer investigation methods as well as the use of EnCase software during sophisticated computer examinations get the EnCE certification. A thorough study of a 100 GB hard drive's data can contain over 10,000,000 pages of electronic information and can take anywhere from 15 to 35 hours or more to complete, depending on the size and kind of media.

e.   "Anti-Forensics"

"Anti-Forensics is a bunch of techniques that is utilized as countermeasures to digital forensics analysis. Its goal is to decrease the amount or quality of evidence collected at a crime scene or to make evidence investigation and assessment quite difficult or impossible. Anti-forensics has four basic goals, according to Liu and Brown (Kruse et al.,2002).

Figure 4. Process of digital forensics

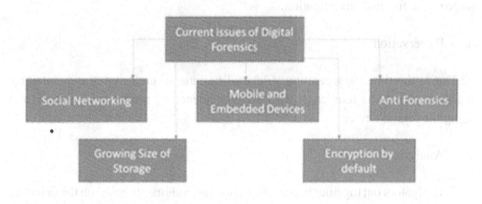

- "Avoiding detention of an occasion."
- "Disrupting the collection of data."
- "Increasing the measure of time that an examiner needs on a case."
- "Casting doubt on a forensics report or testimony."

"Anti-forensics strategies might incorporate information concealing that is encryption or steganography, artifact wiping, trail confusion and evidence eliminating devices."

3. PHASES OF DIGITAL FORENSICS

The progression in digital forensics is done using the following five phases as shown in Figure 3 listed below:

- Identification
- Preservation
- Analysis
- Documentation
- Presentation

a.   Identification

The identification process includes deciding what evidence exists, where it is stored, and in which format it will be stored. Personal computers, smartphones, and digital assistants are examples of electronic storage devices (PDAs). It is the first stage of the forensic investigation.

b.   Preservation

During this time, the separation of data will be done, secured, as well as maintained. It entails preventing tampering with digital evidence by prohibiting persons from using the digital device.

c.   Analysis

Investigators put together information and draw judgments based on the evidence revealed at this level. However, it may take several rounds of inquiry to prove a single criminal situation.

d.   Documentation

This technique demands the production of a record of every observed evidence. It aids in the investigation as well as reconstruction of a scene of the crime. It comprises crime scene photography, drawing, as well as mapping, and also adequate recording of the scene of the crime.

e.   Presentation

The results have been reviewed and presented. It should, however, be described using abstract terminology. The actual information should be cited in all abstracted sentences.

3.1 Different Types of Digital Forensics

a.   Disk Forensics

This technique looks for active, edited, or erased files on a storage media to retrieve information (Wazid et al.,2013).

b.   Networks Forensics

It is a digital forensics sub-discipline. It entails analyzing computer network traffic in order to gather vital information and legal proof. Wireless forensics is a subfield of network forensics. The purpose of wireless forensics is to give the tools needed to collect and analyse data from wireless network traffic.

c.   Email Forensics

Recovers & analyses emails, calendars, and contacts, even those that have been deleted.

d.   Malware Forensics

This branch is in charge of identifying malicious code, as well as analyzing its payload, viruses, & worms.

e.   Database forensics and Memory forensics

It is, in fact, a subset of digital forensics that focuses on evaluating and analyzing databases and the data included inside them. Memory forensics is retrieving raw data from memory space (system registries, caches, and RAM) and sculpting it out of the raw dumps.

f.   Mobile Phone Forensics

It primarily concentrates on testing and analyzing mobile devices. Among other things, it allows you to recover contacts, phone records, incoming and outgoing SMS&MMS, audio files, as well as video content from your phone and SIM card (Jimenez and Lopez,2016).

4. TOOLS FOR DIGITAL FORENSIC ANALYSIS

By automating repetitive procedures and showing data in a graphical user interface to make it easier for the user to find crucial information, digital forensic analysis tools enable effective inspection. Lately, Linux has been employed as a platform for digital evidence investigation, and programmes with user-friendly interfaces like The Sleuth Kit and SMART have been developed. The following commercially accessible digital evidence analysis techniques are covered: Some of the greatest digital forensic software tools are listed below (Dweikat et al.,2020):

- EnCase
- Sleuth Kit

- Forensic Toolkit (FTK)
- Pro Discover Forensic
- CAINE
- SIFT Workstation
- PALADIN
- Google Takeout Convertor
- PDF to Excel Convertor
  a. EnCase

EnCase is a commercial toolset for forensic investigations that is frequently utilised by law enforcement organisations. It can capture data in a form that is forensically sound such that it may be examined by other well-known commercial forensic analysis tools. The program can handle a sizable amount of digital evidence and, when required, deliver evidence files straight to law police or legal counsel. It makes it simple for lawyers to assess the material and helps them to prepare reports quickly. Graphical user interface tools for digital investigations have been launched by the EnCase initiative.

b. Sleuth Kit

The Sleuth Kit is an open-source forensic toolkit that Brian Carrier created as a collection of file system forensic tools for use in forensic analysis and investigation in the Unix environment. Sleuth Kit may be launched from the command line, but many practitioners find it easier to utilize a graphical user interface. It is an open-source forensic toolkit that is extremely portable, adaptable, and practical. Every file system may be supported because Sleuth Kit is open-sourced. Users of the toolkit may add file system support as necessary. It may import the ability to process new image formats from the LibEWF (Expert Witness Format) and AFFLib (Advanced Forensic Format) packages in addition to supporting processing of raw disc images locally.

c. FTK Toolkit

The Forensic Toolkit (FTK) is a court-validated digital investigation platform that delivers computer forensic analysis, decryption and password cracking software all within a spontaneous and customizable interface. It is a commercial forensic software product that supports both 32-bit and 64-bit Windows machines and is easy to use and understand. It has multiple data views that allow users to analyse files in a number of ways and create detailed reports and output them into native format. Recent versions of the FTK include acquisition functionality, index text

Table 1. Showing cost, customization ability of most common forensic tools used for analysis

Software Tool	Scripting	Cost
Sleuth Kit	YES	Free
EnCase	YES	£2450
FTK	NO	£2700

to produce instant search results, data recovery from a file system, email recovery from the leading email services and products, and file filtering shown in Table 1.

5. CYBERCRIME DIGITAL FORENSICS TOOLS

Because of the variety of cybercrimes, different tools are used for digital forensics in cyberspace-related offences. The accompanying subsections momentarily examine the most regularly utilized devices for this reason.

a.   MemGator"

"As the name shows, MemGator is a memory cross examination tool that automates the extraction of data from memory documents and agrees a report on the concentrate information. MemGator unites various memory analysis tools, for example, the Volatility Framework and PT Finder into the one program. Data can be removed comparable to memory details, processes, network connections, malware detection, passwords and encryption keys and the registry."

b.   First on Scene"

"FOS is a scripted code written in visual essential and it works alongside different tools, for example, Logon Sessions, FPort, PromiscDetect, and File Hasher to create an evidence log report. Log report is vital for forensics investigators during the investigation process."

c.   Galleta"

"Galleta tools is accomplished in investigating cookies' documents which are connected to browsing history. These files give a thought on which sites were as of recently visited and where they keep their traces in the form of cookies."

d.   Ethreal"

"Ethreal is network security device utilized for sniffing packet traffic on the organization (approaching and active). Although this tool is useful; be that as it may, it is fragile against encryption codes which Deteriorate its performance."

e.   Pasco"

"Pasco is an instrument utilized extensively in investigating programs' contents and helps in distinguishing the conducted transaction based on the analysed contents. The beginning of the name comes from Latin language where Pasco means browse."

f.   Rifiuti

"This tool plays out its activity on recycle bin of the framework to recuperate any recent deleted files. Rifiuti is an open source delivered under the liberal FreeBSD license."

g.   Network Mapper (Nmap)

"Network Mapper or NMap is an organization security tools that operates based on scanning a remote workstation for tracking down any open ports. NMap has ability to conceal its nature from the source workstation with the goal that it will not cause any ready as a malware attack.
    There are various other digital forensics tools and software kits found in the writing. Notwithstanding, examining them will be beyond the segments of this short paper."

6. USE CASES AND SOFTWARE IMPLICATIONS OF DIGITAL FORENSICS

Investigators can learn information about computer users, locate deleted files, rebuild artefacts, and attempt to collect as much evidence as they can using computer forensics tools and procedures.

a.   **FTK Forensic Toolkit**

The Forensic Toolkit (FTK) enables investigators to conduct comprehensive and successful examinations into a wide range of data carriers and over 270 file types. Mobile phones, desktops, hard discs, registry records, Windows system information files, Apple system files, social networking programs, and more are all supported. If a data carrier uses encryption or password, FTK can decode files and retrieve passwords for over 100 applications. FTK also has comprehensive search capabilities, as well as the ability to filter files inside files. When it relates to search queries, FTK pre-processes as well as pre-indexes information, which saves the time. To aid with this, FTK features a powerful OCR (Operational Character Recognition) engine as well as the ability to automatically undelete files (Scanlon,2009). If desired, data may be tagged and exported by category. Then, using the associated visualization technologies, For example, a digital investigator may show events as timeline, cluster graphs, or geolocations. Finally, the visualization and conclusions may be compiled into a single report that is easily accessible. This scalable software has been authorized by the courts. It comes with a decryption and password cracking application, as well as a user interface that may be customized.

Applications:

- Forensic Images on a Full Hard Drive
- Parse Registry Files
- Locate, organise, and filter mobile data
- Visualization Technology
- Decrypt files and crack passwords

The above picture is an example screenshot of the FTK forensic toolkit software (Jimenez and Lopez,2016). FTK is an all-in-one picture capture, analysis, and reporting tool with the potential to automate typical investigation procedures, comparable to EnCase. The above figure 4 shows one example of this. FTK's ability to employ a database-driven architecture to maintain track of the analysis of a specific disc for distributed analysis is a remarkable feature. This distributed analysis is used for automated data pre-processing, such as restoring lost files and partitions, categorizing files, and so on. Password Recovery Toolkit and Distributed Network Attack are also included.

b. IBM Security QRadar

IBM Security QRadar aids security teams in identifying, assessing, and prioritizing the most critical threats to the organization. The system collects data on assets, services, networks, endpoints, and users, combines it with security and hazard data, and uses sophisticated analytics to detect and monitor the most significant threats as they travel down the kill chain. When a big threat is found, AI-powered investigations enable organizations to up-level their first-line security analysts,

Figure 5. FTK forensic toolkit software screenshot
(Scanlon,2009)

Figure 6. IBM security Q radar software screenshot
("IBM Security QRadar SIEM",2023)

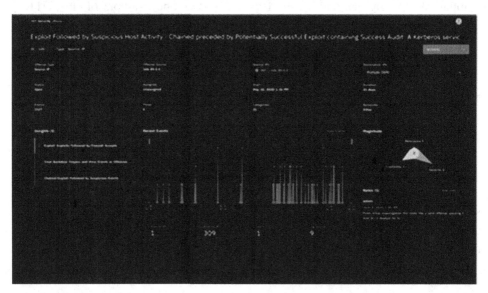

accelerate specialized security procedures, and reduce the incident effect by giving quick intelligence insights into the attack's underlying source and breadth.

IBM QRadar is commercial SIEM (security information as well as event management) software. It collects information in an organization via wireless networks, host assets, computers, applications, as well as vulnerabilities, and other user behavior as well as behaviour's, among other things. The log data & network traffic are then analyzed in real-time by IBM QRadar ("IBM Security Q Radar SIEM",2023) to detect malicious activity and swiftly halt it, averting or reducing harm to the enterprise.

The above figure 5 is the example screenshot of IBM security Q radar software (Frattini et al.,2019). This software collects the data, aggregates, processes, and stores the network's real-time data. Searching event data using particular criteria and displaying events that fit the search criteria in a results list is the software's distinctive feature. It may also run complex searches to filter the presented flows or visually monitor and examine flow data in real-time.

c.   **ExtraHop**

Cybercriminals have the upper hand. ExtraHop is on a mission to help you reclaim it with security that can't be beaten, outwitted, or hacked. Reveal(x) 360, our dynamic cyber protection technology assists companies in detecting and responding to sophisticated threats—before they affect your company. We execute line-rate decryption and behavioral analysis across all infrastructure, workloads, and data-in-flight using cloud-scale AI on petabytes of traffic every day. Enterprises can identify malicious activity, hunt sophisticated threats, and forensically examine any occurrence with confidence because of ExtraHop's comprehensive visibility. IDC, Gartner, Forbes, SC Media, and a slew of other publications have named ExtraHop an industry leader in network detection and response. That's uncompromised security when you don't have to choose between securing your company and taking it ahead (Thethi,2014).

Background:

ExtraHop was formed with a simple mission: to assist enterprises in preventing advanced threats by providing security that cannot be undermined, outsmarted, or hacked, according to the company. ExtraHop was formed with a single mission: to assist enterprises in preventing advanced threats by providing security that cannot be circumvented, outwitted, or compromised.It has roughly 50 artificial intelligence and machine learning patents, as well as 1500+ monthly high-risk threats recognized. The Reveal(x) 360 network detection and response (NDR) solution from Extrahop. It is a SaaS-based network detection and response (NDR) solution that provides unified security across on-premises and cloud environments, 360-degree visibility

Figure 7. ExtraHop software Screenshot
("Extrahoop System User Guide",2023)

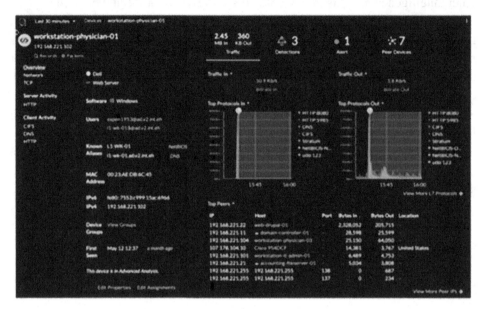

and situational data without friction, and immediate benefit with no management overhead("Extrahop System User Guide",2023). Use cases of Reveal(x) 360 include:

- Compliance & Audit
- Incident response
- Threat Hunting
- Inventory & configuration
- Detect advance threats
- Dependency mapping
- Vulnerability assessment
- Forensic investment
- Monitor sensitive workloads & data

The above figure 6 is the example screenshot of the ExtraHop software. This software leverages cloud-scale machine learning for delivering keen visibility into networks. Threat detection and response, security, and remote site visibility are among the software's distinctive features.

d. **Parrot Security OS**

The company's primary product, Parrot OS, is a Debian-based GNU/Linux operating system with an emphasis on security and privacy. It includes a full mobile lab for all forms of cyber security tasks, such as testing phase, forensic analysis, and ethical hacking, as well as everything you'll need to produce software or safeguard your data. Parrot Security (ParrotSec) is a security distribution geared for the Information Security (InfoSec) industry. It includes a completely mobile laboratory for security and digital forensics experts. For developers, Parrot Security OS is free and open source for researchers, security experts, forensic investigators, as well as privacy-conscious individuals. It is derived from Debian Testing as well as comes pre-installed with MATE as the default desktop environment (Tian et al.,2009). It's a Debian branch that includes Tor, Tor chat, I2P, Anonsurf, Zulu Crypt, and other development, security, and anonymity tools, as well as other Debian forks. It's popular among researchers, security researchers, as well as privacy-conscious consumers. It may be used in virtual environments and docker containers, as well as dual-booting with other operating systems.

It offers a separate "Forensics Mode" and is more covert than the regular mode since it only affects system hard disks or partitions and therefore has no influence on the host system. On the host system, this option is used to undertake forensics activities.

System Basic requirements:

- Processor: x86 architecture with a minimum clock speed of 700 MHz.
- RAM: 256MB minimum for i386 and 320MB minimum for amd64.
- Installation HDD: nearly 16GB
- Architecture: i386, amd64, 486 (legacy x86), armel, and armhf are supported (ARM)
- Legacy boot mode is preferable.

Features:

- Secure- It is constantly updated and published, with a variety of hardening and sandboxing settings.
- Light weight- They are concerned about resource usage, and the system has shown to be highly light and quick, even with outdated hardware or low resources.
- Portable and universal- To make their goods compatible with as many devices as possible, they leverage containerization technologies like Docker or Podman. You may use the Parrot tools on Windows, Mac OS, or any other Linux distribution without changing your habits.

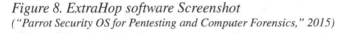

Figure 8. ExtraHop software Screenshot
("Parrot Security OS for Pentesting and Computer Forensics," 2015)

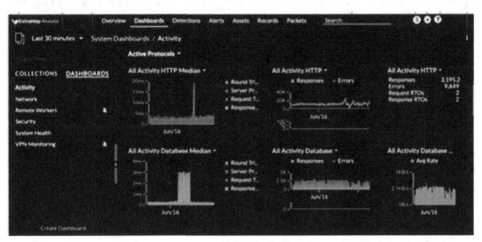

The above figure 7 is an example screenshot of the parrot security OS ("Parrot Security Os for Pentesting and Computer Forensics",2015). When compared to other applications, this will have more advanced functionality and customizability. Its efficiency is that it consumes more resources than MATE and performs best on powerful and contemporary hardware.

e.   Sleuth Kit (+Autopsy)

Sleuth Kit (+Autopsy) is a Windows-based program that is a straightforward, graphical user interface-based application for quickly analyzing hard drives and mobile devices which makes computer forensic investigation easier. You can use this tool to analyse your hard drive and smartphone. It has a plug-in architecture that lets you search for and install add-on modules as well as create custom Java/ Python modules. The Sleuth Kit is a collection of command-line utilities as well as a C library for image analysis and file retrieval. It's used in Autopsy as well as a variety of other free source as well as commercial forensics applications. These programs, which included community-based e-mail lists as well as forums, are used by thousands of individuals all over the world. Basis Technology provides commercial training, support, and bespoke development services.

The Sleuth Kit (TSK) is a UNIX as well as Windows-based library as well a suite of apps for extracting data about disc drives as well as other storage devices to help in computer forensics ("Open-source digital forensics", 2023). It serves as the basis for Autopsy, a more well-known program. The Sleuth Kit includes a user interface for command line tools. The GPL, the CPL, as well as the IPL safeguard the collection as open source. The programme is under ongoing development and is backed up by a development team. Brian Carrier did the original development. The Coroner's Toolkit was the inspiration for this piece. It is the platform's official replacement. The Sleuth Kit can parse NTFS, FAT/ExFAT, UFS 1/2, Ext2, Ext3, Ext4, HFS, ISO 9660, as well as YAFFS2 file systems independently or inside raw (dd), Expert Witness, or AFF disc images . The Sleuth Kit can examine most Microsoft Windows, Apple Macintosh OSX, many Linux, and some other UNIX platforms. The Sleuth Kit can be used as a command line tool or as a library within a different digital forensic application like Autopsy or log2timeline/plaso.

Applications

The Sleuth Kit may be utilized in a variety of ways

- How to figure out even though the system software has destroyed all meta data, what information is on a hard disk drive
- For restoring picture files that have been accidentally erased.
- List of all deleted files are compiled.
- Use the file name or a keyword to find files.

Features

- You can efficiently detect activities using a graphical interface.
- This program does email analysis.
- You may search for all documents or photos by grouping files by type.
- It shows a thumbnail of the photos so you can view them quickly.
- You may name your files anything you like.
- The Sleuth Kit can extract data using call records, SMS, contacts, as well as other sources.
- It may be used to label files and folders based on their path as well as name.

The above figure 8 is an example screenshot of the sleuth kit. Its effectiveness is simplifying the collecting and evaluation of digital evidence by automating several procedures. Few of the commands included in The Sleuth Kit include:

- The **ils command** shows all metadata entries, such as an Inode.
- The application **blkls** displays data blocks in a file system (formerly called dls).
- In a file system, **fls** displays the names of allocated and unallocated files.
- **fsstat** is a command that displays statistics about a file system, such as an image or a storage device.
- **ffind** searches for file names that refer to a certain piece of metadata.

Figure 9. Sleuth Kit (+Autopsy) software screenshot
("Open Source Digital Forensics",2023)

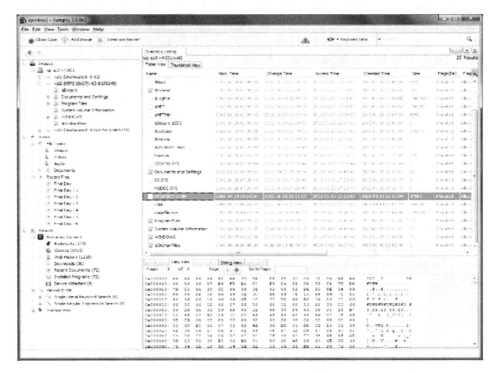

- **mactime** creates a timeline based on all files' MAC timings.
- **disk stat** identifies the presence of a Host Protected Area (now only supported on Linux).

7. DIGITAL FORENSICS CHALLENGES/ADVANTAGES/DISADVANTAGES/APPLICATIONS

7.1 Challenges

To introduce open research questions for the benefit just as improving the exploration in this interesting field is one of the targets of this chapter. In the accompanying sub-areas, various issues yet to investigate and examined ("Kanellis",2006).

a.   Proof Oriented Design"

"Current digital forensics tools were initially intended to analyse the digital evidence irrelevant to the conditions. Implying that, the analysed evidence works with the investigation cycle however doesn't address any issue over the cyberspace. What's more, the greater part of these tool's arrangement with crimes submitted on a computer not against, for instance, a human being."

b.   Data View Inconsistency"

It's usual to notice that the imagined information in cyberspace doesn't always correspond to a comparable saved copy on the disc, resulting in chaos and startlingly erroneous forensic analysis findings." This throws up a window of opportunity for research into how we might resolve this issue and what procedures, tools, upgrades, and other measures may be used to limit the impact of this situation.

c.   Item Interpolation Mechanism"

"It was found a couple of years prior that it is feasible to interpolate a missing piece of a JPEG document. Tragically, this component isn't really found in contemporary digital forensics tools."

d.   Run-Time versus Execution"

"The majority of forensic tools are running in a very sluggish mode execution, the vast majority of them consume a long time of the day to do its job. This might influence the outcomes of the investigation specifically when the time matters."

e.   Digital Forensic Awareness

"One of the key examination regions in this field is the improvement of a complete educational training that can be given to the two investigators as well as judges and persecutors to be illuminated on advanced digital forensics and its application to cybercrime."

f.   Technology Gap"

There is an undeniable innovation gap between cybercriminals and combatting tools and programming units, and it regrettably favours cybercriminals." This emphasises the importance of researching security programming methodologies and tools to close this insurmountable gap.

g.   Technology versus tools"

"The gap between the arising smart innovations and forensic tools. This offers an exceptionally rich area of research in creating tools that are viable with current smart devices and can be accommodated public use."

7.2 Pros of Digital Forensics

"Here, are geniuses/advantages of Digital legal sciences (Nance and Ryan,2011):

- It assists enterprises in detecting sensitive data if their computer systems or organizations are hacked."
- Finds cybercrime from anywhere in the world quickly."
- Assists in the safekeeping of the association's funds and valuable time. Guarantee the integrity of the computer system."
- Allows to concentrate, process, and interpret the real proof, so it demonstrates the cybercrimes activity's in the court."
- To present evidence in the court that will lead to the offender's penalty or punishment."

7.3 Cons of Digital Forensics

"There are few cons/drawbacks of utilizing Digital Forensic (Rittinghouse et al., 2003):"

- Need to create genuine and convincing evidence."
- Digital evidence acknowledged into court. In any case, it is should be demonstrated that there is no changes and manipulations.
- If the tool utilized for digital forensic isn't as per determined principles, then, at that point, in the courtroom, the evidence can be disapproved by justice.
- Legal experts should have extensive computer knowledge.
- Lack of specialized information by the examining official probably might not offer the ideal outcome.
- "Producing electronic records and storing them is a very costly affair."

7.4 Applications of Digital Forensics

"Digital Forensics is a branch of the modern discipline that deals with digital evidence in the prosecution of a crime under legal rules." With the widespread availability and use of various digital media and devices, as well as online media, there are numerous sections of various digital crime such as portable crime scene investigation, network criminology, data set criminology, email legal sciences, and so on. With the rise of digital crime in every industry, digital forensics has a wide range of applications (Kenney,2012). "The significant uses of digital forensics are: -

- "Crime Detection-There are different malware and pernicious exercises that occur over digital media and organizations, for example, phishing, spoofing, ransomware, and so forth"
- "Crime Prevention-There are different cyber-crimes that occur because of lack of safety or existing unknown vulnerabilities, for example, zero-day vulnerability. Thus, cyber forensics helps in discovering these vulnerabilities and keeping away from such crimes to happen."
- Analysis of Crime -This is the fundamental utilization of digital forensics (Electronic Evidence Guide",2013). It includes:
- "Preservation-This interaction includes ensuring the crime location and the digital forensics or arrangement from additional manipulation and photographing and video graphing the crime location, for future reference. Additionally, this cycle includes stopping any continuous order that might be connected to the crime."
- "Identification-This interaction includes recognizing the digital media and devices that can fill in as the potential evidence."
- "Extraction-This interaction includes the imaging of the digital evidence, (to keep up with the authenticity of the Application of Digital Forensics."

8. LEGITIMATE CONSIDERATIONS

One area of digital forensics where courts have yet to make a decision is an individual's right to privacy. The US Electronic Communications Privacy Act restricts the ability of law enforcement or civil investigations to obtain and suppress evidence. The demonstration distinguishes between stored communication (email archives, for example) and transferred communication (like VOIP) (Carrier,2002). The last alternative, which is seen as much more of a privacy invasion, is to obtain a warrant more eagerly. The ECPA also has an impact on businesses' ability to examine their workers' computers and communications, a topic that is still up for discussion in terms of the extent to which such surveillance can be done.

- The ECPA contains comparable privacy laws and controls the processing and sharing of person information both inside the EU and with other nations, according to Article 5 of the European Convention on Human Rights. The UK statute requiring advanced criminology testing is governed by the Regulation of Investigatory Powers Act.

- "In the United Kingdom, comparable regulations governing computer offences can also impact legal professionals." The 1990 Computer Misuse Act prohibits illegal access to computer data; this is a particular problem for civil investigators, who face more restrictions than police enforcement.""

8.1 Legal Consideration

"National and international regulation governs the evaluation of digital media." In the case of civil investigations, legislation may limit investigators' ability to conduct assessments. During its early days in the field, the "Worldwide Organization on Computer Evidence" (IOCE) was one institution that tried to establish sustainable worldwide guidelines for the capture of proof (Bass et al.,1990) computerized. Network monitoring and the review of individual correspondences are usually limited.

- "One area of sophisticated crime scene analysis where courts are still uncertain is whether a person has more right than wrong to security. The US Electronic Communications Privacy Act restricts law enforcement and ordinary experts' ability to prohibit and obtain proof. The demonstration distinguishes between stored communication (such as email files) and correspondence that has been dispatched (like VOIP). The final alternative, which is considered more of a security incursion, is to get a warrant more thoroughly. The ECPA also has an impact on businesses' ability to monitor their employees' computers and communications, a topic that is still up for debate as to the extent to which such monitoring can be done."

Public regulations limit the amount of data that can be seized during a criminal investigation. In the United Kingdom, for example, the PACE act governs the capturing of proof through legal implementation.

- "In the UK similar laws covering computer crimes can likewise influence criminological specialists. The 1990 computer misuse act enacts against unapproved admittance to PC material; this is a specific concern for civil investigations who have a bigger number of constraints than law authorization."
- "Article 5 of the European Convention on Human Rights states comparative security restrictions to the ECPA and limits the handling and sharing of individual information both inside the EU and with outer nations (Palmer,2001). The capacity of UK law requirement to lead computerized

legal sciences examinations is enacted by the Regulation of Investigatory Powers Act."

9. ARTIFICIAL INTELLIGENCE AND ITS APPLICATION IN DIGITAL FORENSICS

"Artificial intelligence (AI) is a well-established field that deals with the management of computationally complex and huge problems. Existing analytical approaches can't expand to meet the increasing demand, so practitioners are looking to artificial intelligence to automate some of the jobs in digital evidence processing (Reddy,2023). As the course of sophisticated criminology necessitates the dissection of a large amount of perplexing data, AI is seen as an appropriate methodology for dealing with a few concerns and difficulties that are now present in computerised legal sciences. The primary notions in many AI frameworks are connected to cosmology, depiction, and information organisation. A chronology is created for each investigation, and with the help of smart technology that provide data, detectives may be able to gather considerably more detailed data points than in a standard case (Gehlot et al.,2022). Artificial intelligence has the potential to provide critical capabilities and aid in the normalisation, management, and exchange of a large amount of information, data, and knowledge in the scientific field. Many tools and methodologies are gradually incorporating machine learning and artificial intelligence into their reasoning (Casino et al., 2022). (Nayerifarda,2023) reveal that, when compared to other forensic domains, image forensics has reaped the highest benefit from the use of machine learning approaches. Furthermore, CNN-based models are the most essential machine learning technologies that are progressively being applied in digital forensics. The current advanced criminological frameworks are ineffective at saving and storing this plethora of various configurations of information, and they are inadequate to deal with such vast and complex data. As a result, they require human interaction, which implies the possibility of postponement and errors (Carrier,2006). However, with the development of AI, this event of mistake or deferral can be forestalled. The framework is planned such that it can assist with identifying mistakes yet at a lot quicker pace and with exactness. Several types of research have highlighted the role of various AI technologies and their advantages in providing a system for storing and breaking down computerized proof. Among these AI tactics are AI (ML), NLP, discourse, and image identification recognition, each with its own set of advantages. For example, ML provides a framework with the ability to learn and improve without being explicitly modified, as in image processing and clinical determination. Furthermore, NLP algorithms aid in the extraction of data from textual material, such as during the file fragmentation process (Olivier et al.,2009).

10. CONCLUSION

Digital forensics is a rapidly evolving discipline with numerous obstacles and crosswinds. Legal consideration plays an important aspect in modern law enforcement investigations as it enquires about how data is collected, studied, analysed, and stored. Legal issues for reviewing digital evidence regarding the committed crime to be used as legal proof in a court of law are an essential aspect of digital forensic inquiry. With the advancement of technology, it is becoming a more advanced issue, as well as an important subject that frequently needs the analysis and extraction of a large quantity of complex data from a crime scene. AI, on the other hand, is offering a viable solution to such complicated and huge data concerns. In the recovery phase, existing forensic technologies are crucial. Each tool has its own set of disadvantages and constraints. These tools and procedures must be advanced and improved in order for computer forensics to be a complete success and legally valid in court. Computer forensics has a promising future. The profession will continue to expand as technology advances, bringing with it new benefits and difficulties.

CONFLICT OF INTEREST

There is no conflict of interest.

CONSENT OF PUBLICATION

There is consent of Publication by all authors.

ACKNOWLEDGEMENT

Non Declare.

REFERENCES

Bass, W. M., Gill, G. W., Jantz, R., Locard, E., Owsley, D. W., Tardieu, A. A., & Vucetich, J. (1980). Digital forensics. *History (London)*, *1*, 1990s.

Carrier, B. (2002). *Open Source Digital Forensic Tools: The Legal Argument (PDF)*. @stake Research Report.

Carrier Brian, D. (2006, February). *Communications of the ACM, 49*(2), 56–61. doi:10.1145/1113034.1113069

Casino, F., Dasaklis, T. K., Spathoulas, G. P., Anagnostopoulos, M., Ghosal, A., Borocz, I., Solanas, A., Conti, M., & Patsakis, C. (2022). Research trends, challenges, and emerging topics in digital forensics: A review of reviews. *IEEE Access: Practical Innovations, Open Solutions, 10*, 25464–25493. doi:10.1109/ACCESS.2022.3154059

Council of Europe. (2013). *Electronic Evidence Guide*. Council of Europe.

Cyber, C. I. P. (2015). Parrot Security OS for Pentesting and Computer Forensics [Online]. *EHacking.* https://www.ehacking.net/2015/06/parrot-security-os-for-pent esting-and.html

Extrahop. (2023). Extrahop System User Guide [Online]. *Extrahop.* https://docs. extrahop.com/8.9/eh-system-user-guide/

Frattini, F., Giordano, U., & Conti, V. (2019, September). Facing cyber-physical security threats by PSIM-SIEM integration. In *2019 15th European Dependable Computing Conference (EDCC)* (pp. 83-88). IEEE. 10.1109/EDCC.2019.00026

Gehlot, A., Singh, R., Singh, J., & Sharma, N. R. (Eds.). (2022). *Digital Forensics and Internet of Things: Impact and Challenges*. John Wiley & Sons. doi:10.1002/9781119769057

Hoelz, B. W., Ralha, C. G., & Geeverghese, R. (2009, March). Artificial intelligence applied to computer forensics. In *Proceedings of the 2009 ACM symposium on Applied Computing* (pp. 883-888). ACM. 10.1145/1529282.1529471

IBM. (2023). *IBM Security QRadar SIEM [Online]*. IBM. https://www.ibm.com/ products/qradar-siem

Kanellis, P. (Ed.). (2006). *Digital crime and forensic science in cyberspace*. IGI Global. doi:10.4018/978-1-59140-872-7

Kenney, D. (2012). Firearm Microstamp Technology: Failing Daubert and Federal Rules of Evidence 702. *Rutgers Computer & Technology Law Journal, 38*, 199.

Kruse, W. G. II, & Heiser, J. G. (2001). *Computer forensics: incident response essentials*. Pearson Education.

Naick, B. D., & Bachalla, N. (2016). Application of Digital Forensics in Digital Libraries. [IJLIS]. *International Journal of Library and Information Science, 5*(2), 89–94.

Nance, K., & Ryan, D. J. (2011, January). Legal aspects of digital forensics: a research agenda. In *2011 44th hawaii international conference on system sciences* (pp. 1-6). IEEE. 10.1109/HICSS.2011.282

Nayerifard, T., Amintoosi, H., Bafghi, A. G., & Dehghantanha, A. (2023). Machine Learning in Digital Forensics: A Systematic Literature Review. *arXiv preprint arXiv:2306.04965*.

Olivier, M. S. (2009, March). On metadata context in database forensics. *Digital Investigation, 5*(3-4), 115–123. doi:10.1016/j.diin.2008.10.001

Palmer, G. (2001). *A road map for digital forensics research-report from the first Digital Forensics Research Workshop (DFRWS)*. Utica, New York.

Reedy, P. (2023). Artificial intelligence in digital forensics. In *Encyclopedia of Forensic Sciences* (3rd ed., pp. 170–192). Elsevier. doi:10.1016/B978-0-12-823677-2.00236-1

Rittinghouse, J., Hancock, W. M., & CISSP, C. (2003). *Cybersecurity operations handbook*. Digital Press.

Scanlon, M. (2009). *Enabling the remote acquisition of digital forensic evidence through secure data transmission and verification*. University College Dublin.

Sleuthkit. (2023). Open-Source Digital Forensics [Online]. Sleuthkit. https://www.sleuthkit.org/autopsy/

Solomon, M. G., Rudolph, K., Tittel, E., Broom, N., & Barrett, D. (2011). *Computer forensics jumpstart*. John Wiley & Sons.

Thethi, N., & Keane, A. (2014, February). Digital forensics investigations in the cloud. In *2014 IEEE international advance computing conference (IACC)* (pp. 1475-1480). IEEE. doi:10.1109/IAdCC.2014.6779543

Tian, K., Zhang, B., Mouftah, H., Zhao, Z., & Ma, J. (2009, June). Destination-driven on-demand multicast routing protocol for wireless ad hoc networks. In *2009 IEEE International Conference on Communications* (pp. 1-5). IEEE. 10.1109/ICC.2009.5198907

Wazid, M., Katal, A., Goudar, R. H., & Rao, S. (2013, April). Hacktivism trends, digital forensic tools and challenges: A survey. In *2013 IEEE Conference on Information & Communication Technologies* (pp. 138-144). IEEE.

Chapter 2
Attacks on Web Applications

Ayushi Malik
University of Petroleum and Energy Studies, India

Shagun Gehlot
University of Petroleum and Energy Studies, India

Ambika Aggarwal
University of Petroleum and Energy Studies, India

ABSTRACT

Online applications hold sensitive data that is valuable, and hackers strive to identify weaknesses and exploit them in order to steal data, pose as users, or disrupt the application. Web applications are increasingly exposed to dangers and attack vectors that are more advanced. Additionally, theft of private information—such as user credentials or billing information for credit cards—occurs frequently. Attackers initially concentrated on obtaining personal information that was accidentally exposed through poorly built or poorly protected web apps. Insecure design, security misconfiguration, vulnerable and outdated components, identification and authentication failures, etc. are some of the most prominent web application vulnerabilities that will be covered in this chapter. The authors are using countermeasures including fuzz testing, source code review, and encoding approaches to get around these vulnerabilities. As a result, this chapter offers information on the various attacks that website visitors who use web applications encounter.

1. INTRODUCTION

The evolution of Internet and Web technologies, combined with rapidly increasing Internet connectivity, defines the new business landscape. A key element of internet

DOI: 10.4018/978-1-6684-8218-6.ch002

commerce is web apps. Everyone connected to the Internet utilises a plethora of web apps for a wide range of activities, such as social networking, online shopping, email, and chats. Web applications are now becoming exposed to more complex threats and attack methods. Hackers may misuse private information, which not only violates users' privacy but also opens the door to user impersonation. This module will familiarize you with various web applications, web attack vectors, and how to protect an organization's information resources from them. It describes the general web application [1] hacking methodology that most attackers use to exploit a target system. Ethical hackers can use this methodology to assess their organization's security against web application attacks. Thus, this module also presents several tools that are helpful at different stages of web application security assessment. This chapter includes several web application security tools to prevent the web from attacks that are familiar nowadays and also provides a brief knowledge about the attack and their countermeasures.

1.1 Evolution of the Web

The World Wide Web is a network of free websites that can be viewed online. It is sometimes referred to as the Web, the WWW, or just the Web. The Web is not the same as the Internet; it is one of several applications developed on top of the Internet. The evolution of the Web may be understood by dividing it into three waves: (1) read-only, (2) read/write Web, and (3) programmable Web. These waves aren't always separated by time, but rather by the introduction of new functions; as a result, they might overlap and coexist at times.

Figure 1. Phases of web evolution

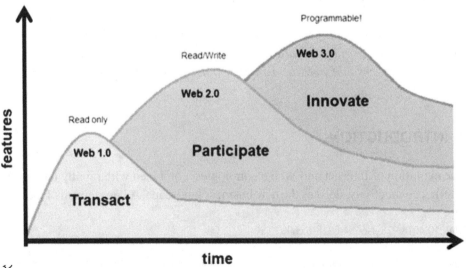

The phrase "Web 1.0" [1] refers to the initial iteration of the Web (read-only Web), which includes programs that can provide information in one direction but have limited communication and user interaction capabilities. Applications that fall under this first wave include search engines and e-commerce platforms because they make it possible to perform transactions involving both physical objects and digital data. The first generation of the web was a time of static sites and solely content delivery. The earlier web made it possible to read and search for information in another universe. There was hardly any user input or original stuff. The term "Web 2.0" [2] refers to the read/write Web's second wave, which places a strong emphasis on participation, teamwork, and co-creation as means of fostering community connection. Illustrations of this current wave include social networking platforms, blogs, and other platforms. Interactive, collaborative, and distributed behaviors—key components of Web 2.0—allow for the online conduct of both formal and informal daily activities. Anyone can create a new application or service utilizing the infrastructure that the web provides thanks to a feature known as Web 3.0 [2], the 3rd wave of the Web (programmable Web). The emergence of cloud computing, which enables the Web to serve as a platform for an ecosystem of people, apps, services, and even things (the Internet of Things—IoT), is what is driving this wave. The fundamental idea of 3.0 [3] is to create data structures and link them together to improve application discovery, automation, integration, and reuse.

1.2 Web 1.0, Web 2.0, and Web 3.0 Distinctions

The primary distinction between Web 1.0, Web 2.0, and Web 3.0 is that the former is thought of as a read-only website that emphasizes the creators' originality in their

Figure 2. Web 1.0 vs. Web 2.0 vs. Web 3.0

Parameters	Web 3.0	Web 2.0	Web 1.0
Year	2016	2006	1996
Basics	Personal and transportable	Read/Write	Mostly Read-Only
Primary Focus	The Individual	Community Focus	Company Focus
Example of Content	Live-streams/Waves	Blogs/Wikis	Home Pages
Focus of Content	Consolidating Content	Sharing Content	Owning Content
Example of Interaction	Smart Applications	Web Applications	Web Forms
Interaction Based On	User Behaviour	Tagging	Directories
Monetization Strategy	User Engagement	Cost Per Click	Page Views
Types of Advertising	Behavioural Advertising	Interactive Advertising	Banner Advertising
Website Example	The Semantic Web	Wikipedia	Britannica Online
Languages and Backlinks	RDF/RDFS/OWL	XML/RSS	HTML/Portals

material. Web 3.0 focuses on connected huge datasets, whereas Web 2.0 emphasizes user and producer content creativity. Following is a list of the minor difference.

2. WEB APPLICATIONS

These are computer programs that work with web browsers to connect web pages to web servers. Through the use of a user-friendly graphical user interface (GUI) [3], they enable users in create requests for, submit, and retrieve data from databases via the Internet. Depending on the device being used to navigate the web applications, users can insert data using a keyboard, mouse, or touch interface. Web applications employ various programming languages like SQL in conjunction with web programming languages like JavaScript, HTML, and CSS to fetch data from databases. Online applications, which enable users to interact with servers via server-side scripting, have contributed to the dynamic nature of websites. They enable the user to carry out certain operations, including browsing, emailing, chatting, online shopping, and monitoring and tracking. Users have access to all desktop apps, giving them the ability to work with the Internet as well. Sun One, AOL/Netscape Enterprise Server, and Microsoft IIS [3] are a few prominent web servers. The development of web applications and their popularity has accelerated due to the rise in internet and online business usage. The variety of functions that web apps provide is a major element in their popularity for commercial use. Additionally, they provide superior services compared to many computer-based software programs, are simple to install and maintain, secure, and quick to upgrade.

The benefits of web applications are:

- Development and troubleshooting are made simple and affordable by operating-system independence.
- They may be accessed from any location at any time with a computer and an Internet connection.
- The user interface is adaptable, making updates simple.
- Any device with a Web browser, such as Personal digital assistants, phones, and so forth., can be used by users to access them.
- All the web application data is stored on dedicated servers that are watched over and managed by knowledgeable server managers, allowing the developers to improve workload capacity.
- In addition to improving security measures, having many server facilities makes it simpler to keep track of thousands of computers running the application.

- They make use of fundamental technologies that are adaptable and support portable platforms, such as Java Server Pages (JSP) [4], Active Server Pages, Servlets, Structured Query Language (SQL) Server, and .NET programming languages.

3. WORKING OF WEB APPLICATIONS

The primary purpose is to retrieve data from a database that has been requested by a user. The web application instantly shows the relevant web pages in the browsers once a user clicks or inputs the URL into it.

This process involves the following sequential steps:

- The user begins by entering the URL or domain name of the website into the browser. The user's request is granted by the Hypertext Transfer Protocol (HTTP) [4] server.
- After receiving the request, the web server checks the file format:
  i)   If the user requests a basic web page with a HyperText Markup Language (HTML) [4] extension, the web server processes the request and sends the file to the browser.
  ii)   The web application server must handle the request if the user submits a request for a web page with extensions.
      - The web server sends the client requests to the web server, which manages it, as a result.
      - To complete the required task, the web application servers enter the databases and update or retrieve the data inside.
      - The user's browser receives the results once the website server has processed the request and delivered them to the web server.

4. ARCHITECTURE OF WEB APPLICATION

Web browsers run web applications by combining client-side scripts with server-side scripts (such as ASP, PHP, etc) [5]. The structure of the web application, which consists of hardware and software that handle operations like reading the request, searching, gathering, and presenting the necessary data, determines how well it will function.

To run the online application, the website's architecture consists of a variety of hardware, web browsers, and external web services. There are three different levels in the structure of web applications:

Figure 3. The architecture of web application

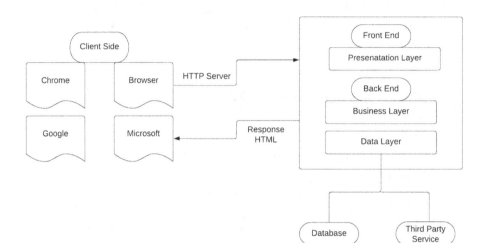

1.   User interface or presentation layer
2.   A layer of business logic
3.   Database Layer

Every physical device used on the client side, such as laptops, smartphones, and PCs, is included within the consumer or presentation layer. These devices have operating systems and browsers that are compatible, allowing users to send requests for necessary online apps. By typing a URL into the browser, a user requests a website, and the request is sent to the web server. The application presents this answer in the browser in the form of a web page once the web server replies to the request and retrieves the requested data.

The web-server logic layer and the business logic layer make up the "business logic" layer itself. The web-server logic layer is made up of a hardware element like a server, a proxy caching server, a firewall, an HTTP [6] request parser, an authentication and login handler, and a service handler. It has a firewall that protects the content, an HTTP [6] request parser that processes client requests and sends back replies, as well as a resource handler that can handle several requests at once. All of the codes that read data from the browser and provide the results are contained in the browser logic layer. The operational logic of the web application is implemented at the business logic layer utilizing tools like .NET, Java, and "middleware" technologies. It specifies the data flow, after which the programmer uses programming languages to create the application. The business logic layer combines historical applications with the

Figure 4. Security threats to web applications

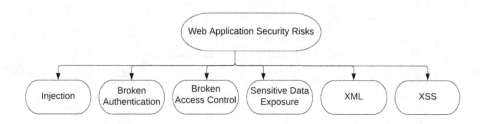

most recent capabilities of the application while storing application data. To access user-requested data from the server's database, a special protocol is required; this layer also houses the software and specifies how to search for and retrieve the data. Cloud services, a business-to-business level that stores all business transactions, and a database server that provides organized records to an organization make up the database layer (e.g., MS SQL Server, MySQL server).

5. WEB APPLICATION SECURITY RISKS

SQL injection, cross-site scripting (XSS) [7] attacks, session management, and flawed authentication will be the main areas of concern. An independent source of security knowledge about web applications is the Open Web Application Security Project (OWASP) [7], a non-profit organization, not because the problems are becoming more common, but rather because the topic of flawed authentication and session management is being examined more closely, claims OWASP [7]. Although online applications follow several security guidelines, they are nevertheless susceptible to attacks like SQL injection, cross-site scripting, session hijacking, etc. Online 2.0 and other web technologies increase the attack surface available for web application manipulation. Business-critical operations include supply chain management (SCM) [8], customer relationship management (CRM), and others. are always supported by Web 2.0 technology and web apps, which boost business productivity.

5.1 Injection Attacks

Injection problems like SQL, LDAP, and command injection occur when untrusted data is sent together with a command or query to an interpreter. The hostile data under assault may cause the interpreter to issue unauthorized instructions or gain unauthorized access to data. "Injection holes" in web applications allow unwelcome

Figure 5. Types of injection attacks

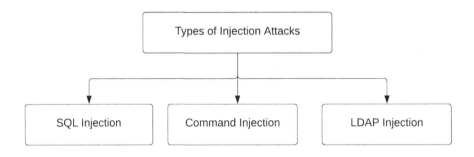

data to be processed and supplied as an element of a request or query. Attackers construct malicious queries or instructions to exploit injection flaws and result in data failure or damage, and a lack of accountability. Application vulnerability scanners may quickly find injection problems in older code, which are frequently present in SQL, LDAP [8], and XPath queries, among other types of queries. Attackers compromise online programs by injecting malicious code, instructions, or scripts into the input gates so that the apps interpret and operate with the newly given malicious input, allowing them to collect sensitive data. Attackers can simply read, write, remove, and update any data by taking advantage of injection weaknesses in online applications.

5.1.1 SQL Injection Attack

To directly alter the database, SQL injection attacks use a series of destructive SQL queries or statements. SQL statements are widely used by applications to connect to external data sources, examine roles and access levels, save, retrieve information for the program and customer, and verify identity to the application. This attack is successful because the program does not sufficiently validate input before submitting it to a SQL query. For instance, in the subsequent SQL [9] query,

select * from tablename where employeeid=3211

becomes the subsequent event after a straightforward SQL injection attack:

Select * from tablename where employeeid=3211 OR 1=1

- Access the program without providing the correct login information
- run queries to the database's data, frequently even data that the application wouldn't typically have access to
- Change the database's contents or remove it altogether
- Access other databases by using the established trust connections between the web application components.

5.1.2 Command Injection Attack

Attackers can use web apps to spread malicious malware to various platforms thanks to command injection issues. System calls to the operating system, the use of outside programs in place of shell commands, and SQL calls to the server-side data are all examples of attacks. The poor web programs are executed and inserted using scripting languages such as Perl [10], Python, and others. Any time a web application uses an interpreter, malicious code might be added by attackers to do damage.

Web applications need to make use of operating system capabilities and third-party software to carry out their tasks. Send mail is a regularly used program even though many other applications invoke outside. Scrub a program thoroughly before sending data through an HTTP external request. Otherwise, information can be altered by attackers by adding special characters, malicious instructions, and command modifiers. The external system is then instructed to execute these characters via the web application. Being a command injection, inserting SQL is a risky activity that is rather common. Attacks using command injection are easy to use and locate, but they are difficult to understand.

The following list of command injection attack types is provided:

- Shell Injection
  1) To access a web server's shell, an attacker tries to create a special input string.
  2) Injection system (), StartProcess, Java. Lang. Runtime. exec (), System. Diagnostics. Process. Start (), and similar APIs are examples of shell functions.
- Html Embedding
  1. Virtual website defacing is accomplished using this style of assault. This attack involves the addition of additional HTML-based content to the weak web application.
  2. User input for a web script is inserted into the output HTML in HTML embedding attacks without being verified for HTML code or scripting.

5.1.3 LDAP Injection Attacks

LDAP Directory stores and groups data according to its properties. A tree of directory entries represents the information's hierarchically ordered structure. Clients can use filters to search directory entries using the client-server approach of LDAP (Lightweight Directory Access Protocol). Similar to a SQL injection attack, an LDAP injection attack uses user parameters to produce LDAP queries. It is an open-standard protocol for interacting with and querying directory services, and it operates on an Internet transport protocol like TCP. Immediate connection to databases hidden behind an LDAP tree is possible by circumventing the LDAP filters used for Directory Services searches and using an LDAP [11] injection method. This technique takes use of non-validated web application input flaws.

Utilizing a local proxy, LDAP attacks take advantage of web-based applications built using LDAP statements. User input may be used by web applications to build unique LDAP statements for requests to dynamic web pages. LDAP injection attacks are frequently used by attackers against web applications that require user input to produce LDAP queries. The underlying LDAP query structure can be found by the attackers using search filter attributes. Using this structure, the attacker adds extra characteristics to user-supplied data to check for LDAP injection vulnerabilities and assesses the output of the web application.

Depending on the target's implementation, attackers can use LDAP injection to

1)   Login evasion
2)   Information sharing
3)   Escalation of privilege
4)   Modification of information

Example:

To check if the application is susceptible to LDAP code injection, send a request to the server with erroneous input. A fault the LDAP server returns can be taken advantage of via code injection techniques. The URL string becomes (&(USER=certifiedhacker)(&)) if an attacker provides the legitimate user name "certifiedhacker" and injects certifiedhacker)(&)). (PASS=bIah)) The LDAP server only processes the first filter, which is the query (&(USER=certifiedhacker)(&)). Since this is a constant, the attacker signs in to the system without using a proper password.

Filtering all LDAP inputs is a crucial protection against such assaults; otherwise, LDAP's flaws enable the execution of unauthorized queries or the alteration of its contents. The process runs with the same permissions as the web application component that performed the command when the attacker alters the LDAP declarations.

5.2 Broken Authentication

Authentication and session management application features are frequently built improperly, enabling intruders to gain access to passwords, keys, sessions, or credentials or to take advantage of other implementation defects to impersonate other users. Management of active sessions and all facets of identity verification fall under authentication. Nowadays, weak identity operations like "modify my password," "forgot my password," "recall my password," "account updating," and so forth cause web apps that use strong authentication to fail. Users must therefore exercise the utmost caution while implementing authentication mechanisms securely. Strong authentication techniques utilizing biometrics or software platforms and hardware-based cryptographic credentials are always preferable. Attackers can impersonate users by exploiting weaknesses in authentication or session management methods [12].

- Session ID in Uniform Resource Locator (URLs)- Example: When a user enters http://certifiedhackershop.com, the web application creates a session ID for the relevant login. The session ID can be obtained by an intruder by tricking the user or by utilizing a sniffer to acquire the cookie containing the session ID. The attacker at this time has the following URL: http://certifiedhackershop.com/sale/saleitems=304;jsessionid =120MTOlDPXMOOQ .This takes him to the victim's already-logged-in page. The attacker can pass for the victim. types of the following URL into the address bar of their browser.
- Exploiting credentials-Due to inadequate hashing methods, attackers can determine passwords kept in databases. If user credentials are not encrypted, attackers can intercept the credential database of the online application and use every user's password to their advantage.
- Exploiting timeouts - Sessions would last until the time provided in the session timeout if a platform's session timeouts are configured for a longer period of time. If the session out time gets set for a lengthy time, the session will be active for a long time. The hacker uses the same browser later to carry out the attack when a user just closes the tab without logging out of websites they've visited on a public computer because session IDs may still be active and take advantage of the user's capabilities.

5.3 Sensitive Data Exposure

Web applications must store sensitive data either in a database or on a filesystem, including credentials, credit card numbers, account information, or other authentication data. The software may be in danger if users don't keep their storage areas secure enough to prevent hackers from accessing them and using the data they contain for malicious purposes. Numerous web apps fail to adequately safeguard their sensitive information from unauthorized users. When transferring sensitive information from the server to the client or the other way around, web applications utilize cryptographic methods to secure their data. Due to weaknesses like information leakage and unsafe cryptographic storage, sensitive information is exposed. Although the data is encrypted, some cryptographic encryption techniques have inherent flaws that allow attackers to use them to their advantage and steal the data. Insecure encryption and storage of sensitive data by an application using a badly written encryption code database, the attacker can quickly take advantage of this issue and acquire or modify sensitive information that isn't adequately encrypted or hashed, such as credit card data, Social Security Numbers (SSNs), and other authentication credentials, to commit identity fraud, credit card fraud, etc. By encrypting critical data with the right algorithms, developers can prevent such assaults. The cryptographic keys must be securely stored by developers at all times. Attackers can readily access these keys and decrypt the sensitive data if they are kept in insecure locations. The attacker can also access the web application as a valid user due to insecure keys, certificates, and password storage. Sensitive data exposure might result in significant losses for a business [13]. Therefore, businesses must use appropriate content-filtering technologies to prevent information leakage from all sources, including systems and other network resources.

5.4 XML External Entity Attack (XEE)

A server-side request forgery (SSRF) attack known as the "XML External Entity Attack" occurs when a misconfigured XML [14] parser allows an application to parse XML data from an untrusted source. In this attack, the attacker sends a fraudulent XML input to the target web application with a reference to an external agency. This malicious input gives the attacker access to secure documents, including services from servers or associated networks, when it is processed by a target web application's badly configured XML parser. As XML characteristics are widespread, the intruder takes advantage of them to dynamically construct documents or files during processing. Attackers frequently take advantage of this vulnerability since it can be used to get private information, perform denial-of-service attacks, expose sensitive data via HTTP(s), and, in the worst-case scenario, result in remote code execution or launch Cross-Site Request Forgery (CSRF) [14] attacks against any vulnerable service. XML employs an entity, which is frequently described as a storage unit." Entities are XML's unique features that allow access to local or distant resources and are thus defined anywhere in a system using system identifiers. Since the entities might also come from external systems, it is not essential for them to be components of an XML document. To parse entities, the XML [14] processor uses system IDs that serve as universal resource identifiers (URIs). Since the XML parsing procedure replaces these entities with their actual data, only the attacker can exploit this weakness by compelling the XML [14] parser to retrieve a file or its contents that he has specified. Being a trusted application, this attack could end up being more damaging.

5.5 Broken Access Control

The concept of access control describes how a web application permits select privileged users to create, amend, and delete any record or piece of material while limiting the access of other users. A method known as "broken access control" allows an attacker to infiltrate a network by finding an access control hole, getting around authentication, and finally bypassing access control altogether. Access control flaws are frequently found as a result of application developers' ineffective functional testing and the absence of automated detection. It enables an attacker to create, access, change, or remove every record while pretending to be users or authorities with privileged rights.

According to the OWASP [15] 2017 R2 edition, broken access control is the result of missing function-level access control and insecure direct object references.

- **Insecure Direct Object References:** An insecure direct object reference is created when developers make internal implementation objects, such as documents, directories, database entries, or key-through references, publicly accessible. For instance, if a primary key is a bank account number, there is a potential that the application will be hacked by attackers using such references.

- **Absence of Function Level Access Control:** In some online applications, function level protection is controlled through configuration. Attackers take advantage of these deficiencies in function-level access control to get access to restricted functionality. In this scenario, the administrative functions will be the attackers' primary targets. To stop such attacks, developers must implement appropriate code checks. An attacker can easily find these issues, but it is much more challenging to pinpoint the susceptible web pages or functionality that can be attacked.

5.6 Security Misconfigurations

Programmers and system administrators should ensure that the entire application stack is configured properly to prevent security misconfiguration at any layer of the stack, including the platform, web-based application, application server, framework, and custom code. For example, if the developer fails to configure the server correctly, this may lead to several issues that compromise site security. Unvalidated inputs, form and parameter manipulation, inappropriate error handling, inadequate transport layer protection, etc. are all issues that can result in such situations.

5.7 Cross-Site Scripting (XSS)

These attacks take advantage of flaws in dynamically produced web pages, giving nefarious attackers the ability to insert client-side script into websites that other users are seeing. It occurs when erroneous input data is sent along with dynamic content for rendering to a user's web browser. Attackers conceal harmful JavaScript, VBScript, ActiveX, HTML, or Flash within valid requests before injecting it for execution on a victim's system. Attackers get through client-ID security measures to obtain access, insert harmful scripts onto particular websites, and then exit. Even HTML website content can be rewritten by this malicious software.

The exploits that can be carried out using XSS [15] attacks are listed below-

- Execution of a malicious script
- Sending people to a bad website
- Utilizing user rights improperly
- Pop-up ads and ads in IFRAMES that aren't visible
- Manipulating data
- Session Hijacking
- Cracking passwords with brute force
- Theft of data
- Internet exploration
- Remote monitoring and keylogging

6. WEB APPLICATION THREAT

Attacks employing the URL and port 80 are not the only threats to web applications. Despite using ports, protocols, and the Open Systems Interconnection, vendors must be able to manage all attack vectors to protect the integrity of mission-critical programs from any future attacks.

Figure 6. CIA Triad

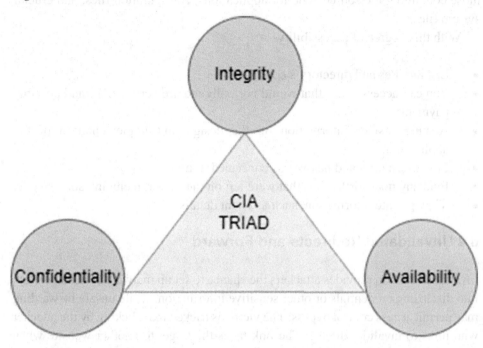

Web applications are now an essential component of the internet and frequently include sensitive data that has to be adequately secured and protected. Three things must be protected to secure a web application: the availability, confidentiality, and integrity of the data. This chapter will provide an overview of the different attacks that affect these three areas and how the web application's vulnerability is used.

6.1 Directory Traversal

Attackers exploit HTTP to execute commands outside of the root directory of the web server by using directory traversal to go to restricted directories. It is possible for inadvertent information exposure or modification to occur when access is granted outside of a defined application. Multiple folders that serve as the application's data and parts are configured for complex applications. The valid parts of an application can be found and executed by a program by navigating through these numerous directories. The directory architecture of an application, as well as frequently the core web server and operating system, are exposed by "directory traversal" or "forceful browsing" attacks. Attackers can use directory traversal to run commands outside of the root directory of the web server and access restricted directories, such as those containing the source code for applications, configuration files, and crucial system files.

With this degree of accessibility-

- List the files and directory's contents
- You can access pages that would normally require verification (and possibly payment).
- Acquire insider information on the design and implementation of the application.
- Find usernames and passwords concealed in files.
- Find any noteworthy files that were left on the server, including source code.
- View private information, such as client details.

6.2 Unvalidated Redirects and Forward

Invalid redirection provides attackers the chance to set up malware or deceive users into disclosing credentials or other sensitive information, while unsafe forwarding may permit access control bypass. The victim is tricked into clicking by the attacker who links to invalid redirects. The link takes the target to another website when they click on it under the impression that it is a legitimate website. Such reroutes cause the setup of malware and may even trick users into disclosing passwords or other private information. Unsafe forwarding is a target for an attacker looking to avoid security tests.

Unsecure forwarding may enable access control bypass, which could result in:

- **Session Fixation Attack**-An explicit session ID value is used by the attacker in a session fixation attack to lure the user into accessing a trustworthy web server.
- **Security Management Exploits**-In order to change or disable security enforcement, some malicious actor's system security, either on networks or at the application layer. An attacker that takes advantage of security management can directly change protection rules, change resources, applications, and system data, as well as remove or add existing policies.
- **Failure to Restrict URL Access**-The display of links or URLs is frequently blocked by an application to preserve or protect sensitive functionality. These links or URLs are directly accessed by attackers, who then carry out illegal actions.
- **Malicious File Execution**–Most apps include flaws that allow malicious file execution. Uncontrolled input into a web server is what led to this vulnerability. Attackers are then able to process and run files on a remote

server, launch remote code execution, set up the rootkit from a distance, and—in certain cases, at least—take complete control of systems.

6.3 Watering Hole Attack

During a watering hole attack, the attacker researches the kind of websites that the target firm or person commonly visits and then checks those specific websites for any potential security flaws. Injecting malicious script or code into the web application after discovering the website's vulnerabilities allows the attacker to reroute the webpage and download malware to the victim's computer. The intruder waits for the target to browse the compromised web application after infecting the weak web application. This attack is known as a "watering hole" because the perpetrator waits for the target to walk into the trap, much like a lion would wait for its prey to enter a waterhole to drink. The website redirects when the victim visits it, which causes malware to download onto the victim's computer. The attacker determines the kind of websites a target firm or individual frequently visits and checks those specific websites to look for any potential security flaws. When an attacker discovers a weakness in a website, the attacker injects malicious script or code into the web application, which can cause the webpage to be redirected and download malware onto the target PC. As the attackers wait for the victim to enter the trap, much like a lion waiting for its prey to enter a waterhole to drink water, this attack is called a "watering hole attack". When a user visits a rogue website and connects to a malicious server, malware is downloaded onto the victim's computer, which compromises both the machine and the network/organization.

6.4 Cross-Site Request Forgery (CSRF) Attack

This, often known as a "one-click attack," takes place when a hacker instructs a user's web browser to send a request to a vulnerable website via a malicious website. CSRF vulnerabilities are frequently found on websites with a financial focus. CSRF is one of the strategies used to enter these networks because external attackers typically

Figure 7. Watering hole attack

cannot access corporate intranets. Web applications are vulnerable to CSRF attacks because they are unable to distinguish between requests sent with malicious code and legitimate requests. Attacks known as Cross-Site Request Forgery (CSRF) take advantage of flaws in web pages to force a user's browser to submit requests that it was not intended to send. The victim user is actively using a reliable website while concurrently visiting a dangerous website,

In this case, the attacker creates a malicious script and stores it on a hostile web server. When a user accesses the website, the malicious script launches, giving the attacker access to the victim's browser.

Functioning of CSRF Attacks- In this attack, the intruder waits until the user is connected to a reliable server before tricking them into clicking on a malicious link that executes arbitrary code. Any arbitrary code is run on the secured server when the user clicks the link.

6.5 Cookie/Session Poisoning

Cookies are widely used to sustain sessions between online applications and users; as a result, cookies can need the periodic transmission of sensitive credentials. The attacker can easily alter the cookie information to increase access or pretend to be another user. Sessions typically serve to connect each user specifically to the online applications they are viewing. By poisoning cookies and session data, an attacker may be able to add malicious content, change the customer's online experience, and gather undesired information. Website designers frequently encrypt cookies to safeguard them. Many people who see cookies have a false sense of protection because of the simplicity of encoding techniques like Base64 and ROT 13 [15] (which rotates the alphabet's letters thirteen 13 times).

6.5.1 Threats

If cookies and sessions are compromised, allowing the attacker access to the user credentials, the attacker may get access to an account and assume the identities of other users of an application. By assuming a user's online identity, attackers can examine their previous purchases, make new purchases, take advantage of services, and gain access to a weak web application. Making direct use of the cookie for authentication is one of the simplest examples. By displaying cookie data and/or adding new user IDs or other session identifiers to the cookie, a proxy is used to rewrite the session data in a different way. This is another technique for poisoning cookies and sessions. Cookies come in both secure and unsecured varieties and can be either persistent or not. These four options are all possible. Persistent cookies are stored on a disc as opposed to non-persistent cookies, which are stored in memory.

Web applications only communicate secure cookies through Secure Sockets Layer (SSL) connections.

6.5.2 The Process of Cookie Poisoning

Web applications employ cookies to simulate a stateful user browsing experience based on the end customer and identification of the application server of web application components. This method involves altering a cookie's value at the client before a request is sent to the server. Any response to the given string and command can be used by a web server to deliver a set cookie. Cookies are a common method of identifying users and are kept on user computers. Once configured, the web server gets all requests made by cookies. Cookies permit modification and analysis by JavaScript to provide the application with more capabilities. This method involves the hacker sniffing the user's cookies, changing the cookie parameters, and then sending the new cookie parameters to the web server. The attacker submits a request, which the server accepts and processes.

6.6 Denial of Service (Dos)

It targets a service's availability by reducing, limiting, or preventing legitimate users from accessing system resources. Users may lose time and money if a website for a banking or email service is down for numerous hours or even days. DoS attacks aim to decrease, restrict, or completely block legitimate users' access to system resources on a computer or network. In this attack, Attackers flood a victim's system with unauthorized service requests or traffic to exhaust its resources, taking the system offline, leaving the victim's website inaccessible, or at the very least noticeably degrading the victim's system and network performance. Instead of gaining unauthorized access to the system or corrupting its data, a DoS attack seeks to stop authorized users from using it.

The examples of different DoS attacks are as follows:

- Overburdening the victim's computer system with traffic
- Overloading a service with much more events than it can manage, such as internet relay chat (IRC).
- Sending faulty packets to cause a transmission control protocol (TCP)/ internet protocol (IP) stack crash
- Connecting with a system in a way that causes it to crash
- Forcing a system to become stuck by making it enter an endless loop

DoS attacks can take many different shapes and aim at different kinds of services. The following may result from the attacks:

- Use of limited and non-renewable resources
- Utilization of the CPU, storage space, bandwidth, or data structures
- Actual damage to or modification of network components
- Destruction of computer software applications and documents

DoS attacks often target internet connectivity or bandwidth. By exploiting current network resources to overburden the network with a large volume of traffic, bandwidth assaults deprive legitimate users of resources. Attacks on connectivity overwhelm a computer with numerous connections that consume all of the OS's resources, preventing it from handling valid user requests.

Consider a pizza delivery service that does a large portion of its business over the phone. If an attacker intended to stop this firm in its tracks, he could find a way to block the phone lines and prevent it from operating. The attacker exhausts all available connections to the system, rendering all normal business operations impossible.

Dos attacks are a type of security breach that typically do not lead to data theft. These assaults, however, may cause the target to lose time and resources. Failure, however, could result in the discontinuation of a service like an email. In a worst-case scenario, a DoS assault could result in the loss of millions of users' files and programs who happened to be browsing the Internet at the time of the attack.

6.7 Hidden Field Manipulation Attack

Since most e-commerce websites include hidden sections in their cost and discount specifications, attackers target these sites with hidden field manipulation attacks. Developers employ unknown columns to store client data, such as product pricing and discount percentages, in each client session. Developers of such systems may believe that all of their apps are secure while they are still in development, but hackers may adjust the product's price and even carry out transactions using those altered prices. User decisions on HTML pages are typically kept as form field values and sent to the software as an HTTP request (GET or POST). HTML also can keep data items as secret fields, which the browser does not render on the screen but instead collects and sends as form submission parameters. Attackers can modify post requests to the server by looking into the HTML code of the webpage and changing the hidden field values.

Example: On an e-commerce platform, a specific mobile phone might be listed for $1000, but the hacker might be able to get it for $10 by changing some of the concealed languages in the price box. Even though they may be utilizing the most recent anti-virus software, IDS, routers, and other tools to safeguard their networks from intrusions, such attacks cause significant losses for website owners. In addition to monetary losses, owners risk losing their market reputation.

6.8 Web-Based Timing Attack

A web-based timing attack is a form of side-channel attack used by attackers to obtain passwords and other sensitive information from web applications by monitoring the server's response time. These assaults take advantage of side-channel leakage and calculate the processing time for private key operations.

- Direct Timing Attack: These attacks are conducted by estimating the time it takes the server to complete a POST request, which allows the attackers to determine whether a username is present. Similar to this, attackers examine passwords character by character and use timing information to identify the place where the password match failed. Attackers then utilize this information to ascertain the password of the target user.
- Cross-site Timing attack: In contrast to a direct timing attack, which involves the attacker directly passing the request to the target website, this timing attack uses JavaScript to transmit specially crafted route requests to the target website. After that, the attacker evaluates how long it took the victim to download the data request.
- Browser-based Timing Attacks: These assaults are highly skilled side-channel assaults. Attackers use a browser's side-channel leaks to estimate how long it will take to execute the requested resources, as opposed to relying on the erratic download time. In this instance, time estimation starts as soon as a resource is downloaded and ends when processing is complete.
- Cache Storage Timing Attack: Developers have access to the entire cache (memory) using the Cache API[15] interface, which is used to load, retrieve, and delete any answers. Depending on the size of the resource, loading it from the disc requires some time. Attackers can estimate the corresponding response size by estimating how long it will take the browser to complete this task.

6.9 MarioNet Attack

It is a browser-based assault that injects malicious code into the browser, and the infection endures even after closing the malicious website that the virus was propagated through or switching to another one. The majority of the most recent web browsers

Figure 8. Illustration of MarioNet attack

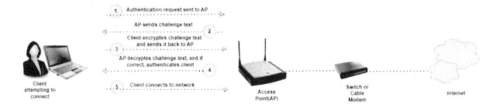

are compatible with a brand-new API called Service Workers, which enables websites to separate actions that produce web pages from expensive computing chores to prevent UI freezing when processing massive volumes of data.

Through a website they control, intruders register and enable the Service Workers API. The Service Workers API immediately starts up when the target browses that website and can continue to operate in the surroundings even if the user is not actively doing so. Attackers take advantage of the Service Workers SyncManager [16] interface to maintain the Service Workers API.

It can therefore withstand any tab breakdowns and power outages, increasing the likelihood that the browser will be attacked. It relies on formerly accessible HTML5 APIs and makes use of JavaScript's capabilities. It can be used to start a botnet and conduct nefarious operations including distributed password cracking, DDoS, click fraud, and crypto-jacking.

Additionally, this exploit enables attackers to temporarily inject executable payload into busy websites, obtain private data like user credentials, and ultimately take control of the exploited browsers from a centralized server.

6.10 ClickJacking Attack

When the destination webpage is loaded into an iframe element that is concealed by a web page element that seems authentic, a clickjacking attack is conducted. The victim is duped into hitting on any harmful web page element that has been placed invisibly at the top of any trustworthy web page by the attacker, who carries out this assault. Attackers use numerous attack routes and tactics known as UI redress attacks in addition to clickjacking, which is not a single approach. They carry out these attacks by taking advantage of flaws in HTML [16] iframes, bad configuration, or the X-Frame-Options header. Attacks known as "clickjacking" come in a variety of forms, including "likejacking" and "cursorjacking". Attackers use social media, email, or any other media to provide the victim with a link to the infected website to carry out these attacks. When doing clickjacking, the attacker loads the victim page inside an iframe with low opacity. The attacker then creates a page with all the clickable elements—such as buttons—exactly as they are on the chosen target website. Now, the victim is duped into clicking on the covert controls or the deceptive UI elements, which instantly set off several malicious actions like injecting malware, fetching malicious websites, obtaining private data like credit card numbers, transferring money from the attacker's account, and making online purchases.

The various clickjacking techniques employed by attackers are listed below:

- Complete transparent overlay: This method overlays the previously created malicious page with the transparent, legitimate page or tool page. It is then imported into an anonymous iframe and given a higher z-index value to position it above the rest.
- Cropping: This method just overlays the controls that the user has chosen from the transparent page. Depending on the attack's objective, this technique could involve hiding buttons behind links and text tags with fake information, altering button labels to read incorrect commands, and covering the entire legitimate website with false information while leaving only one legitimate button exposed.
- Hidden overlay: Using this method, the attacker builds an iframe made of IXI pixels with malicious content hidden behind the mouse cursor. Although the

Figure 9. Illustration of clickjacking attack

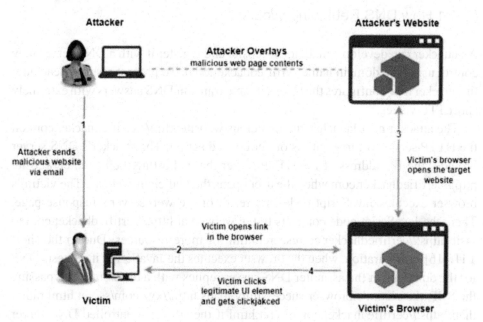

cursor hides the harmful text, if the customers click on it, it will be recorded on the infected page.

- Click event dropping: Using this method, a malicious page might be concealed behind a trustworthy page. Additionally, it can be used to set the top's CSS pointer-events attribute to none. This may result in click events "dropping" through the fake page and just registering the malicious page.

- Rapid content replacement: With this method, opaque overlays are placed over the targeted controls and are only temporarily removed to register a click. When utilizing this method, the attacker must correctly forecast how long it will take the target to browse the website.

6.11 DNS Rebinding Attack

Attackers circumvent the same-origin policy's security restrictions by using the DNS rebinding technique, enabling the malicious website to interact with or make extreme requests to local domains. If a customer is a member of an organization, for instance, he or she often utilizes the private network. The same-origin policy prevents access to any external resources inside of that private network (SOP). As a result, due to limitations in the SOP, attackers are unable to interact directly with the local network. They, therefore, work around this SOP's network security by using the DNS rebinding method.

6.11.1 How DNS Rebinding Works

An attacker can develop a malicious website and register it with a DNS server they control using the domain name certifiedhacker.com. To prevent response caching, the attacker now configures the DNS server to transmit DNS answers with extremely short TTL values.

The attacker next launches the malicious website http://certifiedhacker.com on the HTTP server to carry out his or her desired action. The attacker's DNS server provides the IP Address of the HTTP server that is hosting the assaulter website http://certifiedhacker.com when the user opens the malicious website. The victim's browser executes JavaScript code as a result of the web server's response page. Then, the JavaScript code connects to the website at http://certifiedhacker.com to visit https://certifiedhacker.com/secret.html for more resources. Due to the short TTL [16] configuration, when the browser executes the JavaScript, it requests DNS for the domain, but the assaulter DNS server replies with a different IP. Bypassing the SOP, the victim's browser successfully loads http://xyz.com/secret.html rather than https://certifiedhacker.com/secret.html if the attacker-controlled DNS server returns the private or internal IP of xyz.com.

Figure 10. Demonstration of DNS rebinding attack

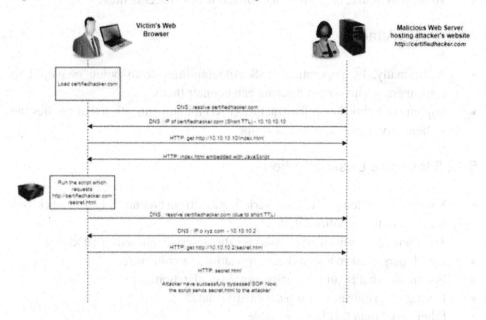

6.12 Countermeasures

6.12.1 Sensitive Data Exposure

- Avoid developing or utilizing flimsy cryptographic algorithms.
- Offline encryption key generation and safekeeping
- Make sure encrypted data on discs is difficult to decrypt.

6.12.2 XML External Entity

- Avoid using an XML parser with a poor configuration to process XML input containing references to external entities.
- XML unmarshaller configuration needs to be secure.
- Use a secure parser to decode the document

6.12.3 Broken Access Control

- Before sending the authorized user to the requested resource, perform access control checks. Avoid using weak IDs to avoid attackers from figuring them out.
- Create a mechanism for session timeouts.

- To prevent abuse, only allow authorized users to access files.

6.12.4 XSS Attacks

- Additionally, by preventing XSS vulnerabilities from being conveyed to consumers, script output filtering can counter them.
- Implement public key infrastructure (PKI) for authentication that verifies the authenticity of any introduced scripts.

6.12.5 Insecure Deserialization

- Validate untrusted input before serialization to ensure that only trusted classes are used in the serialized data.
- Trust boundaries must be crossed for the deserialization of trusted data.
- Developers must redesign their applications' architecture.
- For classes that require security, avoid serialization.
- During deserialization, protect sensitive data.
- Filter serial data that is not reliable.

6.12.6 Using Components with Known Vulnerabilities

- Check the dependencies and client- and server-side components' versions regularly.
- Check for vulnerabilities in your components frequently using resources like the national vulnerability database (NVB).
- Apply security updates frequently.
- Security scanners should be used often to check the components.
- Enforce security guidelines and recommended component usage standards

6.12.7 Insufficient Logging and Monitoring

- Define the assets covered by the log's scope to incorporate business-critical areas.
- Establish a minimal standard for logging and make sure that all assets adhere to it.
- Make sure that user context is recorded in logs so that they can be traced back to specific users.
- Ascertain what to capture and where to access logs for early event detection.
- Sanitize all event data to stave against attempts at log injection.

6.12.8 Directory Traversal

- Define the restrictions on who can access the website's secure parts.
- Implement security measures and hotfixes to stop exploitation Many flaws like Unicode [17] could impact directory traversal
- Security fixes should be updated regularly on web servers.

6.12.9 Unvalidated Redirects and Forwards

- Avoid utilizing forwards and redirects.
- If you can't eliminate destination parameters, make sure the value you supply is correct and permitted for the user.

6.12.10 Watering Hole Attack

- Apply software updates often to fix any vulnerabilities
- Observe the network traffic
- Protect the DNS server to stop hackers from rerouting the website
- Investigate user behavior
- Examine well-known websites

6.12.11 Cross-Site Request Forgery

- When processing a POST, look at the HTTP Referrer header and disregard URL parameters.
- Use CSRF tokens, like nonce tokens, that are provided through the hidden form field to block unauthorized access. Refer to headers, like the HttpOnly [17] setting that transmits a special X-Requested-With header with jQuery

6.12.12 Cookie/Session Poisoning

- Credentials for cookie authentication ought to be connected to an IP address.
- Make logout options accessible.
- Verify all cookie values to make sure they are accurate and well-formed.
- To prevent dangerous scripts from stealing cookies from the browser, use virus and malware detection software.
- frequent clearing of the browser's cookies.
- When a user submits a request, use cookie randomization to alter the website's or a service's cookie.

- To prevent session sniffing, use a VPN that uses traffic routing and high-grade encryption.

6.12.13 Web Service Attack

- WSDL Access Control configuration Access control permissions for all WSDL [18]-based SOAP message types
- Use SAML-based document-centric authentication credentials. Use several different types of security credentials, such as the SAML [18] assertions, X.509 Cert, and WS-Security
- Install firewalls that can execute web services. filtering at the SOAP and ISAPI levels
- Set up firewalls and lDS systems to identify web service anomalies and signatures.
- Firewalls and lDS systems should be set up to filter out inappropriate SOAP and XML syntax.
- Consolidate in-line requests and answer schema validation.
- When dereferencing URLs, prevent external references and use previously downloaded content.
- Updating and maintaining a safe repository of XML [18] schemas
- Use X.509 certificates, password digests, or Kerberos tickets as authentication in SOAP headers.
- Maintain the integrity of the messages by signing with a digital signature at the recipient's end.
- Updating and maintaining a safe repository of XML schemas
- Use X.509 certificates, password digests, or Kerberos tickets as authentication in SOAP headers.
- Maintain the integrity of the messages by signing with a digital signature at the recipient's end.
- To deny access to the web service file, use URL authorization (.asmx)
- Utilize NTFS permissions to grant users access to WSDL files.
- To stop the dynamic production of WSDL, disable the documentation protocols.

6.12.14 ClickJacking Attack

- Never employ client-side techniques like Framebusting or Framebreaking since they can be readily circumvented. Instead, use a server-side technique like the X-Frame-Parameters header and its options SAMEORIGIN, DENY, and ALLOW-FROM URI.

- Mask the HTML document and only reveal it once you've made sure the page isn't being framed
- Utilize Content-Security-Policy (CSP) HTTP header [19] to define sources in complex installations since it offers a lot of flexibility.

6.12.15 JavaScript Hijacking

- To automatically encrypt the text in JavaScript, use.innerText rather than. innerHTML.
- Due to the function's vulnerability, eval should not be used.
- Write serialization code not
- To protect the properties and data elements, use the encoding library instead of creating XML dynamically.
- Instead of using client-side code for encryption, use SSL/TLS for secure communication.
- Avoid creating XML [19] manually by utilizing any suitable framework and instead to return JSON containing an external object, use the format "result": ["object":" within array"].

6.12.16 Username Enumeration

- Make sure that user identities in inputs generate simply generic error messages in outputs.
- Instead of consecutive numbers, utilize randomly generated information for usernames.
- Use appropriate SQL [20] injection and XSS safeguards to stop dumpable user enumeration.
- Deploy CAPTCHA to all input-accepting pages at all times to prevent automated data collection.

6.17 Protect Yourself Against Web Application Assaults

The beforementioned techniques can be used to prevent web application assaults. To protect the web server, you can employ packet filtering, LDS, or a WAF firewall. To defend the server against attackers, you should also often upgrade the server's software using patches. Sanitize and filter user input, check the source code for SQL [20] injection, and use third-party software as little as possible to safeguard online applications. Furthermore, you can eliminate verbose error messages that could provide attackers access to data by using stored procedures and parameter queries when retrieving data. Personalized error pages can be used to safeguard

Figure 11. Defend against web applications attacks

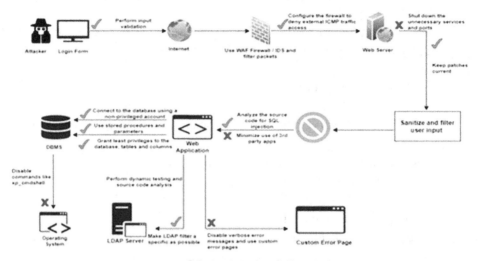

Defend against web applications attacks

web applications. Connect with a non-privileged account and provide tables, the database, and columns with the fewest privileges possible to prevent SQL injection into the database. Disable OS-impacting commands.

REFERENCES

Agarwal, A., Bora, N., & Arora, N. (2013). Goodput enhanced digital image watermarking scheme based on DWT and SVD. *International Journal of Application or Innovation in Engineering & Management*, *2*(9), 36–41.

Almeida, F. (2017). Concept and dimensions of web 4.0. *International Journal of Computers and Technology*, *16*(7), 7040–7046. doi:10.24297/ijct.v16i7.6446

Chaudhary, R., Singh, P., & Agarwal, A. (2012). A security solution for the transmission of confidential data and efficient file authentication based on DES, AES, DSS and RSA. *International Journal of Innovative Technology and Exploring Engineering*, *1*(3), 5–11.

Hou, X. Y., Zhao, X. L., Wu, M. J., Ma, R., & Chen, Y. P. (2018, April). A dynamic detection technique for XSS vulnerabilities. In *2018 4th Annual International Conference on Network and Information Systems for Computers (ICNISC)* (pp. 34-43). IEEE. 10.1109/ICNISC.2018.00016

Kaur, G., Pande, B., Bhardwaj, A., Bhagat, G., & Gupta, S. (2018, January). Defense against HTML5 XSS attack vectors: a nested context-aware sanitization technique. In *2018 8th International Conference on Cloud Computing, Data Science & Engineering (Confluence)* (pp. 442-446). IEEE. 10.1109/CONFLUENCE.2018.8442855

Khanzode, C. A., & Sarode, R. D. (2016). Evolution of the world wide web: from web 1.0 to 6.0. *International journal of Digital Library services, 6*(2), 1-11.

Król, K. (2020). Evolution of online mapping: From Web 1.0 to Web 6.0. *Geomatics, Landmanagement and Landscape, 1*, 33–51. doi:10.15576/GLL/2020.1.33

OWASP. (2018). *Owasp top 10 application security risks - 2017, 2018*. OWASP. https://www.owasp.org/index.php/Top_10-2017_Top_10

OWASP. (2018). *net antixss library*. OWASP. https://www.owasp.org/index.php/.NET_AntiXSS_Library

OWASP. (2018). *Proyecto owasp api de seguridad empresarial (esapi)*. OWASP. https://www.owasp.org/index.php/Category:OWASP_Enterprise_Security_API/es

Patil, K. (2021). A Study of Web 1.0 to 3.0. AGPE THE ROYAL GONDWANA RESEARCH JOURNAL OF HISTORY, SCIENCE. *ECONOMIC, POLITICAL AND SOCIAL SCIENCE, 2*(2), 40–45.

Praseed, A., & Thilagam, P. S. (2018). DDoS attacks at the application layer: Challenges and research perspectives for safeguarding web applications. *IEEE Communications Surveys and Tutorials, 21*(1), 661–685. doi:10.1109/COMST.2018.2870658

Register. (2018). Gits Club GitHub Code Tub With Record-Breaking 1.35Tbps DDoS Drub. [Online]. *The Register.* https://www.theregister. co.uk/2018/03/01/github_ddos_biggest_ever/

Rodríguez, G. E., Benavides, D. E., Torres, J., Flores, P., & Fuertes, W. (2018, January). Cookie scout: An analytic model for prevention of cross-site scripting (XSS) using a cookie classifier. In *International Conference on Information Technology & Systems* (pp. 497-507). Springer, Cham. 10.1007/978-3-319-73450-7_47

Rodríguez, G. E., Torres, J. G., Flores, P., & Benavides, D. E. (2020). Cross-site scripting (XSS) attacks and mitigation: A survey. *Computer Networks, 166*, 106960. doi:10.1016/j.comnet.2019.106960

Singh, H. A. (2020). EVOLUTION OF WEB 1.0 TO WEB 3.0.

Sriramya, P., Kalaiarasi, S., & Bharathi, N. (2020, December). Anomaly Based Detection of Cross Site Scripting Attack in Web Applications Using Gradient Boosting Classifier. In *International Conference on Advanced Informatics for Computing Research* (pp. 243-252). Springer, Singapore.

Taha, T. A., & Karabatak, M. (2018, March). A proposed approach for preventing cross-site scripting. In *2018 6th International Symposium on Digital Forensic and Security (ISDFS)* (pp. 1-4). IEEE. 10.1109/ISDFS.2018.8355356

Tian, Z., Luo, C., Qiu, J., Du, X., & Guizani, M. (2019). A distributed deep learning system for web attack detection on edge devices. *IEEE Transactions on Industrial Informatics*, *16*(3), 1963–1971. doi:10.1109/TII.2019.2938778

Zubarev, D., & Skarga-Bandurova, I. (2019, June). Cross-site scripting for graphic data: vulnerabilities and prevention. In *2019 10th International Conference on Dependable Systems, Services and Technologies (DESSERT)* (pp. 154-160). IEEE. 10.1109/DESSERT.2019.8770043

Chapter 3
Demystifying Cyber Crimes

Kritika
Government of India, India

ABSTRACT

With the National Cybercrime Reporting portal witnessing an increase of 15.3% increase in the cyber cases in second quarter of 2022, the post pandemic world has seen a tremendous rise in the number of cybercrime cases. The cases have increased many folds, 125% from 2021, and still continues beyond. The targets of crime are not limited to any age group and innocent children have not been spared. The exposure to online classes, conferences, meetings, etc. has opened the door to criminal activities in a humongous way. Thus, there comes the need for each and every one of us to be well aware of the recent practices that these criminals use and not fall into the clutches of these nefarious cyber criminals.

1. INTRODUCTION TO CYBER CRIME

One of the most significant advancements that has been made in the 21st century which has impacted our lives to a great extent is the advent of internet. Way back from the use of ARPANET in 1969 to the furtherance of Web 3.0, knowledge and technical maneuvering has lived up to it. The way we communicate, play games, work, shop, pay bills without having to wait in long queues, greet pals on their special occasions from far off lands, from landline to video calls, anything and everything we do has made our lives better and comfortable.

But this comes along with a dark side, a much deeper side of the world that many of us are not aware about. With the inception of internet, it has undergone due changes from web 1.0 to latest induction of web 3.0 widely used for many purposes legit and illegit. The modern day technology has made easy access to bundle of

DOI: 10.4018/978-1-6684-8218-6.ch003

knowledge from various sources all across the globe that learning of new skills has become tip of the iceberg. The recent advancements has not only reaped upper hand on technocrats but has also whipped hand on criminal mindset as well.

The methods of crimes have ascended from traditional ways like theft, bank robbery, bribing, blackmailing to modern day crimes *aka* cyber crimes. The indiscriminate exposure to web has augmented these crimes ranging from phishing to ransomware, vishing and many more.

Cyber crime is a term which is most widely used in 21ˢᵗ century defined as any crime administered through the use of computer or any other devices linked to communication like phone, email etc. with the intent to cause fear, anxiety, stress or blackmailing to publish on social media. The severity of crimes has mounted many folds with the increase in daily number of cases being reported on *cybercrime.gov* portal despite conducting a large of awareness programmes for children, youth, parents and senior citizens through various mediums online and offline. The psychology behind ushering these crimes is either greed, revenge or adventure. The Figure 1 represents the comparison between traditional as well as modern day cyber crimes.

Figure 1. Traditional vs. cyber crimes comparison

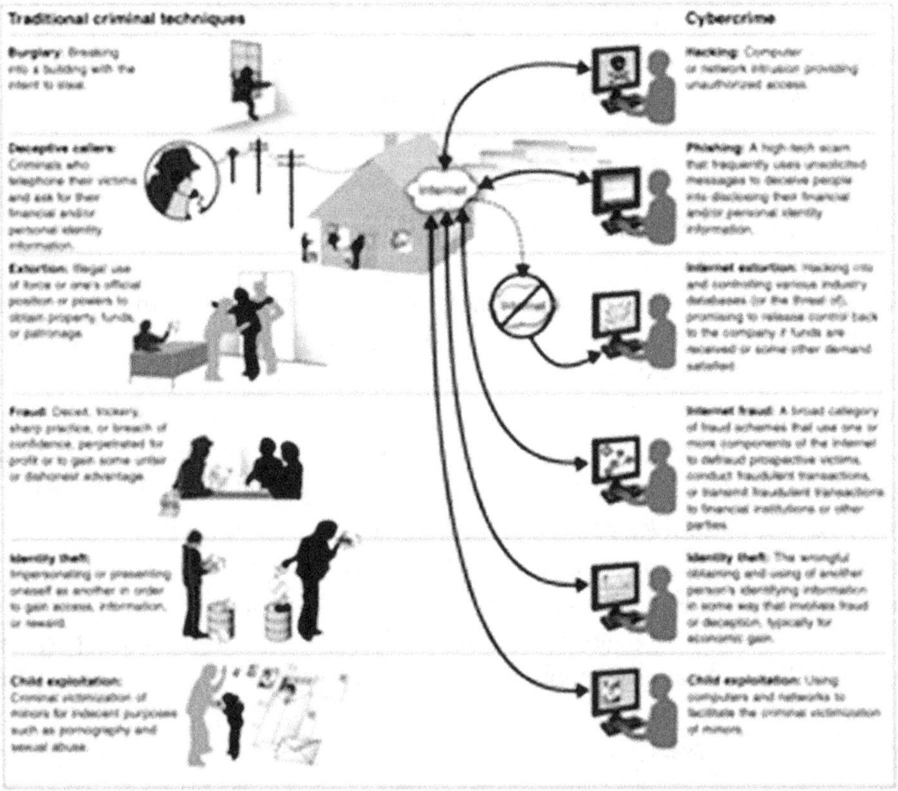

Figure 2. Types of cyber criminals

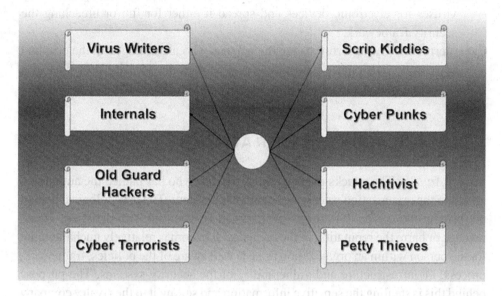

2. TAXOLOGY OF CYBER CRIMINALS

The Figure 2 represents the types of cyber criminals which are detailed as:

Scrip Kiddies: Scrip kiddies are defined as the novice or immature hackers with restricted programming skills and computer knowledge and perform the hacking function with the help of toolkits only. They often tend to perform such functions only for some social media attention.

Cyber Punks: They are advanced technical skills with the competence of creating their own software and are well aware about the telecommunication system they are engaging an attack with.

Internals: They are usually ex-employees or employees engaged in the criminal activity often to take revenge or personal benefit, steal and leak the information to other rivalry agencies.

Petty Thieves: An astute computer professional who takes supremacy of the organisation's poor security infrastructure for personal monetary benefits.

Old Guard Hackers: People who are more interested in opprobrium of a personal property without the pretention of indulging in any sort of criminal activity.

Hachtivist: The criminals who legitimize themselves on account of moral behaviour like associated with political issues, whether the issue is genuinely a political one or not is difficult to assess as it is generally full of revenge, power, greed or public attention.

Virus writers: The people who are engaged in the process of writing malicious viruses for electronic devices and spread it either for fun or breaching the security framework.

Cyber terrorists: Cyber terrorists are the persons who are indulged in the criminal activity on cyberspace who threaten or violate the security protocols of a particular nation and breach their security and blackmail them or spread fear.

3. CLASSIFICATION OF CYBER ATTACKS

Insider Attacks: The attacks made by those persons who has gained the authorized access to the organisation's network or computer system. The psychological behind performing these types of attacks of a person is when he is dissatisfied, greed, revenge or harm the reputation of the company. It becomes relatively unchallenging for the person within an organization as he is well aware of the policies, framework, process, IT infrastructure and vulnerability of the security systems. The purpose behind this is stealing the sensitive information and selling it to the rivalry company for monetary benefits. One of the most effective ways to prevent these types of attack is by installing Intrusion Detection Systems in an organization. The Figure 3 showcases the types of insider attackers.

Figure 3. Types of insider attackers

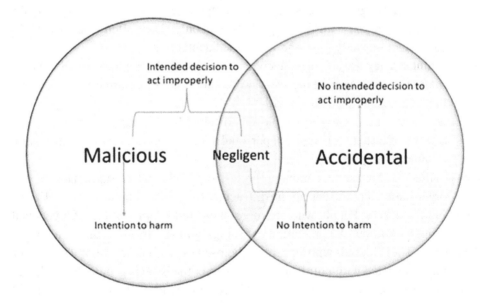

Figure 4. Types of external threat actors

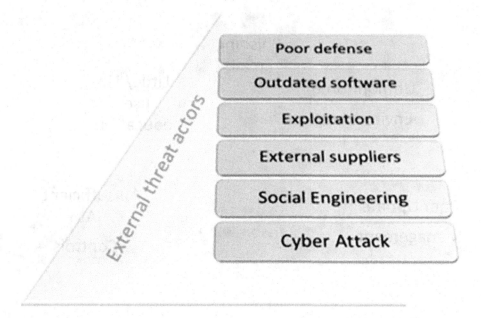

External Attacks: These attacks occur when the assailant is either hired by internal or external member of the organization. The company that becomes the victim to the attack not only faces financial losses but also loss of reputation as the aggressor is someone who is from outside, scanned and gained access to the network security system of the organization which shows a lack of essential security infrastructure as well as manpower. It can be prevented by carefully analyzing the firewall logs of the network. The Figure 4 represents the types of external threat actors.

4. TYPES OF CYBER CRIMES

4.1 Ransomware

An organization or user's access to files on their computer is restricted by encrypting these files and asking for a ransom in return to decrypt it which is the least expensive and quickest way to make money by putting the company's confidential data at stake.

The ransomware attack is on the increase post pandemic period with the tremendous augmentation of internet and mobile banking, online share trading, dissipating of shares and securities and a large number of lay offs which became inevitable in the

Figure 5. Causes of ransomware

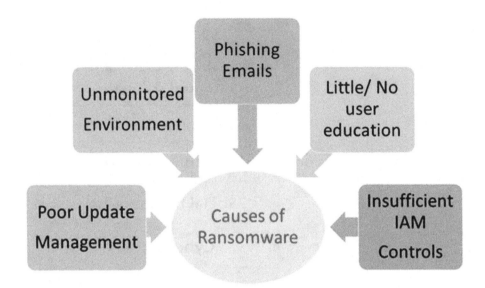

period or cut in salaries due to economy shut down. The recession period has forced people to adopt illegit methods of earning prompt capital. The Figure 5 displays various causes of ransomware.

Example- The Hyderabad Police arrested an unemployed youth for stealing credit card details of the persons lodging in a five star hotel who then used it to make online purchases on various e-market websites.

There are numerable ways to protect from the ransomware attack. Some of them include:

- Back up Data- It is always advisable to follow 3-2-1 rule while backing up the sensitive data. Three separate copies to be considered with two different storage devices and one offline copy with utmost care to be taken to backup data once a day before ending work.
- Keeping all systems up-to-date- It is always suggested to keep the systems or any software being actively used in latest version. The updated version of the software always comes with modern day vulnerability removal and can detect in case of any malicious threat tries to enter system through emails, or pirated software devices. It is always advisable to download the original software from the authentic sites.
- Installation of Anti-virus or Firewalls- Though the modern day devices be it windows, mac or linux comes with an in-built antivirus or firewall system to

Figure 6. Steps to be taken in case of ransomware attack

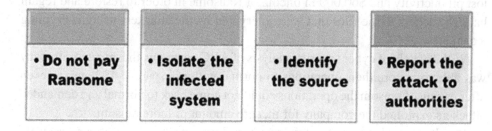

prevent the devices from attack of exploiters. Keeping these up to date will help recover any losses in data if incurred unintentionally also. Eg-Windows Defender

- Network Segmentation- The ransomware attack can spread like a fire through the network usage. Always restrict the number of users and never share the wifi password with every stranger or guest. It is always preferrable to have a guest account built in case to be used by non-authorised persons so that attackers might not get access to the network traffic of the authorized persons(family members).

- Email Protection- The malware attacks through email phishing is the fastest way to spread malware and affect the electronic devices and steal confidential data and cause breach of the same. The site named *"Virus Total"* can be used to scan the links or any documents received via email with the details of sender address and the if it is sent from authentic source or not.

- Application Whitelisting- The appropriate way to find out the authentication or originality of the application to be downloaded. The applications which are listed under whitelist are considered to be dangerous and will be restricted or blocked in case of person so accidentally downloads it.

- Security Awareness Training- The end users and the organisation's employees being the most vulnerable targets of the attackers, a regular routine based training on latest trends and new types of attacks possible must be addressed.

The Figure 6 advices the steps to be taken in case of a ransomware attack.
Illustration:

A construction company XYZ suffered a ransomware attack that infected their backups and internal work stations, completely incapacitating the employees nearly 35 in number for a total of 15 days. The attackers were able to penetrate into the internal security system of the company because of the lack of proper maintenance security infrastructure. The company was fooled of having a back up of the data on the daily basis with multiple secure locations. The attack resulted into $100,000 in lost productivity and $60,000 in bitcoin of Ransome in order to restore and regain back the access to the files that were encrypted by the attackers using encrypting techniques.

With the help of cyber security analysts of ABC company, the infected company was able to restore their operations to normal within a span of 48 hours as even after paying of ransom the operations could not come back to normal and demanded attackers exploited the company for meagre amount of more ransom.

Results were that the operations went back to normal after 48 hours, security system of the company got improved and a more reliable automated data recovery plan was proposed and implemented and training to the professionals of the company was provided after on regular basis.

4.2 Cyber Bullying

Cyber bullying is often termed as e-publicising of servile messages about a person in general children or girls anonymously with the mal-intention to petrify, threaten, mortify, exasperate or aim minor children or girls. This is generally carried out using process of morphing of images or videos and then extorting them in demand of ransom.

It is also a type of "*unlicensed use*" or "*unlicensed access to the facilities*" being provided to the individuals at every level be it home, school or work place. Parents take such crimes to be of trivial nature but can lead to fatal consequences of committing suicides. When a person becomes victim of such crimes, they remain to be forever even if the crime comes to an end but the memory, the fear, the scars rests lifelong.

Example: A boy child aged 14 years was bullied by some seniors in the school due to his shy nature and one of the senior students from the same school published his morphed pictures on the social media due to which he faced a lot of embarrassment and committed suicide.

The Figure 7 represents different forms of cyber bullying prevalent in modern day world.

- *Persistent Cyberbullying:* The widespread accessibility to social media and texting services even to children of 10 years and persistent nature of which making them fall prey to the nostalgic mind of the tormentor gradually resulting in deep depression and anxiety due to constant occurring of such instances.

Figure 7. Forms of cyber bullying

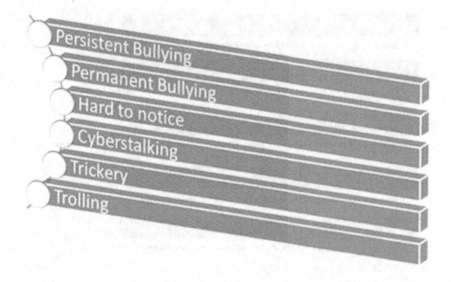

- *Permanent Bullying:* This version of crime occurs on online platforms, creating a lasting digital trail that never disappears from internet platforms haunting the victim in the longer run as it gets spread from one form of social media to another like a fire.
- *Hard to Notice:* The type of cyber bullying which does not attract a lot of intention, making a large majority of professors and parents unaware and unable to identity it timely leading to distant edging of victim. It makes the individual slip to depression or turning into an aggressive behaviour.
- *Cyberstalking:* It is a type of crime in which victim is often stalked either using internet or in-person by deliberately sending e-mails, text messages, social media posts or other mediums with the intention of making wrong accusations, spreading of vindicative rumours. The tasks can be carried out by tracking locations, breaching of confidential data, hate speech etc.
- *Trickery:* The act of duping someone into leaking their personal information without knowledge and publishing it on social media in demand of bribery.
- *Trolling:* The mental harming of an individual by intentionally posting images, comments or posts on social media with no intimate connection with victim.

The Figure 8 represents the ways to protect from cyber bullying.
Illustration:

Figure 8. Ways to protect from cyber bullying

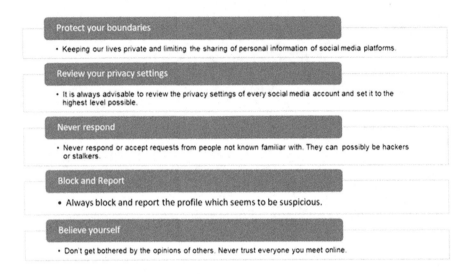

A primary school girl of the age of 10 years during one of the conversations with her teacher confessed that she met a random guy on some social media and added him to her social media account and started chatting. Initially, the boy was behaving nicely and mannerly but later when they got comfortable talking, he forced her to send pictures of her and even to meet up in person. Due to which she got little annoyed and even stated that she had not told her parents yet stating that "as they are not familiar to the use of social media and won't understand the situation" which is one of the biggest mistakes which the children of today's generation make.

Sensing the seriousness of the current situation, the teacher informed the parents of the girl and advised her to share any such incidence if it occurs in future as well. The teacher well on time saved the girl from becoming the victim of cyber bullying and also awakened other girl students of the class as well and conducted an awareness session regarding the same.

The result comes out to be that children should not be given the liberty to use social media account at an early age. If given, pros and cons of the same must be addressed thoroughly and the activities of the child must be monitored on regular basis without making the child feel uncomfortable. Parents are the first teacher to any children and they should take up their responsibility properly.

4.3 Identity Theft

Identity theft is the crime of acquiring personal of monetary information of another person in order to use their identity as an imposter to gain outlawed access for making use of paranoid information for transactions and purchases.

There are two ways in which an identity theft can be performed. First, is the traditional way and second is the modern way using technology. The traditional way of stealing an identity depends on either dumpster diving or shoulder surfing in public like in the organisations working, cafes, ATMs or eavesdropping the conversation between the two individuals. The modern way of stealing the identity of an individual is using the modern day techniques and technology like skimmers for stealing debit or credit card details from the magnetic strip attached to them as it copies the entire details of the person which is electronically stored in that strip. The personal computer plays a very vital role in performing a crime when handed over to a stranger for carrying out his tasks without knowing the real intention of that person who might install a malware using his pen drive. The next time when the person uses his personal device, it might leak all his activities including his banking transactions and a financial fraud can be carried out illegitimately or may share his credentials on social media and use it later for the purpose of black mailing.

Example: An airlines got duped of 17 crore through online booking fraud which was carried out using various credit cards bookings. The details regarding various credit cards were obtained by the hacker from restaurants, shopping malls and petrol pumps employees who swiped their card for transaction at various intervals of the day. The information was misused as the slips were discarded away carelessly by the people whose information got leaked.

The Figure 9 showcases ways to protect from identity theft.
The Figure 10 classifies the types of identity theft.
Illustration:

A scam related to one of the social media sites which is popular among the young generation made the person victim of the cyber crime called identity theft. The attacker hijacked the victim's account and started sending fake messages as dire need of money due to emergency conditions at home as his close family member met with an accident. The people in his friend list started sending money to the link shared by the impersonator without knowing the fact that the scenario created was false. When one of his close friends asked enquired about the same, the person found out that the account has been hacked and by the time he filed a complaint

Figure 9. Ways to protect from identity theft

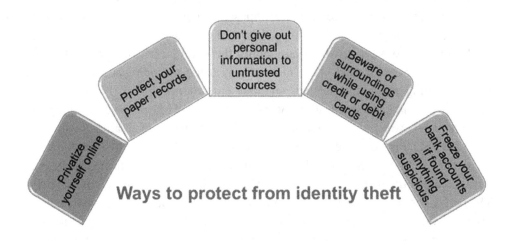

Ways to protect from identity theft

Figure 10. Types of identity theft

Financial Identity Theft	• The use of other person's identity to carry out finance related crimes which includes stealing credit card details, online banking fraud or QR code related scams.
Social Security Identity Theft	• In order to obtain various monetary benefits like medical, credit card, loans, disability, sensitive information is breached.
Medical Identity Theft	• Posing as imposter in order to obtain free medical benefits.
Synthetic Identity Theft	• Combining real and fake information in order to create a new identity to carry on illegit activities like stealing credit card money.
Child Identity Theft	• Slithering of child identity in order to obtain a residence, find employment, obtain loans or any other outstanding warrants.
Tax Identity Theft	• Imposting as someone in order to collect tax return benefits on their behalf.
Criminal Identity Theft	• Criminal posing as someone in order to avoid arrest, summons, or conviction records.

against the same, the impersonator has already de-activated the account and has obtained lakhs of money.

All the incidence took place because of large amount of personal information being displayed on the social media website and the privacy settings which were set to bare minimum exposing all his whereabouts and his friends list.

As a result, the precaution to be taken while using any social media is that share as minimal as possible and don't add strangers to the account, they might be one of the attackers without your knowledge. Always remove the image description like location, time, date, device used before uploading any image to the social media account.

4.4 Email Spoofing

Creating and catapulting an email using a counterfeit sender's address in such a way that the recipients believe the email to be from a trusted source. The intention of the attacker is to gain access to the sensitive information such as bank data or social security numbers or leaking of company's crucial data which has become a frequent practice these days.

Reasons for performing email spoofing:

- **Scamming:** The email containing pretentious and luring particulars like discounts, free tickets, lottery etc. with recipients believing it to be from a trustworthy source falls into a trap in anticipation of getting offers.
- **Injecting Malware:** The easiest and the quickest way to inject malicious software in victim's device in order to steal credentials without the knowledge of a person through files attacked to email.
- **Phishing:** The hackers pretending to be sending emails from a registered banks or organisations to get access to the user's financial statements and pilfer money.

The Figure 11 represents the reasons for email spoofing.

The process of carrying out email spoofing is carried using a fake website(a cloned website of the original one) whose link is hidden in a phishing mail which when clicked by a victim directs him to the fake website which looks authentic at first go and when the data is entered gets stored into the database which is set up by the attacker.

The Figure 12 showcases the process of carrying out email spoofing.

Figure 11. Reasons for email spoofing

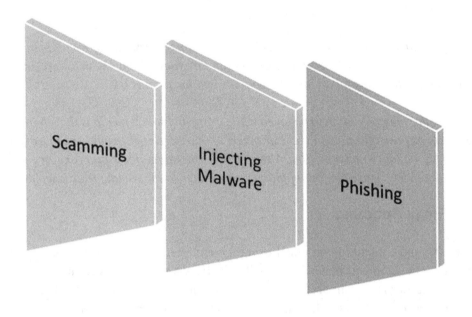

Figure 12. Process of carrying out email spoofing

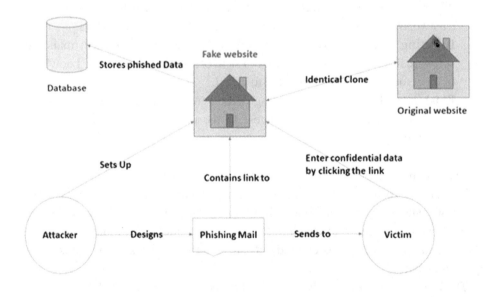

Figure 13. Ways to prevent email spoofing

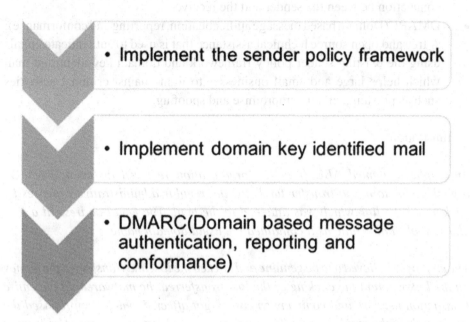

- **Implement the sender policy framework**

- **Implement domain key identified mail**

- **DMARC(Domain based message authentication, reporting and conformance)**

Example: A branch of the bank company suffered email spoofing when a large of bank account holders turned out to withdrew their cash and closed their accounts from the corresponding branch of the bank as they received an email stated that the bank is very bad financial condition and could close its operations in no time. This spoofed mail forced many of the account holders to take the urgent step of withdrawing their money creating chaos in and outside the branch office of the bank.

The Figure 13 represents ways to prevent email spoofing.
There are three ways to protect yourself from the email spoofing:

- Sender Policy Framework: It is an email substantiation mechanism designed to specify which email servers can send mails on behalf of the domain we specify. In order to carry out this process, the individual needs to authorize the machine which can send emails on his behalf and also include few additional information to the existing DNS information.
- Domain key identified mail: The process of encrypting email messages using public key cryptography technique to protect against forgery. It is a type of digital signature that has been attached to the concerned mail sent in order to

boost confidentiality and integrity of sender and build a stronger trustworthy connection between the sender and the receiver.

- DMARC(Domain based message authentication, reporting and conformance): A free and open source technical assistance that is used to authenticate emails using the existing sender policy framework and domain key identified mail which helps large and small businesses to fight against criminal activities such as phishing, email compromise and spoofing.

Illustration:

An employee named ABC from 'X' organization received an email from 'Y' organization stating to transfer funds for the regular administrative expenses by clicking on the link which was attached to the mail. The email address at a first glance looked authentic as it contained the usual domain name.

The employee believed it to be genuine and clicked the link and transferred the money on the link wherein the clicking of the link transferred the malware and the entire transaction network and company's network got affected which compromised the credentials of the employees working in that organization. It also posed a threat to the other network contractors connected with the agency undermining official and public confidence in its reality.

The lesson learnt is that do not attempt to click the links attached to mail luring with offers such as lottery or anything even related to the official matters without checking the credibility of the link and without authorizing the sender's details.

Also, the information posted to contact should be as minimal as possible asking to send the email regarding the reasons stating for why and what for the credentials of the person or organization is required and cross verification should always be carried out before taking further action.

4.5 Forgery

Forgery is a process of producing the false documents or papers that looks to be genuine at first such as visas, passports, cheques, or academic certificates which are generated by the criminals using electronic devices, scanners and sophisticated printers without leaving the trace behind.

Digital forgery is much easier and quicker to carry out than traditional forgery of signature or producing fake documents by stealing as the information in digital form gets stored in either files in electronic devices or on network backup. With the

Figure 14. Types of forgery

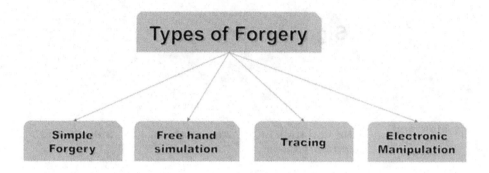

use of more and more technology, people tend to save their transaction passwords, email passwords and other crucial information like adhaar card number, PAN number etc. on auto fill mode which proves to be dangerous whether stored on public or personal devices. The Figure 14 displays types of forgery.

Simple Forgery: Also known as traditional forgery, one of the easiest ways of committing forgery crime as by performing an illegal sign of the authority by copying the signature style or handwriting by passing it as original and authentic.

Free hand simulation: It is the process of collecting a number of samples of the signatures and style of handwriting and recreating the same using free hand technique.

Tracing: The most prominent and effective way of carrying out the forgery by using the tracing technique which exactly matches with the native one making it difficult for the forensic department to exactly rule out the difference between the native and forged signature or document carried out using the process.

Electronic Manipulation: The use of modern day technology such as photoshop had made the forgery of documents or scanning of signatures a walk in the park with even altering the digital texts.

Example: An ex-service personnel of DEF organization forged the signature of the finance manager of the organization in case of release of huge amount of benefits rendered to him after achieving the age of superannuation. Since, he was in close relation to the finance manager of the organization, we forged his signature on a cheque double the amount due to him.

The Figure 15 shows different signs of forgery.

Slow and methodical strokes: When a signature is carried out by the authentic person, it is generally fast paced and stroke free errors. When a signature is carried

Figure 15. Different signs of forgery

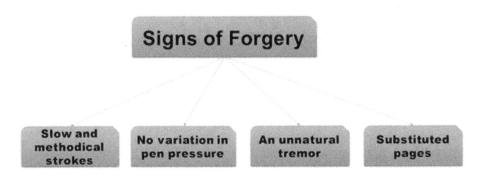

out by the imposter, it is generally slow and there are chances of committing an error stroke on the document he is trying to forge.

No variation in pen pressure: The pen pressure vary with accordance to the speed of writing. The authentic person though signing quickly, the pressure put by the pen is quite less. In the other case, the pressure on the pen put will be likely to be large in order effectuate the forged signature.

An unnatural tremor: Tremors happen when the person is hysterical or is perpetuating for long. A tremor can be caught during the forensics while comparing it with the bona fide due to fear, stress, anxiety or several other factors.

Substituted pages: The characteristic of the substituted page can be revealed by looking at the letters, thickness, size or colour of the page. The substituted page in most of the cases does not tend to match with the pages of the genuine document.

Illustration:

An assistant of the banking firm was caught with the sharing of the huge number of digital certificates of counterfeit representation that looked authentic at first resulting in the transfer of huge sums of money in his bank account with different bank account holder's name. Since, he was the assistant in the accounts division, the digital signature forgery of the highest authority was facile for him to effectuate and commit the crime.

Preventive measures to be carried out to avoid forgery:

- Don't let anyone or everyone access to your confidential data such as Pan card, digital signature etc.
- Use safe browsing measures
- Destroy the confidential papers to avoid it from misuse.

- Secure your devices and don't compromise the security by letting it out to anyone.
- Use secure connections while entering passwords or performing transactions. Avoid using public networks.

4.6 Child Pornography

Child pornography has become one of the humongous pursuits on the cyberspace. There are still thousands of websites present in the cyber world even after banning millions of them. The content whether displayed in adult or child pornography is illegal in most of the nations.

The sharing, downloading and watching of these contents is a heinous crime committed by the nefarious people in order to earn few legal tenders. In the current scenario, since the introduction of cryptocurrency, there is no trace of who sends to whom with the anonymity of the block being created and most widely on dark web.

One of the prime reasons for unwavering such crime is the sexual attraction towards the children of different age groups. They consider themselves as pedophile or hebephile. The other concern regarding this is money laundering or child trafficking.

The Covid-19 pandemic has made children more vulnerable to such crimes as the use of cyberspace has increased multifold due to online classes and online assignments. As a result of which children are more prone to using their electronic devices and hiding it out from parents or guardians as online molesters threaten them in various ways. Urban and sub-urban cities were red flagged and marked as hotspots of child pornography.

Example: A person X was imprisoned for a period of 15 years due to his indulgence in child pornography and was alleged of selling 300 plus videos on different sites. He shared the link using different mediums of social media in order to make few bucks and was caught when one of the individuals filed a complaint against him.

There is a minuscule difference between child abuse and child pornography where child abuse is the act of physically, mentally, psychologically or sexually maltreating the child mostly by their parents or house help. Whereas child pornography refers to personating of the sexual content including images or videos for the purpose of sexual arousal of an individual which results in rape, sexual assault even by the juveniles. Since, the content is freely available to be accessed by anyone from anywhere without confirming the age. The youth pornography is also a crime since it leads to more heinous crimes and sometimes to the situation of suicide.

During the pandemic, there was 95% increase in the traffic on these sites and more than 20% increment in the porn content than before.

Figure 16. Ways to protect your child

Be involved in child's life

Encourage them to speak

Be more vigilant towards your child's activities

Educate them about the good and bad touch

Teach them about the boundaries

Be a friend, guide and philosopher to your child

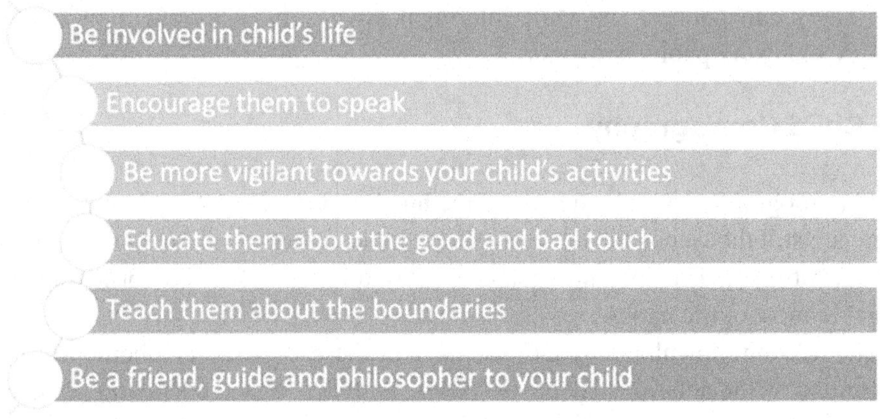

The effects of such disparaging acts would result in low self-esteem, anxiety, depression, lack of confidence, difficulty in making new relationships, fear, prefer isolation than outside world in children. This will make a child become dull, aggressive and will face difficulties in trusting other people.

The Figure 16 shows ways to protect your child from pornography abuse.

Few of the ways to protect your child from abuse or pornography are:

Getting involved in child's life- From the time he gets up till the time he/she enters the bed. Build a comfortable environment for the child and choose caregivers carefully after self-satisfaction and thorough police investigation.

Encourage them to speak- There might be instances when a child refuses to share the things happening in child's life due to fear of getting scolded or being misunderstood which often makes child go into deep depression or anxiety and lack of self-confidence.

Be Vigilant- Always keep a check on the day to day activities of your child. From monitoring his online activities to the child's friend circle and the people he/she might be visiting on daily basis without knowledge. This helps you as a parent to guide your child in right direction and can help him not fall victim to the abusers.

Good and bad touch- Talk to your child about their bodies more often. Tell the difference between the good touch and the bad touch of someone and steps they should take in case they feel uneasy with someone's touch. The first form of child abuse starts with the bad touch which leads to sexual abuse then to a more heinous crime of pornography.

Limit boundaries- Teach your child about limiting the boundaries. Talk about the social media restrictions they need to put. Don't ask to befriend anyone who they meet on social media or even in public. Educate them to stay away from strangers and not accept anything from them or get persuaded by the things they offer.

Be a friend, guide and philosopher- Build a friendly environment for your child where he can easily share matter bothering him and guide him the right path and be available for him in times of need. Take utmost care of child's mental and physical health.

Illustration:

A girl aged 12 years under the peer pressure created a social media account hiding it from parents. In the initial days, the girl added only the people she knew but as the addiction grew, she started adding unknown people and indulging in chat with them. In this scenario, a person aged between 30-32 years sent a request to girl which she accepted and they started having conversations and the girl unknowingly leaked the important credentials like address, school name, place of stay etc. to that person.

Since, the conversation was going on from a long time, the stranger was able to build trust and one day called the girl to meet. She refused at first and then told the entire incident to her parents. The girl luckily escaped from becoming the victim of child pornography.

The parents were shocked to know when she told her the entire incident and they immediately filed the complaint and the person got life time imprisonment and made her aware of such nefarious people in the world.

This would not have happened if the parents would have created a friendly environment and talked to the child on the daily basis about the people she meets and the activities she indulges in. Since, parents are the first teacher to any child, make sure you start as early as possible since the advancement of technology has opened more doors to such criminals.

4.7 Cyber Defamation

Defamation is a term used to cause damage to a person's prestige. The word cyber defamation is more or less the same with the difference that is implemented in a cyberspace. The term cyber defamation correlates to the fact of using someone else's personal information with the false interpretation of it in order to cause harm, injury, damage, inconvenience, obstruction to the person's repute using an electronic device such as computer or mobile phone and uploading or distributing the same on social media to cause harassment, defamation or false acquisition on the individual.

There are two types of defamation which can be caused such as libel(permanent type) i.e. defamation which is jotted down in the form whether printed or digitally or slander(transient type) i.e. defamation which is uttered aloud.

There exists a very minute difference between cyber defamation and freedom of speech as mentioned in Article 19(1)(a) of the Constitution of India. The latter is more implicit than the former which is subject to reasonable restrictions.

There are times when a criminal makes a partially true statement in order to escape from the charges of defamation but that does not guarantee him the right to be escaped as even if a little damage caused to the person's repute, he shall be guilty of the charges of the defamation whether in cyber world or contemporary world.

Example: Publishing of misleading information on social media about someone else in order to cause him shame. A former businessman posted defamatory statement about his colleague stating the latter killed his own child in his drunken state which was fallacious to the obvious statement that his child was killed some 15 years ago in a car accident.

The Figure 17 represents the ways to avoid cyber defamation.

Be aware of your surroundings- Make sure of the place where you are and person with whom are sharing the talks. People often tend to misinterpret and manipulate the way something is said and done. Even a good deed can be represented as taking advantage of the situation.

Control the meaning- Be clear with your words and avoid using ambiguous words or phrases especially when at a senior most position or in the field of media. Words can easily be twisted.

Use of language of opinion- While expressing the opinion, use the defensive mechanism i.e. honest opinion with accurate facts and figures. Instead of directing pointing out anyone use indirect figure of speech like "I think", "she reckons", "they claim" etc. to least open the doors towards false allegations.

Put them together- It is always advisable to put all the facts and figures at one place before making any comment or statement in the public which could otherwise result in false allegations towards a person in order to defame him resulting in the criminal activity.

Act ethically- Always stick to moral code of conduct and ethical conduct when dealing in public at large. Avoid using de meaning words or phrases.

Put the right tier of meaning- When in court use the right level of words. It can be misleading or misinterpreted. Choose the tier of words during an argument to avoid it from being manipulated.

Illustration:

Figure 17. Ways to avoid cyber defamation

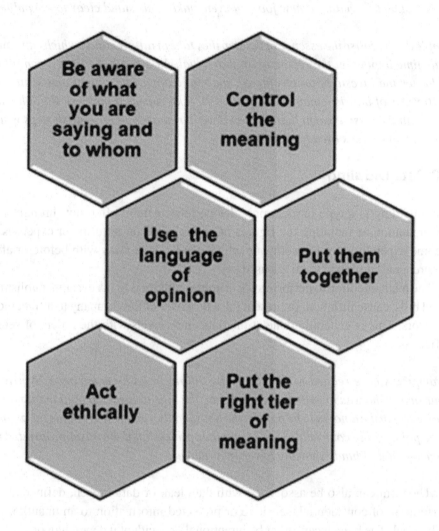

A case related to the rivalry between the corporate team mates led to the defamation of one person towards other in hatred. The two team mates were working on project and seeing the one getting more recognition than other, jealously arouse between the them. The employee A caught hold of some of the old pictures of the employee B which employee A morphed and used to defame him.

Employee A posted the pictures on social media and circulated in the common group of the organization due to which employee B faced a lot of humiliation and *in fact was voluntarily made to resign from the organization. Employee B faced a lot of troubles in getting a new job and even making his stand clear to the public.*

People don't realise the essence of the step they take in little jealousy which can cause a life time impact on the repute of an individual. As a result, he filed a complaint and after the investigation completed, the real culprit was caught and sentenced to 10 years of imprisonment and a fine of 10 lakh rupees. Employee B got his job back with the lesson learnt that don't believe in rumors and take harsh steps which can ruin someone's bread and butter.

4.8 Data Diddling

Data diddling is a type of electronic device related fraud often put through with the pretention of mutating the figures of an organization in terms of expenses or income expenditure of the company which can be done away with before or after the preparation of data to be released.

The main intention to perform such criminal activity is to tax evasion, publishing fraud bills, cause inflation, or present false turn over of the company to attract more investors. This is generally done at the data entry levels with the orders of senior officials.

Example: A case related to the electricity billing fraud by company X. A private contractor who was to collect and distribute the bills to his local area manipulated the figures of the amount to be paid by the households, raising it to double the amount to be paid. As the contractor was a computer professional, he misappropriated the huge amount of funds showing lower remittances.

The term can also be associated with data leak or data breach, defined as the perforation of confidential, sensitive or protected information to an unauthorized individual. The breach can either be intentional or accidental depending on the way the data has been breached from the organization. These are the potential threat actors responsible for dribbling of the information for few bucks or to meet the extraordinary expenses that cannot be met with the existing source of income. To gain little monetary advances risks their life as well as company's organization in hands of nefarious criminals. The Figure 18 represents various threat actors.

Accidental insider: The employees at present or in past, colleagues, business partners or contractors can pose a threat to the insightful information of the organization. The Figure 19 showcases different types of accidental insiders.

Figure 18. Various threat actors

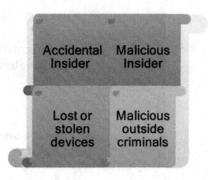

Figure 19. Types of accidental insider threat actors

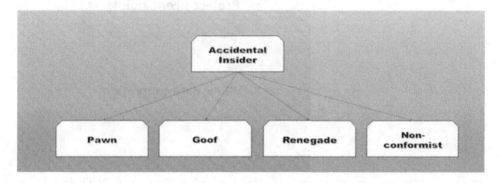

- Pawn: The employees which in literal terms are pawns in the hands of others, who are tricked into performing malicious activities or blow the lid off to the chiselers through social engineering or spear phishing.
- Goof: The employees who believe themselves to be absolved from the security protocols of the organization. Out of comfort or ineptitude, they actively wear out the security breaches and leave vulnerable data in the hands of chiselers.
- Renegade: Renegade are the people from outside the organization such as contractors, cooperators or collaborators who indulge in committing a crime by using their access to decamp intellectual property or client for mere monetary or personal gain.
- Non-Conformist: These are the people who for monetary gains misappropriate the information of the company for their profit. They are more dangerous when highly skilled with access to system administration or database management related tasks.

Figure 20. Ways to protect from data diddling

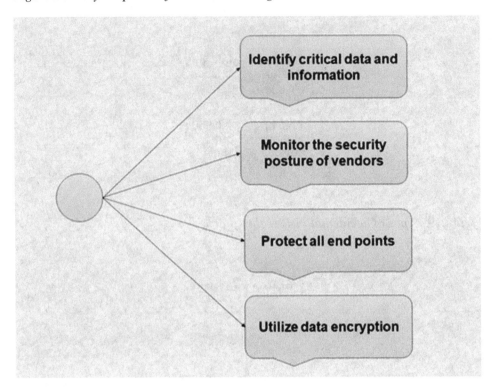

Malicious Insider: The individual intentionally shares the confidential information of the organization to the outsiders or arch rivalries causing harm to company's repute and finances. Such persons generally have the authorized access to the legitimate information of the company which is often used illegitimately.

Lost or Stolen Devices: Any type of electronic device be it laptop, computer, pen drive, hard disk or even the mobile phone, if left in the unlocked state with crucial particulars of whether the organization or personal can be misused by the people with criminal mindset.

Malicious outside criminals: These are the people who use various castigate trackers in order to gather particulars or facts and figures of the company through various sources from lower level of individuals to bribing the senior officials with luxurious and luring offers. The Figure 20 represents ways to protect from data diddling.

Ways to protect from data diddling are:

• Identify critical data and information: The first and foremost step to protect the confidential information from getting breached is premediating where and who has the access to that data. Only authorize people who can be trusted with the information and keep a close watch on their activities.

- Monitor the security posture of vendors: When working with different types of vendors at different levels, always ensure that they follow the best security protection protocols since their security infrastructure is yours as well. The data diddling will not only affect them but you as well.
- Protect all end points: It is always advisable to protect the end points or last location of the data transfer or stored. With the increase in the internet activity, there are multiple end points possible at the same time. Make sure to encrypt all the end points and restrict their usage in case of sharing or transfer of data related to the organization.
- Utilize data encryption: Adopt the latest encryption technique available in the store. Train your employees with the latest updates and make sure all the software used are up to date. It is always advisable to make multiple copies of the data and encrypt it using different encryption-decryption techniques.

Illustration:

A case related to data diddling was reported few months back in which an employee of the organization was responsible for the breach of the sensitive information to the arch rivalry of the organization for few bucks because of which he staked the repute of the organization.

The employee happened to be the head of the system administration of the organization with full access to all the information of the company from salaries to confidential data. Being an expert in his area and with latest updation of his technical skills he was able to decrypt the confidential particulars of the company in no time and for his own profit, sold the information to the arch rival of the company.

A complaint was filed against the breach and was employee found guilty lost his job as well as fined with 15 lakhs of rupees for causing harm to the company's repute and selling the confidential data for few bucks.

The lesson learnt is even the smallest greed can lead the individual to criminal activities, the consequences of which is unpredictable.

4.9 Juice Jacking

Juice jacking is a type of cyber attack in which the USB enabled devices like laptop, computer or mobile phones are compromised. The hackers attain all the information

Figure 21. Types of juice jacking

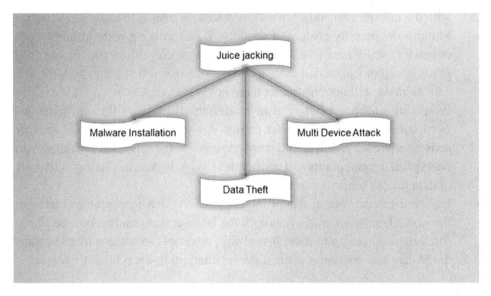

stored in that device when the device is connected to the charging ports. It is more prevalent in airports, metro stations or public places where free charging ports are made available to charge your device.

The features of juice jacking are:

- Simple to use but very effective.
- Does not require any sort of request or permission from the user as the attacker has already pre-installed all the apps in the device.
- No additional software is required to be installed on the devices.
- The users are often not as much aware of the malware attack that takes place in their devices.
- It is not dependent on any single type of platform, compatible with android as well IOS devices.

The Figure 21 shows types of juice jacking techniques.

Malware Installation: The malware gets transferred to the device once the connection is established successfully and remains installed till the time user gets to know about its installation and remove it. The cyber criminals use this method to install various types of malwares such as adware, ransomware, spyware or trojans.

Data Theft: The installation of malware using the juice jacking technique enables the hackers to gain access to the personal credentials of an individual including bank account, passwords, e-mails or security version of device. They can clone all the apps present in the device and can use it for own benefit.

Multi Device Attack: The way of attacking multiple devices at the same time. Once the device is charged by the affected cables, can act as a carrier of the virus to other devices if they come in touch with the affected device.

Example: A person was travelling from United Kingdom to Canada for a business trip. Since, he was continuously using his laptop, it got discharged and in dire need to finish his presentation, he connected his laptop to one of the charging ports available at the airport while waiting for the flight. As soon as he connected his device to the charging port, all his important credentials got breached at the hands of the attacker including his business quotation and the device got cloned without his knowledge due to which he lost the deal.

Ways to protect from juice jacking:

- Always carry power banks for charging mobile phones or laptop. Avoid using public ports for charging.
- Switch off the mobile phone or the device before charging, if there exists an urgent to charge.
- Use specialized cables which are available in the market only meant for charging purposes and not data transfer.
- Use USB condoms which acts as a data blocker.
- Install an anti-virus always.

Illustration:

In the scenario where work from home is common nowadays, people often tend to move to nearby cafes or outlets where they could work as well as spent time meeting new people or gossip over a cup of coffee. These outlets often provide with the free charging ports along with the option to carry out through personal chargers.

An individual forgot to bring his charger as well as charge his phone to the fullest. While using hotspot on his device, his mobile phone ran out of battery, to which he used the free port for charging his device.

As soon as he connected his mobile to the usb charging port, unaware of the juice jacking attacking, within few minutes his phone got cloned and affected with malware. All his credential information including bank related apps, passwords and access to emails all were in the custody of the attacker.

The person came to know about the attack when he lost few bucks from his bank account without his transferring and filed a complaint against the same. Lack of awareness can also be the cause of becoming victim to cyber crime.

4.10 Intellectual Property Crimes

Intellectual Property is the term related to the proprietorship of contemplation or depiction by one who came up with at first. The term connotes in many ways- patent, trademark, layout, trade secrets, copyrights, or industrial designs.

In terms of cyber space, intellectual property crimes are often carried out by breaching the copyright infringement, leaking the designs of one intellect to the other, signing consent without the owner's permission using digital signatures.

Example: Company X sued a person for using the similar domain name to their company deceptively, breaching the trade mark act. The person defended his act by saying that it is only applicable to the goods and not the websites. Since, the case was tilted towards the company, the person was made to pay compensation for the same.

5. PSYCHOLOGY OF CYBER CRIMINALS

The psychology behind the mind of criminals whether in cyberspace or world at large depends largely on scary childhood experiences, mentally stressors, monetary issues, peer pressure, fun time activity or to wreck someone of their assets for proving a degree of oneself.

The ideological perspectives of the motives behind the cyber crime can be correlated to the two studies, namely, Routine Activity Theory(RAT) and Social Learning Theory(SLT). Routine activity theory can broadly correspond to different cyber topologies via, cyber enabled, cyber dependent and cyber related crimes.

Cyber dependent attacks are the ones classified as the attacks with the help of internet technology. Eg- malware, phishing, ransomware etc. Cyber enabled attacks are classified as the integration of traditional attacks along with the use of internet technology. Eg- online fraud, online pornography etc. Cyber related attacks are classified as the content related crimes. Eg- cyber violence, cyber obscenity, cyber bullying etc.

Routine activities theory(RAT) is an addendum of lifestyle exposure theory which makes some people prone to crimes than others. It may include both vocational as well as recreational activities where it is argued that an individual whose daily activities are linked to potential offenders have higher chances of being victimized. It is occupied with two factors viz., actuating offenders and proficient guardianship.

Traditionally, it was used in terms of crimes related to burglary, vandalism and assault. With the advent and advancement of technology, it can be closely related to cyberspace though arduous in terms of contiguity and ephemerality.

In cybercrime, the potentiality to offenders sometimes is determined by the number of friends on social media and ratio of strangers to actual friends. Alluring goals has also been associated with the construct of risk of victimization permeated with the elements of observability, enervation, readiness and assay. *Observation* is defined as the person or object that is in range to the offender being more susceptible. *Readiness* is accessing the aim and escaping the crime scene. *Assay* represent the right assessment of the target either for contentment, marketing or steadfastness. *Enervation* refers to the innate opposition to its elimination.

The other dimension of RAT is the guardianship. Guardianship can be classified in terms of technical, social, behavioral and personal guardianship. *Technical guardianship* refers to the act of using protective software such as anti-virus, firewalls, filtering and blocking software. *Social guardianship* refers to the circle we are surrounded by be it in personal space or professional work force. *Behavioral guardianship* refers to the enlisting of changing passwords and the act of creating passwords. *Personal guardianship* refers to the act of protecting himself/herself from being the victim to the hands of the nefarious people.

The three principles of victimization of RAT highly depends on, namely, in range of the inspirational offender, observability of the favourable aim and incompetent guardianship plays a very vital role in determining the nature of criminality of an individual.

Social learning theory(SLT) refers to the high level swotting conforming to the interactions, social structure and circumstances. The theory suggests the examination of cybercriminal behaviour which is influenced under the dimensions of social control, grasping, industrialization, psychopathologies, class strains and physiological deficiencies.

The circumstances under which an individual may indulge in committing a crime can be broadly categorized as: first, altercations in the relativeness of two individuals who underprop, obligate or template the legit and community-based norms. Secondly, an individual is more prone to digressive models than acknowledging models. Thirdly, the paradoxical behaviour of the individual is held up and fortified at the levy of existing norms.

With the exposure to these environments, offenders acquire delineation by pronouncing, schooling and simulation. The aggressive of these criminals will be enhanced by the unbiasedness of the positive engagement and punishment where the criminal is more prone towards committing the crime and escaping the punishment.

One of the major attributes of committing crimes is the social environment or the peer pressure. Under the pressure, a person is more inclined towards committing crimes such as cyber harassment, cyber stalking, phishing etc.

Example: A rise in the phishing activities being reported from a particular village in state of Jharkhand. The people of that village portray themselves as the bank employees or the volunteers to aware people and trick young and old into leaking their sensitive information like debit card details, bank account details, passwords, adhaar card details etc. Every second person in the village in this profession and in order to prove themselves to be better than the other indulge in more heinous crimes.

The profile of the cyber criminals can be postulated under the following:

- Technical Knowhow: Associated with the level of computer skills and knowledge a person is competent with.
- Personal Traits: The psychological traits of a person inherited from his surroundings.
- Social Characteristics: The social environment influencing the social behaviour of the individual.
- Motivating factors: The inspiration and motivational aspects which turn them into cyber criminals.

REFERENCES

de Jackson, C. (2021). A comprehensive analysis of social learning theory linked to criminal and deviant behaviour. American International Journal of Contemporary Research, 11.

Jeetendra, P. (2017). Introduction to Cyber Security Ahmad Rahayu(2022). A systematic literature review of routine activity theory's applicability in cybercrime. *Journal of Cyber Security and Mobility, 11.*

Prashant, M. (2019). *A Textbook on cybercrimes and penalties.* Snow White India.

Chapter 4
Early Detection of Security Holes in the Network

N. Ambika

https://orcid.org/0000-0003-4452-5514
St. Francis College, India

ABSTRACT

The previous method that has been suggested is to increase the automatic closing of security holes in networks that are vulnerable. This process is the amalgamation of various phases, which begin with collecting information and end with mitigating vulnerabilities. The network's internal domain name is considered input in the proposed method for internal audit purposes. The operational services collect live IPs. Exploits are typically created to gain access to a system, enable the acquisition of administrative privileges, or launch denial-of-service attacks. Each exploit is configured and executed after the list of exploits for each service has been obtained. Whether the endeavor can effectively approach the framework, the moderation step is conjured, checking the sort of access got. The suggestion evaluates the incoming data using the knowledge set stored in its memory. It maintains a table detailing the IP address incomings. The guilty detection is increased by 27.6%, and the security is increased by 36.8% compared to previous work.

INTRODUCTION

5G will entertain users and devices with high mobility in an ultrareliable and affordable manner, enable connectivity of many machines as part of the Internet of Things, and provide ubiquitous broadband services. IP-based communication in 4G has already contributed to creating new business opportunities. 5G is regarded

DOI: 10.4018/978-1-6684-8218-6.ch004

as a brand-new ecosystem that connects nearly all aspects of society, automobiles, household appliances, health care, businesses, etc., to the internet. However, a new set of security flaws and threats (Ambika N., 2020) (Ambika N., 2022) will be introduced due to this development, posing a significant threat to both current and future networks.

The life cycle has four stages (represented in Figure 1)-

- Preparation stage - This phase focuses on the preparation, design, creation, and modification of the network slices. An arrangement of elements and their configuration is referred to as a slice. Content exposure, data leakage, injected malware, and other attacks could result from errors in the network slice template. These attacks could compromise the network's confidentiality, integrity, and authenticity by allowing access to unencrypted channels and user data leakage from the databases. Security measures like encrypting and decrypting the slice template and performing real-time security analysis are necessary to stop these kinds of attacks.

- Installation, configuration and activation stage - Installation of the slices onto the network, the configuration of the services following the request, and activation of the pieces—that is, as ready-to-use software or service— are all part of the second phase of the life cycle. The actual danger in this stage is the production of phony cuts and re-designing the cuts during or before the last actuation. These attacks focus on APIs, which can ultimately impact installation and configuration and result in an activation error in a slice. Safety efforts should be taken to get APIs by giving functional and availability freedoms to the approved individuals and using TLS or O-Auth for confirmation and approval.

- Run time stage - This phase lets you know that the slice is being used and lets you change its requirements, configuration, resource allocation, deallocation, and network functions. The objectives of assaults in this stage are regulators, hypervisors, the general cloud framework, control channels, and bringing together control components. However, attacks continue to primarily target APIs. Performance attacks, privacy breaches, and data exposure are all types of attacks. The safety efforts that should be taken care of validation and uprightness of the organization cuts to forestall counterfeit solicitations, cut seclusion to forestall DoS and DDoS assaults, and secure-5G displaying to forestall availability of unapproved and pernicious solicitations. Additionally, dynamic NFV can be utilized, which provides a security mechanism on demand.

- Deactivation stage - It is the last period of the organization's cut life cycle, in which the assets and organization's capabilities feel better. The slice

Figure 1. Life cycle of slice and the associated threats
(Dangi, et al., 2022)

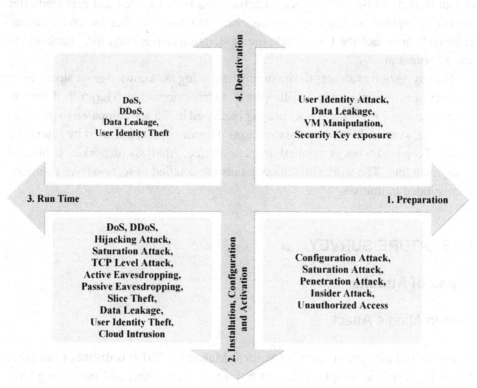

has been removed from use. Even after decommissioning the portion, the most common threats that can harm the network are improper resource and network function use and slice handling. User information databases, cloud storage, and centralized control elements are the targets of these attacks. To avoid attacks during this phase, resources and network functions no longer in use should be appropriately redirected, and sensitive data no longer required should be deleted.

The method (Filiol, Mercaldo, & Santone, 2021) that has been suggested is to increase the automatic closing of security holes in networks that are vulnerable. This process is the amalgamation of various phases, which begin with collecting information and end with mitigating vulnerabilities. The network's internal domain name is considered as input in the proposed method for internal audit purposes. A scan of all IPs in the class of the network domain is carried out to obtain IP addresses that are still active once the internal domain name that it will analyze has been received. The operational services collect live IPs. Exploits are typically created to gain access to a system, enable the acquisition of administrative privileges, or launch denial-of-service attacks. Each exploit is configured and executed after the list of exploits for each service has been obtained. Whether the endeavor can effectively approach the framework, the moderation step is conjured, checking the sort of access got.

The suggestion evaluates the incoming data using the knowledge set stored in its memory. It maintains a table detailing the IP address incomings. The guilty detection is increased by 27.6% and the security is increased by 36.8% compared to previous work. The work is divided into six sections. Introduction is followed by Literature survey. Proposed work is detailed in section three. Analysis of work is explained in segment four. The work simulation details are detailed in section five. The work is concluded in unit six.

LITERATURE SURVEY

Types of Attacks

Man-in-Middle Attack

The proposed mitigation method (Amin & Mahamud, 2019) is distinct from other forms because it is simple, does not rely on cryptography, and saves long-term memory in a different file. When an ARP poisoning attack occurs, the user will

Figure 2. Man-in-middle attack
(Liu, Crespo, & Martínez, 2020)

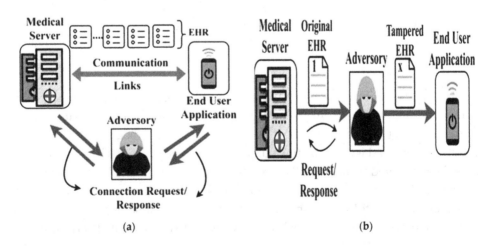

(a)   (b)

receive an alert from the bash script. The mitigation procedure will require two other scripts. The first script will run once and save the default gateway's MAC address and IP address in a user-defined file. The script will end after saving the data to the desired file. In the background, the second script will always be running. After requesting a file containing the default gateway IP and MAC for matching purposes at the beginning of script 2, the remainder of the process will proceed.

The partial key for communicating pairs is generated by a single centralized server, Scheme's "Partial Key Generator." (Pal, Saxena, Saquib, & Menezes, 2011) Using the master key and the node's identity, the First Partial Key Generator generates the secret key for any node. The shared key between PKG and the node for which this private key is generated is the generated personal key. It manually distributes the secret keys to the appropriate nodes, following each node's remote key generation. It communicates with the generator. Each node stores its private key. It never keeps any node's secret key. The generator calculates the secret key at runtime using the node's identity and its master key if any node requires it. The random number sent by the generator to create the shared key is the partial key of any node. Before shipping, it encrypts the partial key using the secret key of the receiving node.

A Wireless Public Key Infrastructure (WPKI)-based Security Enhanced Authentication and Key Agreement (SE-EPS AKA) (Li & Wang, 2011) is the basis for the recommendation. During the subscriber's first registration, network access, or when the SN cannot resume the IMSI using the temporary mobile subscriber identity, the SN must request the UE to send the IMSI in plain text. As a result, the IMSI risks being leaked, which could lead to attacks like "man-in-the-middle attack," business tracking, and subscriber location, among others. The authentication vector (AV) sent between HSS and MME in plain text can be easily intercepted due to the wired link's lack of protection. To accomplish mutual authentication between HSS and UE and to generate the session cipher key, the protocol continues to be based on a symmetric key cryptosystem. While the security of the long-term shared cipher key is dependent on the network and susceptible to leakage, the subsequent local communication's security is significantly compromised. The protocol ignores the service network identity (SNID) protection. Because the SNID is transmitted in plain text over both wired and air interfaces, an attacker can easily eavesdrop on the legal SNID to positively initiate attacks such as network fraud or pseudo base station. It cannot provide incontestable business because the protocol does not support digital signatures.

Brute Force Attack

A brute force attack is a trial-and-error attempt to crack a username or password. The essential tools are dictionaries. Brute-force attacks work better when a Central Processing Unit and a Graphics Processing Unit are used together.

Five network entities comprise the work (Ferrag & Maglaras, DeliveryCoin: An IDS and Blockchain-Based Delivery Framework for Drone-Delivered Services., 2019): Autonomous vehicle, Macro eNB, package buyer, package vendor, and package delivery service. It distinguishes self-driving organization assaults as well as bogus exchanges between self-driving hubs. The Strong Diffie–Hellman assumption in bilinear groups is absent from the proposed scheme, which uses short signatures, hash functions, and no random oracles. Within the blockchain-based delivery platform, the consensus is reached by a UAV-aided forwarding mechanism.

Denial-of-service attack

OpenStack is a well-known open-source cloud platform used (Köksal, Dalveren, Maiga, & Kara, 2021) as an NFVI–VIM component and meets the requirements of 5G NFV networks. OpenStack creates VMs for every VNF. Two distinct organizations are made on OpenStack to interconnect different VMs together. Traffic generators come in two varieties. The first is a standard DNS traffic generator for making typical DNS queries that can be made in the 5G network. It produces a scaled measure of

Figure 3. Threat model
(Ferrag & Maglaras, DeliveryCoin: An IDS and Blockchain-Based Delivery Framework for Drone-Delivered Services., 2019)

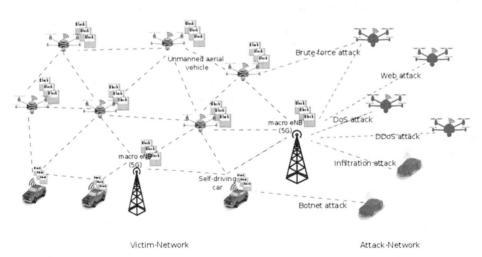

Figure 4. Proposed work
(Ferrag, Shu, Djallel, & Choo, 2021)

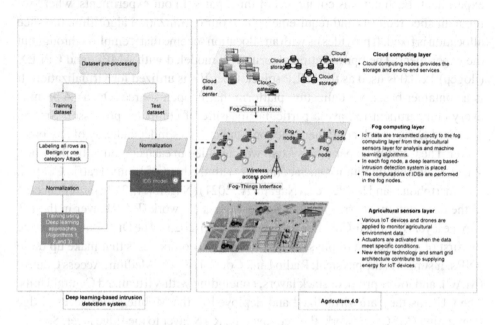

traffic that can go through a solitary server in genuine 5G organizations. Here, the DNSPerf instrument is utilized to make such traffic. The second is a DDoS traffic generator that aims to consume VNF resources by producing bad DNS traffic. Using the NetSniff networking toolkit, bad DNS traffic is generated. Both tools utilize Packet Capture API (PCAP) files containing numerous DNS queries. The creators of these tools create these PCAP files. The DNS server will not respond to the actual DNS queries because it is unable to respond to all of these queries. The return of DNS query responses to the client is yet another load balancer function. The IP addresses and port numbers of IPS VMs are stored in the load balancer. In this case, the NGINX tool is used to load balance the DNS queries. NGINX can load balance UDP packets, which are generally the basis of DNS queries. The goal of a load balancer is always to maintain the same load on all IPS VMs. Concerning the test environment, the number of packets of regular incoming traffic and DDoS attack traffic has been reduced.

A mathematical model is used in the proposal (Sattar & Matrawy, 2019) to guarantee end-to-end delay for 5G core network slices and on-demand slice isolation. It uses interslice separation to give greater unwavering quality and better accessibility, between cut disengagement for giving segregation between cuts. A six-server testbed was developed to evaluate the work. The following are the hardware requirements for PS1, PS2, PS3, A1, and A2: 8GB of RAM, a 2.50GHz (4 cores) Intel Xeon(R)

CPU E5420, and 100Mb/s network bandwidth. We use 12 core network slices in each experiment. Each cut was comprised of three parts. In our experiments, where we input all slice requests and requirements into our optimization algorithm, the slice allocation is fixed. It provides us with an allocation scheme that it employs throughout the experiment. The optimization algorithm is modeled with AMPL, and CPLEX (ilogcp) 12.8.0 is used as the MILP solver. OpenVZ is utilized for virtualization. It is a container-based virtualization platform that is open source. OpenVZ permits every compartment to have a particular measure of computer processor, Smash, and Hard Drive. Each container functions and operates independently of any other server. The CentOS 6 operating system was installed in each container by the work. We allocated bandwidth for each container with the help of Linux Traffic Control.

Smartphones and IoT devices (Nagaraj A., 2021) (Nagaraj A., 2022) are connected to the base stations' corresponding Radio Access Network (RAN) layer in the IoT Device layer (Benlloch-Caballero, Wang, & Calero, 2023). The Distributed Units are the first of a collection of physical and virtualized components that make up these gNBs. It supports the physical, Radio Link Control (RLC), Medium Access Control (MAC), and lower protocol stack layers, concluding with Virtualized Central Units. The CUs, as they are virtualized and deployed in the Mobile/Multi-access Edge Computing (MEC) Network, then connect the RAN layer to the Edge layer. Session Management Function (SMF), User Plane Function (UPF), and Access Management Function (AMF) are all provided by the 5G/6G Core layer. Virtualization has been applied to these tasks. As in the Edge layer, cloudified functions in the Core layer can provide Cloud as a Service application that runs in the infrastructure's core network and is centralized regarding resources and computational power. Either a DSP or an ISP can execute individual shut control circles with practically no human mediation to completely robotize the reaction against digital assaults in their separate regulatory spaces. The DUs have been responsible for encapsulating the raw data from the devices using the Common Public Radio Interface (CPRI) protocol so that it can be correctly routed by the ISP's physical devices. as soon as traffic reaches the Edge Computing layer. This paper's self-protection close control loop is made up of various software components that are architecturally chained together to form a tight circle. Each of these parts has a particular undertaking, and the blend of every one of them brings about the precise identification of an assault, the investigation of its goal, and the resulting activity against the danger. Message Bus software is responsible for facilitating communication between each component. Information about the topology of all network devices, ports, and connections between ports and devices that are accessible on each machine is provided by the Resource Inventory Agent (RIA) component. The primary objective of developing the SMA component has been to enhance and extend the capabilities of a conventional IDS. This Snort IDS and a 5G multi-tenant traffic classifier that we developed make up the SMA, which enables the generation of concrete, granular, and effective alerts and the collection of pertinent information about the tenants of 5G networks. The Analyser is the first component to be used in the management layer. The metrics reported by the previous segment (SMA) to the Metrics Exchange are the subject of its analysis. The component in charge of making a decision is called the Decision Maker. The creation of a plan is the responsibility of the Planner component. The information provided by the Decision is extended in this plan to produce a set of actions that can be put into the existing system. The orchestrator is in charge of putting the plan the planner made up earlier into action. The FCA component is an agent whose primary purpose is to make network traffic control visible. An FCA agent is associated with and installed on each of the infrastructure's computers, allowing for control of network traffic between physical and virtual machines.

Malware

The work (Liu, Li, Long, & Bilal, 2023) is another graphical brain network model Organization Traffic Chart Organization Brain (NT-GNN), for malware identification that integrates network traffic. To gather the organization traffic made by the execution of the harmful program, we introduce and execute the Apk application in a genuine climate and catch the organization traffic it makes after some time with projects

Figure 5. Proposed work
(Liu, Li, Long, & Bilal, 2023)

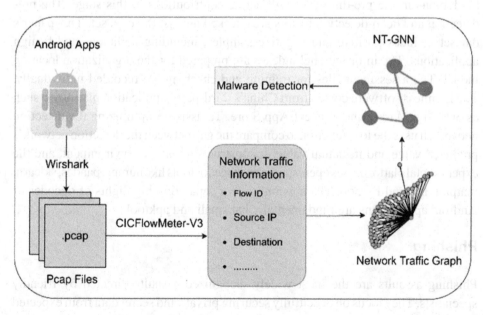

like CICFlowMeter-V3. A feature vector that describes the time-varying network traffic between two endpoints is attached to each flow F. We select the entire data set to input into the model after cleaning the data by removing zeros and nulls. When the edges represent the methods of communication between these endpoints and the nodes of the traffic graph, which represent the network endpoints, the traffic graph can be extracted from each pcap traffic packet's numerous pieces of traffic. After training in a supervised anomaly detection environment and learning relevant concepts from the training data, it can classify new graphs with binary class labels. Based on the unique properties of the node, the messaging–passing phase creates and transmits information following the network's structure. Amassing neighbor data and refreshing state data are the following two assignments in this stage. To deploy NT-GNN, it utilized a Windows 10-based operating system, an Intel(R) i7-11700 CPU, 32 GB of RAM, and a GeForce RTXTM 3090 Ti GPU. Because the neural network model has a lot of data, the GPU is used to speed up the training process. The CICAndMal2017 dataset was created by running malicious and benign smartphone applications. It allowed us to avoid the behavior of more advanced malware samples, which change their behavior to produce incorrect results when they recognize the emulator environment at runtime. Figure 5 portrays the same.

There are two parts to DLAMD (Lu, et al., 2021). Pre-detection is the first step, and its purpose is to exclude malware with apparent characteristics from the initial rapid detection of various Android applications. Deep detection takes a lot less time and costs less to compute, and the framework is more effective at detection as a result. Deep detection is the second phase. Applications that are decided to be dubious in the pre-discovery will act as contributions to this stage. The pre-detection and deep-detection results are checked against the data set. This paper's dataset includes positive and negative samples, including malicious and benign applications. The informational indexes are prepared for the organization location model. The assessment files for training and checking are recorded individually. The harmless software comes from China's third-party application platforms, such as 360 App Market and Tencent App Store. To assist in optimizing the detection network, it uses the loss function to compare the gap between the detection network's predicted value and its actual value. It constructs a realistic environment, and the experimental platform uses open-source packages or tools like numpy, pandas, sklearn, matplobtlib, and Python. The apparatuses for separating highlights of opcodes of Android applications are fundamentally baksmali and apktool.

Phishing

Phishing assaults are the most widely recognized assaults directed by friendly specialists. They focus on deceitfully securing private and secret data from expected

Figure 6. Phishing attacks
(Salahdine & Kaabouch, 2019)

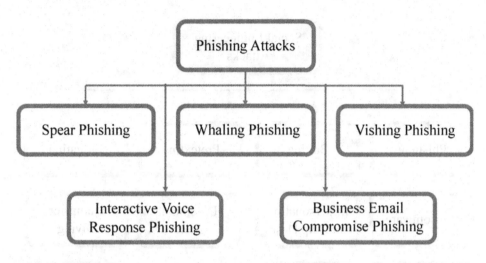

targets utilizing calls or messages. Assailants misdirect casualties to acquire delicate and personal data. Fake websites, emails, advertisements, anti-virus software, scareware, PayPal websites, awards, and freebies are among them.

Social Engineering Attack

In social-based attacks, perpetrators exploit the victims' psychological and emotional states by developing relationships. These assaults are the most challenging and effective as they include human collaborations. Social engineering attacks may combine the previously discussed aspects, such as: human, PC, specialized, social, and physical-based. Depending on different points of view, they can be divided into several categories. They can be divided into two groups based on the involved entity: software or a person. Depending on how the attack is carried out, they can be broken down into three categories: social, specialized, and physical-based assaults. Through breaking down the different existing arrangements of the social designing assaults, likewise group these assaults into two principal classifications: both direct and indirect. The first kind of attacks use direct contact between the attacker and the victim to carry out their attacks. They allude to assaults performed using actual contact, eye-to-eye connections, or voice collaborations. They might also need the attacker in the victim's working area to carry out the attack.

Figure 7. Social engineering attacks
(Salahdine & Kaabouch, 2019)

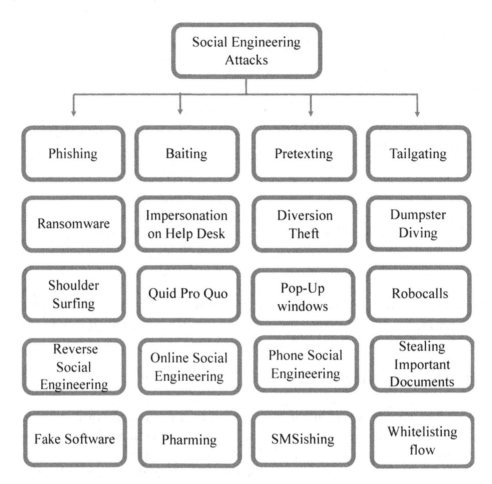

PROPOSED WORK

Table 1. Notations used in the work

Notations	Description
N	Network under consideration
D_i	Device
L_i	Location of the device
S	Server
id_i	Identity of Device D_i

Assumptions

- The devices maintain the knowledge hub encompassing a set of characteristics and behaviour of normal and compromised IP's.
- The location of the devices mapping is maintained in the evaluation nodes (server).
- The devices maintain their observations w.r.t the input received from various devices. The server examines the collected observations and distributes on positive outcome.

Registration Phase

The device registers itself by sharing its information with the subscriber (server). In the equation (1), Device D_i is sharing its identification id_i and Location information L_i with the server S.

$$D_i : id_i \parallel L_i \rightarrow S \tag{1}$$

Knowledge Updating Phase

The server updates the device with the knowledge set. In the equation (2), the server S is updating knowledge K_i to the device D_i.

$$S : K_i \rightarrow D_i \tag{2}$$

IP Address Evaluation Phase

The device evaluates the incoming data using the knowledge set stored in its memory. It maintains a table detailing the IP address incomings. Table 2 details the same.

Table 2. Parameters evaluated

Parameters evaluated	Measured quantity (example)
Identification of the communicating device	D_j
Amount of transmitted data	A_i
Time duration of transmission	T_i

ANALYSIS OF WORK

The previous method (Filiol, Mercaldo, & Santone, 2021) that has been suggested is to increase the automatic closing of security holes in networks that are vulnerable. This process is the amalgamation of various phases, which begin with collecting information and end with mitigating vulnerabilities. The network's internal domain name is considered as input in the proposed method for internal audit purposes. A scan of all IPs in the class of the network domain is carried out to obtain IP addresses that are still active once the internal domain name that it will analyze has been received. The operational services collect live IPs. Exploits are typically created to gain access to a system, enable the acquisition of administrative privileges, or launch denial-of-service attacks. Each exploit is configured and executed after the list of exploits for each service has been obtained. Whether the endeavor can effectively approach the framework, the moderation step is conjured, checking the sort of access got. The suggestion evaluates the incoming data using the knowledge set stored in its memory. It maintains a table detailing the IP address incomings.

SIMULATION

The work is simulated using NS2. Table 3 portrays the parameters used in the system.

Table 3. Parameters used in simulation

Parameters used in simulation	Description
Dimension of the network	200m * 200m
No of devices installed	12
No of evaluators	4
Amount of data sent by devices	274 bits
Length of identification of devices	24 bits
Length of location information	32 bits
Length of knowledge bits transmitted	700 bits (with interval)
Time duration of simulation	60m

EARLY DETECTION

The suggestion keeps a track of all the activities of the IP address. Hence the malicious activity can be traced at early stage. The detection is increased by 27.6% compared to previous work. Figure 7 portrays the same.

Figure 8. Early detection of compromised nodes

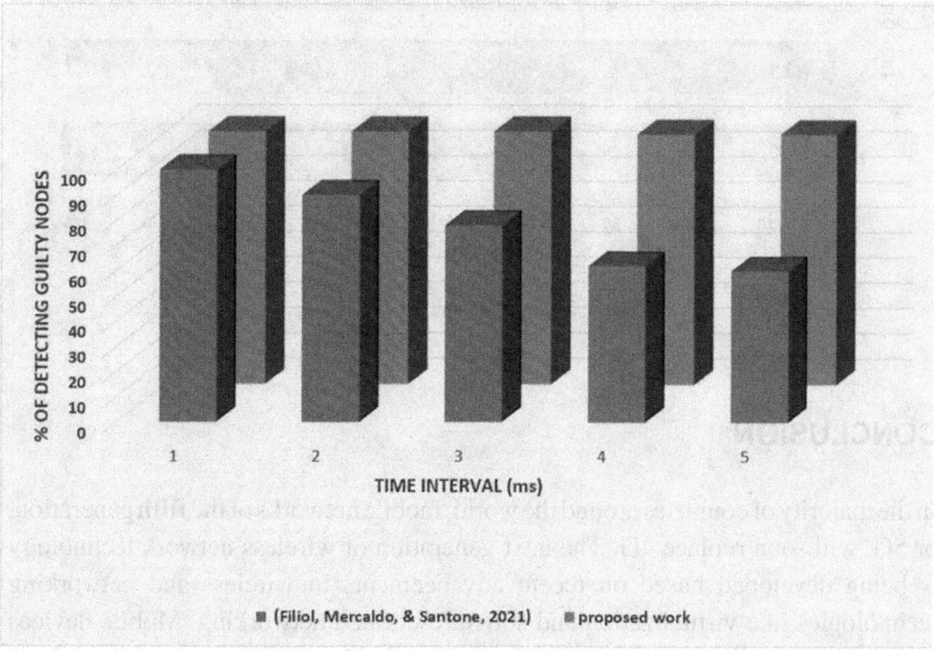

SECURITY

Detection of compromised nodes increase security in the system. As the devices are able to detect the guilty nodes at early stage, security of the system increases. The security is increased by 36.8% compared to previous work. Figure 8 represents the same.

Figure 9. Security in the system

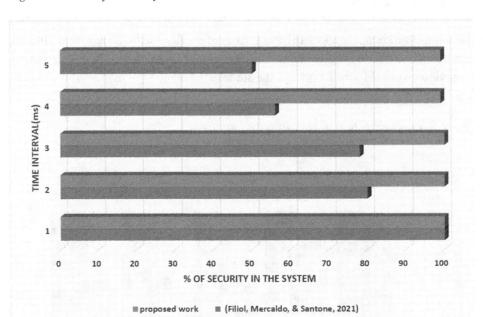

CONCLUSION

In the majority of countries around the world, mobile networks of the fifth generation, or 5G, will soon replace 4G. The next generation of wireless network technology is being developed based on recent advancements in wireless and networking technologies like virtualization and software-defined networking. Mobile devices will be able to switch between providers and technologies to maintain a high level of Quality of Service in a 5G environment where multiple wireless service providers share an IP-based core network. Through vertical handover, mobile devices will acquire a concept of social nodes because they will be constantly connected to the network. These nodes are more susceptible to impersonation, eavesdropping, man-in-the-middle, denial-of-service, replay, and repudiation attacks because they are easier to locate.

The previous method that has been suggested is to increase the automatic closing of security holes in networks that are vulnerable. This process is the amalgamation of various phases, which begin with collecting information and end with mitigating vulnerabilities. The network's internal domain name is considered as input in the proposed method for internal audit purposes. A scan of all IPs in the class of the network domain is carried out to obtain IP addresses that are still active once the internal domain name that it will analyze has been received. The operational services collect live IPs. Exploits are typically created to gain access to a system, enable the acquisition of administrative privileges, or launch denial-of-service attacks. Each exploit is configured and executed after the list of exploits for each service has been obtained. Whether the endeavor can effectively approach the framework, the moderation step is conjured, checking the sort of access got. The suggestion evaluates the incoming data using the knowledge set stored in its memory. It maintains a table detailing the IP address incomings. The guilty detection is increased by 27.6% and the security is increased by 36.8% compared to previous work.

REFERENCES

Ambika, N. (2020). Improved Methodology to Detect Advanced Persistent Threat Attacks. In N. K. Chaubey & B. B. Prajapati (Eds.), *Quantum Cryptography and the Future of Cyber Security* (pp. 184–202). IGI Global. doi:10.4018/978-1-7998-2253-0.ch009

Ambika, N. (2022). Minimum Prediction Error at an Early Stage in Darknet Analysis. In *Dark Web Pattern Recognition and Crime Analysis Using Machine Intelligence* (pp. 18–30). IGI Global. doi:10.4018/978-1-6684-3942-5.ch002

Amin, A. M., & Mahamud, M. S. (2019). An alternative approach of mitigating arp based man-in-the-middle attack using client site bash script. *6th International Conference on Electrical and Electronics Engineering (ICEEE)* (pp. 112-115). Istanbul, Turkey: IEEE. 10.1109/ICEEE2019.2019.00029

Benlloch-Caballero, P., Wang, Q., & Calero, J. M. (2023). Distributed dual-layer autonomous closed loops for self-protection of 5G/6G IoT networks from distributed denial of service attacks. *Computer Networks*, *222*, 109526. doi:10.1016/j.comnet.2022.109526

Dangi, R., Jadhav, A., Choudhary, G., Dragoni, N., Mishra, M., & Lalwani, P. (2022). ML-Based 5G Network Slicing Security: A Comprehensive Survey. *Future Internet*, *14*(4), 116. doi:10.3390/fi14040116

Ferrag, M., & Maglaras, L. (2019). DeliveryCoin: An IDS and Blockchain-Based Delivery Framework for Drone-Delivered Services. *Computers*, *8*(3), 58. doi:10.3390/computers8030058

Ferrag, M., Shu, L., Djallel, H., & Choo, K.-K. (2021). Deep Learning-Based Intrusion Detection for Distributed Denial of Service Attack in Agriculture 4.0. *Electronics (Basel)*, *10*(11), 1257. doi:10.3390/electronics10111257

Filiol, E., Mercaldo, F., & Santone, A. (2021). A method for automatic penetration testing and mitigation: A red hat approach. *Knowledge-Based and Intelligent Information & Engineering Systems: Proceedings of the 25th International Conference KES2021.* 192, pp. 2039-2046. Szczecin, Poland: ELSEVIER.

Köksal, S., Dalveren, Y., Maiga, B., & Kara, A. (2021). Distributed denial-of-service attack mitigation in network functions virtualization-based 5G networks using management and orchestration. *International Journal of Communication Systems, 34*(9), e4825. doi:10.1002/dac.4825

Li, X., & Wang, Y. (2011). Security enhanced authentication and key agreement protocol for LTE/SAE network. *7th International Conference on Wireless Communications, Networking and Mobile Computing* (pp. 1-4). Wuhan, China: IEEE. 10.1109/wicom.2011.6040169

Liu, H., Crespo, R., & Martínez, O. (2020). Enhancing Privacy and Data Security across Healthcare Applications Using Blockchain and Distributed Ledger Concepts. *Health Care, 8*, 243. PMID:32751325

Liu, T., Li, Z., Long, H., & Bilal, A. (2023). NT-GNN: Network Traffic Graph for 5G Mobile IoT Android Malware Detection. *Electronics (Basel), 12*(4), 789. doi:10.3390/electronics12040789

Lu, N., Li, D., Shi, W., Vijayakumar, P., Piccialli, F., & Chang, V. (2021). An efficient combined deep neural network based malware detection framework in 5G environment. *Computer Networks, 189*, 107932. doi:10.1016/j.comnet.2021.107932

Nagaraj, A. (2021). Introduction to Sensors in IoT and Cloud Computing Applications. UAE: Bentham Science Publishers. doi:10.2174/97898114793591210101

Nagaraj, A. (2022). Adapting Blockchain for Energy Constrained IoT in Healthcare Environment. In K. Kaushik, S. Tayal, S. Dahiya, & A. O. Salau (Eds.), *Sustainable and Advanced Applications of Blockchain in Smart Computational Technologies* (p. 103). CRC press. doi:10.1201/9781003193425-7

Pal, O., Saxena, A., Saquib, Z., & Menezes, B. L. (2011). Secure Identity-Based Key Establishment Protocol. *International Conference on Advances in Communication, Network, and Computing* (pp. 618-623). Bangalore, India: Springer, Berlin, Heidelberg.

Salahdine, F., & Kaabouch, N. (2019). Social Engineering Attacks: A Survey. *Future Internet, 11*(4), 89. doi:10.3390/fi11040089

Sattar, D., & Matrawy, A. (2019). Towards secure slicing: Using slice isolation to mitigate DDoS attacks on 5G core network slices. *IEEE Conference on Communications and Network Security (CNS)* (pp. 82-90). Washington, DC, USA: IEEE. 10.1109/CNS.2019.8802852

Chapter 5
Introduction to Dark Web

Qasem Abu Al-Haija
iD https://orcid.org/0000-0003-2422-0297
Princess Sumaya University for Technology, Jordan

Rahmeh Ibrahim
Princess Sumaya University for Technology, Jordan

ABSTRACT

Darknet is an overlay portion of the Internet network that can only be accessed using specific authorization using distinctively tailored communication protocols. Attackers usually exploit the darknet to threaten several world-wile users with different types of attack/intrusion vectors. In this chapter, we shed the light on the darknet network, concepts, elements, structure, and other aspects of darknet utilization. Specifically, this chapter will extend the elaboration on the darknet, the dark web components, the dark web access methods, the anonymity and confidentiality of the dark web, the dark web crimes, the cyber-attacks on the dark web, the malware on the dark web, the internet governance of dark web, the payment on the dark web, and the impacts of the dark web. This chapter will profound the knowledge of the dark web and provides more insights to readers about darknet attacks, malware, and their counter-measures.

1. INTRODUCTION

As a result of technological developments brought on by digitalization, a wide variety of assault types have emerged. As more people rely on the internet to satisfy their needs, web security has become a significant concern. In the late 1990s, Cyberspace flourished, and it arose to transform many aspects of life globally. You

DOI: 10.4018/978-1-6684-8218-6.ch005

Figure 1. The overly network

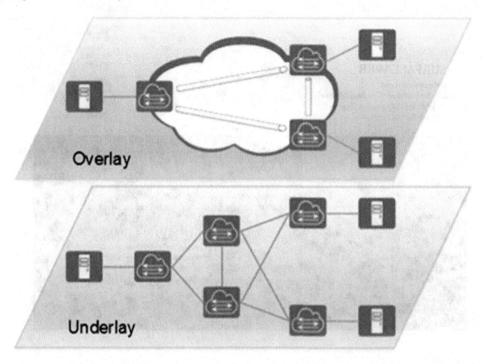

can communicate with anyone online if you have an Internet connection. The main issue is that privacy and anonymity should have been considered when designing the Internet. Therefore, everything can be traced or tracked. One of them is the US Federal Government, which had serious privacy concerns in the middle of the 1990s.

For the US Navy's Naval Research Laboratory, a consortium of statisticians and computer researchers developed the novel technique known as "Onion Routing" (NRL). To enable anonymous, two-way communication, it is possible to build an overlay network (Ibrahim et al., 2022) built over another network, as shown in Figure 7.1. A darknet is a network that employs onion routing. The Dark Web was created by fusing these various darknets. The NRL staff quickly understood that everyone, not just the US government, needed access to the network to be completely anonymous. The Onion Router (TOR) was developed due to the NRL's forced release of its onion routing technology under an Open-Source License (Ciancaglini et al., n.d).

1.1 Internet Structure

Figure 7.1.1 shows the three components of the World Wide Web (www): the dark web, the deep web, and the surface web. Popular web search engines make it simple

Figure 2. The layers of the internet (AL Attack Map, 2020)

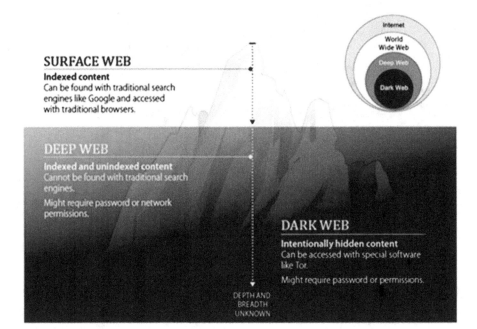

for the general public to access the Surface Web, also called the visible or indexed web. Search engines only return 4% of the results on the web surface.

The deep web (Mirea et al., 2019) is exclusive to certain users, in contrast to the surface web, which is available to every user. The Hidden or Invisible Web is another name for it. An estimated 96% of the Internet comprises the deep web and dark web. It is primarily meant for private intentions. The deep web content can be found in databases, online banking, and anything else that needs a password.

The World Wide Web (WWW) is also referred to as the "Dark Web" (Ibrahim et al., 2022), and its content cannot be accessed using the same browsers as the surface web because it is not a part of the surface web. It started as a way for the US Military to help conceal information using intelligence assets stationed far away. Most unlawful and harmful online activity occurs on the "dark web." It is a hub for fraud/hacking services. Additionally, it is a venue for illegal activities like terrorism, child pornography, and phishing scams.

Several browsers, for instance, The Onion Router (TOR), FreeNet,, Invisible Internet Project (I2P), Whonix, and Rife, permit us to access the Dark Web. A subtype of the Deep Web called the Dark Web additionally offers concealed facilities and is finished with an onion extension.

2. DARK WEB CONCEPT

The term "Dark Web" describes an encrypted Internet network only accessible with specialized software. The Dark Web could be a private network like the I2P network or a peer-to-peer network like Tor. These networks route communication through the encryption layer to preserve user anonymity. The dark web—which is purposefully concealed from the surface web, is thought to have developed from the deep web. It is only accessible through a few browsers, enabling user anonymity and concealment while facilitating easier dark web exploration (Kaur & Randhawa, 2020). Due to their users' anonymity, they can be used in legal and illicit situations. This offers additional protection for those looking to engage in prohibited action on the dark net with little risk for being discovered.

2.1 Dark Web Components and Elements

The development of the Dark Web has involved using various protocols and techniques (Abu Al-Haija & Al Badawi, 2022; Al-Fayoumi et al., 2022; Kaur & Randhawa, 2020). Virtual private networks (VPNs) for data transmission, routing formulas, encryption methods, and browsers make up the Dark Web's essential elements (Butler, 2019). Maintaining anonymity is imperative if you want to use the dark web. It would be best if you used a trustworthy Virtual Private Network in addition to your browser to protect your identity (VPN). There are two paid options: (1) Phantom VPN blocks the tracking and monitoring of Internet activity by ISPs, internet snoopers, and advertisers. (2) Nord-VPN function as a private service provider that provides desktop iOS and Android apps for Windows, Linux, and macOS.

Encryption is used extensively on the Dark Web. The TOR browser uses multiple high-level encryption levels and random routing to safeguard your privacy. Suppose you are utilizing a centralized communication system while on the dark web; a third party may have access to your data. It implies that secret information shouldn't be disclosed, especially if discovered by outsiders, which could lead to problems. Anonymity typically solves this problem. However, since an outsider can retrieve the data sent/received, the problem still exists. Then comes the Pretty Good Privacy (PGP) encryption technique (Butler, 2019). It is a powerful type of encryption used to protect private data or communications. It was intended to deliver security features like non-repudiation, privacy, and authentication.

The fundamental component of PGP is its asymmetric encryption system. In asymmetric encryption, data is encrypted and decrypted using two keys: the encryption/encipherment key, which is made public, and the decryption/decipherment key, which is made private. The encryption key is the one that is made available to the general public. In this kind of cryptography, if someone encodes a message using your encipherment key, then you are the only one who deciphers the message into its readable form (Al-Fayoumi et al., 2022). PGP also supports authentication. PGP functions differently when it comes to authentication. Public key encryption and hashing are both employed. It uses both public key encryption and private key encryption to ensure privacy. To create a digital signature, you need two public-private key pairs, one hashing algorithm, and one secret key. PGP Encryption offers a variety of advantages. The data is always secure because no one can access it or steal it over the Internet. The sharing of data and information online is secure. After being destroyed, sensitive information or messages cannot be recovered. Second, because this encryption method authenticates the source's info to prevent outside parties from intercepting it, hackers cannot change emails or other forms of communication.

2.2 Dark Web Access Methods

Various browsers were designed to get into the Dark Web (Naseem et al., 2016; Rudesill et al., 2015). The most popular Dark Web browser is called TOR (founded in the 1990s). TOR's alpha release, created in C, Python, and Rust, was made available on September 20, 2002. The Tor browser uses the onion routing algorithm (Abu Al-Haija, Krichen, & Abu Elhaija, 2022) to encrypt user data before passing it through several relays (intermediary computers) in the network. As a result, a network with numerous encryption layers is produced. As explained below (Tor Project, n.d), TOR switches connections between three relays:

- *Guard & Middle Relays are also called non-exit relays. A short relay is used to help build the TOR Circuit. Although it behaves as a second host between guard and exit relays, the intermediate relay does not carry out either of those tasks. The guard relay must be efficient and dependable. When a user or client attempts to link up to the TOR Circuit, their IP address may be reached for the first time ().*
- *Exit relay: Last relay in the TOR circuit. The traffic is routed through the relay to reach its destination. Clients won't see their IP addresses; only the IP address of the Exit relay will be displayed. Figure 7.2.2.1 shows that only the predecessor and descendant information is available for each node.*
- *Bridge: TOR users will only use the relay's IP addresses. However, authorities or ISPs can forbid the use of TOR by blocking the IP addresses of the public TOR nodes.*

Tor uses a layered network, not a decentralized one. It relies on several directory authorities and central servers managed by a set of helpers (volunteers) connected

Figure 3. Data flow in tor browser-based onion routing

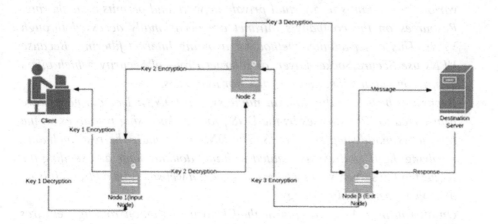

to the Tor Institution. Due to its high levels of centralization, Tor is extremely vulnerable. If the US, German, or Dutch directory authorities are shut down, it is sufficient to shut down five of the ten directory authority servers. As a result, there will be a significant decrease in the newly added relay's ability to communicate with the network, and the Tor network will become very unstable (Tor Project,). Tor only supports TCP, has a very small selection of extensible protocols, and needs to be sufficiently secure to thwart witchcraft attacks. Once sufficient relay nodes are forged, a witch attacker can decrypt all user data if three relay nodes are chosen in order, and the witch attack fakes all three.

2.3 Techniques for Anonymity and Confidentiality on the Dark Web

Two key ideas are at the foundation of the Dark Web: confidentiality and anonymity; several tactics are used, as will be explained below, to preserve anonymity and confidentiality (Kaur & Randhawa, 2020).

- *Proxy: This is a service where client requests are compiled and sent to the destination on the requestors' behalf. After obtaining the responses, the proxy relays the data to the original requester. It serves as a go-between for the source and the destination. These web filtering proxies help filter and avoid web content. In some cases, users' access to particular websites is restricted by proxy servers.*

- *Network tunneling and virtual private networks (VPNs): VPNs are the most popular form of network tunneling. It is a private network that links various participants in a virtual private network and permits data sharing. Resources on the company's intranet are occasionally accessed through VPNs. This is yet another method for avoiding Internet filtering. Because VPNs use Secure Socket Layer, or Internet Protocol Security, which offers secure communication, they are superior to proxies.*
- *Bypassing based on the domain name system (DNS): Domain names are transferred to IP addresses by the DNS process. Accessing resources on the Internet is made simpler by DNS. The DNS will handle the rest, including resolving the IP address associated with that domain name and sending the request to the server, so all we need to access a website is its address. It is an alternative censorship strategy.*
- *Onion routing: A networking method known as "onion routing" ensures content encryption while sending it to the exit node. Additionally, it conceals who is speaking to whom at all times. Anonymous connections can be made. It is distinct from other approaches, as was previously stated. The link travels a long distance over an encrypted chain in the style of an onion from Source A to Destination B.*

3. DARK WEB CRIMES

The exponential rise in anonymizing software has led to a rapid rise in cybercrime, and the dark web has become a haven for illegal activity. The dark web sees much traffic related to illegal activity. Although they might also use it for good reasons, most people who use Tor do so because they want to remain anonymous. A system that protects users' anonymity and keeps an eye on their behavior to ensure they aren't visiting dubious websites simultaneously is impossible to design. But this is different.

Most people use different browsers to gain access to concealed dark websites to view and distribute pictures of child exploitation and purchase unlawful drugs. Later on, Dr. G. Owen and N. Savage (University of Portsmouth) researched the use of Tor and hidden services over six months. As a result, they decided that over 80% of Tor traffic requests to hidden sites encountered throughout the investigation were for common child abuse websites (Sadeghi et al., 2013). Nevertheless, they recognized that taking into account the frequent use of government agencies' computers to repeatedly look for websites for child exploitation pictures as part of their investigations, this information might not be entirely accurate.

The proportion of police action to the tor traffic brought on by criminal activity online cannot be calculated or determined. Even though police activity only accounts for half of the examined child exploitation traffic, there is still a sizable amount of users' payload access on the dark web related to child abuse. The next subsections show quintessential violations committed on the darknet:

3.1 Sex Trafficking and Humanitarian Trafficking

Millions of people are impacted by the global human rights issue of human trafficking under the name of slavery. International labor organizations anticipated that 40.3 million persons would be sufferers of contemporary slavery (O'Neill, 2013). Prostitution and pornography are forms of human trafficking, but it also happens in cafes, bars, the drug trade, street gangs, etc. Over the past ten years, human trafficking has been complicated because it is now carried out digitally and covertly. By connecting to numerous customers, it is simpler to take advantage of multiple victims due to globalization and technology.

Numerous subcategories of human trafficking exist, such as labor/ sex trafficking, baby/ organ trafficking and others. Transplantation tourism, as used in the dark world, is bringing people to the locations of organ recipients to remove organs. Global Financial Integrity estimates that in 2017, the illegal organ trade generated annual profits of between $840 million and $1.7 billion. Another flourishing sector of human trafficking is infant adoption. Young women are forced to work as industrial ovaries in "baby factories" that harvest babies to sell (N & Zhang, 2020). Even though the darknet was developed to safeguard anonymity, such protocols aid individual traffickers and conceal clients of by-law enforcement. Numerous studies and documents show that the Darknet's accessibility to subpar protocols, anonymous IP allotments, and peer-to-peer and imperceptible disbursement transactions benefits criminal activity. Using Bitcoin and other cryptocurrencies to pay for illegal applications on the dark web is simple. The Darknet's criminogenic characteristics give criminals an advantage.

3.2 Child Pornography

Child pornography is frequently distributed on the dark web by pedophiles and other criminals. Most people who access secret child pornographic websites do so through Tor. To host child pornographic content, Freedom Hosting assigned nearly 550 servers in Europe. Additionally, live child abuse is exploited for profit through its video feature applications (Ibrahim et al., 2022). Webcam child prostitution also uses voice-over IP (VoIP). An alarming threat exists when pictures of victims of child sexual abuse are offered for sale online. Following the shutdown of the South

Korean-based child porn website "welcome to video," hundreds of people were detained globally in 2018. According to estimates, 144,000 Britons had access to pornographic material in 2018 (Chertoff, 2017). Terrorism and the sale of weapons Illegal weapon sales and purchases are made easier thanks to the dark web. Despite having a lower volume than other crimes committed on the dark web, trafficking in arms has a much bigger impact on global security. The largest market share for selling weapons on the dark web is in Europe, specifically Germany and Denmark (Chertoff, 2017).

3.3 Terrorism and the Trafficking of Arms

Terrorism and terrorist organizations present serious threats to international security on the dark web. Additionally, they used the dark web to raise money, buy weapons, plan international terrorist attacks, radicalize and inform their members, and recruit new members.

3.4 Drug Trafficking

On the dark web, there are numerous sites for buying and selling prohibited drugs. Two different types of drug markets are commonly available on the dark web: general drug stores and the narcotics market, which sells cocaine, marijuana, psychedelics, and illegal tobacco. The exchange of cryptocurrencies for the purchase of illegal goods is made possible by the dark web. The Silk Road, one of the busiest drug marketplaces, offered drugs worth more than a billion dollars. It has been discontinued in 2013. However, various underground drug marketplaces exist on the dark web (Dan.Is, n.d.). "Grams," the most popular drug-related dark web search engine, has a Google-like logo.

3.5 Malware/Ransomware

Malware and the dark web go hand in hand in many ways, specifically when it comes to the accommodating command-and-control (C&C) structure. The best aspect of Tor or I2P is robust cryptography to conceal server locations. Forensic investigators cannot effectively use traditional investigation techniques like looking at a server's IP address, verifying registration information, etc. Tor is used for command and control by many online criminals. Many well-known malicious software types use Tor for part of their systems by including official Tor clients in their installation files. In line with Trend Micro (a well-known antivirus company), when MEVADE malware switched to Tor-hidden C&C services in 2013, an increase in Tor traffic was attributed to this for the first time.

The VAWTRAK malware, for instance, is a banking Trojan disseminated through phishing emails (N & Zhang, 2020). Every instance interacts through a number of command and control (C&C) servers. To do so, the addresses of such servers can be found by downloading a converted icon file (such as favicon.ico) from websites hosted by Tor (Al-Fayoumi et al., 2022). Although this method conceals the place of a malicious server, hosts who get into it are still at risk. They already have malware on their computers, so this doesn't seem to be a problem. A further malware group that utilizes the dark web is Crypto Locker ransomware.

Crypto Locker is a type of ransomware file that encrypts documents and directs compromised users to a website. As a result, a fee is required for anyone who wants to access these files. Because the payment options and local languages are automatically adjusted, it is intelligently designed. Because the deep web has made it simpler for them to create infrastructures that are more resilient to potential takedowns, it exemplifies why cybercriminals are drawn to it (Tor Project, n.d.). Numerous ransomware recognition engines have established means to recognize affected files in response to the increase in ransomware attacks. However, even if the corrupt file is discovered and removed, its data cannot be recovered (Ferry, 2019).

3.6 Bitcoin and Money Laundry

The digital currency known as Bitcoin was created with anonymity in mind. Even illegal purchases are common and widely accepted in the modern era. Even though Bitcoin transactions are considered anonymous, users must link their identities to their cryptocurrency wallets.

According to the blockchain's architecture, Bitcoin transactions are entirely public and open to scrutiny. As a result, though more effort is needed, tracking money is possible. The system has added several services to increase anonymity and make tracing digital currency more challenging. Usually, to achieve this, bitcoin is "mixed" through a network of microtransactions before being returned to its owner (Ibrahim et al., 2022; Yang, 2019). With this procedure, the owner receives the funds with a lower chance of them being tracked down, and a small fee is taken out. Laundry services facilitate the increased anonymity of money flowing through the bitcoin system. PayPal, ACH, and Western Union are anonymous services constantly being added to the deep web (Tor Project, n.d.).

According to Europol officials, the use of bitcoins for illegal activities has reportedly increased. Since their creation in 2014, Distributed Denial of Service (DDoS) "4" Bitcoin (DD4BC) (Ibrahim et al., 2022), a consortium of DDoS attacks, has targeted more than 140 businesses. This stirs up more organizations, which ultimately results in cyber extortion. Europol officials claim that the DD4BC band first endangered email targets with a DDoS attack if a bitcoin ransom was not paid.

The rise of Bitcoin in the realm of the dark web also helps the growth of online terrorists.

3.7 Proxying

Users of Tor platforms are vulnerable to attacks due to the anonymity these platforms provide. The standard "HTTPS" designation for a secure connection is missing from this site's URL. Users should bookmark at the TOR page to confirm that they are at a trustworthy website. When using a website proxy, the con artist deceives the user into thinking he is on the original page before changing the link and rerouting him to his scam link. When a user purchases cryptocurrency, the money is transferred to the swindler.

3.8 Torture

Red Room websites are those where viewers can watch live streams of rapes, murders, child pornography, and other forms of torture for thousands of dollars. There is, however, no proof that they exist. TOR cannot access them if they exist because it streams live video too slowly. According to reports, visitors to a pedophile website paid a sizable sum of income to look at videos of Scully torturing and abusing a youthful child. The company that Scully owned produced the show No Limits Fun (NLF). Even though Daisy's Destruction features horrifying child abuse and simple sexual assault, it was one of his videos that received much discussion on the forum. It was broadcast on "Hurtcore" pedophile websites, where pedophiles witness the abuse and torture of children and infants (Dan.Is, n.d.).

3.9 The Hidden Wiki

The Hidden Wiki is the primary directory and one of the most frequent locations where new users first arrive on the dark web. Many of its links lead to websites that advertise money laundering, murder commissions, cyberattacks, restricted substance closeness, and how-to guides for making unpredictable. However, the link to the Hidden Wiki is constantly changing (O'Neill, 2013), requiring URL changes to avoid detection by law enforcement.

3.10 Gambling

Despite being a trade that takes place inside and outside the dark web, betting (gambling) on it uses cryptocurrency. Many renowned Bitcoin gambling sites have blocked U.S.-specific IP addresses due to strict regulations and prosecutions from the U.S. government. However, dark web users can gamble at will using their current IP addresses (O'Neill, 2013).

3.11 Exploit Markets

Before patches are applied, exploits can be introduced as malware-based software vulnerabilities. A "zero-day exploit" is a software vulnerability that is undiscovered by vulnerability mitigation experts. It frequently appears in recently released software for which a patch still needs to be released to correct its flaws. Exploit markets serve exploit-based transactions on zero-day exploits. The cost of such exploits differs according to how well-known and challenging the software is to crack (Charlie, 2007).

3.12 Murder

It could be argued that assassination is one of the most disreputable activities on the dark web. The dark web also has marketplaces for assassination because of its untraceable origin. A wager on someone's assassination can be made on the Assassination Market website, with a payout if the date is correctly predicted.

Due to their knowledge of the precise time the assassination occurs, the tangible murderer side or the other involved in the murder gains prevail. The market still makes money even though the underlying structure drives the subject's murder. Additionally, it is very difficult to establish criminal responsibility for assassination because the wager is at the moment of death instead of the murder of a person (Andy, 2013). Other websites, like White Wolves and C'thuthlu (Zane, 2014), don't have these complementary procedures and outright advertise hiring assassins.

4. CYBER ATTACKS ON THE DARK WEB

The DarkWeb can support a range of cyberattacks. The Dark Web's main drawback is its anonymity, which gives attackers more leeway to target their intended victims and increases their confidence.

4.1 Correlation Attacks

It is a pervasive passive-type attack vector. Such attacks enable the attacker to seize control of the opening and end routers of the TOR network and utilize timing and information characteristics to correlate the streams that pass through them, jeopardizing the anonymity of the network. Many government agencies have used correlation attacks in the past to obliterate the privacy of several users.

This attack uses a highly sophisticated mathematical technique, so there is no other way to stop it. Not only is software the target of this kind of attack, but also users. For instance, a dark-market administrator might disclose his age, criminal record, and other details on the website. It will help the authorities track down anyone who connects to the TOR network while the administrator is online and help them keep track of all suspects' online activities. A correlation attack was launched against the TOR network by Carnegie Mellon University. The FBI was then paid $1 million for the TOR user data. Numerous pornographic webpages (such as Silk Road 2.0) were taken down due to the attack. The only defense against this kind of attack is the choice of a trustworthy VPN to block it.

4.2 Congestion Attacks

A clogging attack, also known as monitoring the link among the two knots, creates the route between them. In this case, the connection speed at the victim side should change if the attacker successfully blocks one of the nodes along the target path. It is an all-out, direct assault. Murdoch and Danezis termed an attack against the TOR network in 2005 which can divulge all routers engaged in a TOR track by combining a clogging attack with timing analysis.

Figure 7.4.2 depicts how the congestion attack works on routers with various bandwidths. The fact that the exit router is up and running and we only need to find one node makes this attack successful. The bandwidth multiplication method enables small-scale bandwidth links to use large-scale DoS bandwidth links. It also removes the specific DoS restriction. This kind of attack can be stopped by staying away from the use of fixed path lengths. End-to-end encryption is a possible substitute. Thirdly, this attack can be stopped by blocking JavaScript in client browsers and delaying connection establishment.

4.3 Attacks Based on Traffic-Timing Correlation

These are ongoing and end-to-end attacks (Nasr et al., 2018). These additional de-anonymization attack types exist. The entry and exit relays of the target are changed during this kind of attack. An attacker can identify which server a client is communicating with by examining the traffic patterns between the entry and exit relays. Complex mathematical techniques are not required for deanonymization. For instance, a Harvard University student who used TOR to send phony bomb threats to avoid an exam was detained.

According to FBI data, Guerilla Mail was used to send the emails. Users of the email service Guerilla Mail can make fictitious email addresses. Every email sent includes the sender's IP address. The FBI claims that the student used TOR to send the emails. Through correlation, the FBI located the student. The execution of traffic and timing attacks is straightforward when there are few TOR clients. More sophisticated timing-traffic correlation attacks are used to deanonymize hosts in such a case. TOR uses delaying, packet buffering, and shifting techniques to stop these attacks.

4.4 Attacks Using Traffic Fingerprinting

These uni-target passive attacks examine the traffic flow pattern, making it possible to sniff an entire website without jeopardizing encryption. Two common methods for collecting traffic records using TOR. The attacker's first strategy is called the initial strategy, which involves slowing down traffic at the entrance. However, with this practice, the compromised host is not- obliged to connect to the node of attacker. The second technique enables the hacker to listen in on communications between the compromised host and the TOR circuit's entry host (Internet Service Provider-ISP) by posing as an administrator, such as an ISP. This approach works well. Several defense techniques, such as obfuscated HTTP (HTTPS), also known as secure HTTP, guard host adaptive padding, pipeline randomisation, and traffic mutating, are available to stop this attack.

4.5 Attacks Caused by DDoS

The attacker bombards the target with numerous fraudulent requests to impede the compromised connections or make them unavailable. It doesn't remove user anonymity. One of the greatest mysteries is what happened to the marketplace in Abraxas. It is currently unknown if the Abraxas market was the compromised host of an exit scam or a DDoS attack. Because the market changes mysteriously whenever the price of bitcoin rises, most information points to it being an exit scam. The Abraxas market may have been the target of a DDoS cyberattack. Before the marketplace was disabled, users complained about a slow server and trouble logging in. Second, according to administrators of the Abraxas market, there was a significant DDoS attack, and the site will soon be back online on Reddit (AL Attack Map, 2020; Zillman, 2015). Multivariate threat detection is a quick and efficient way to find DDoS attacks.

4.6 Attacks Using Hidden Services

Through concealed services, users can get into resources without disclosing their identities. The initial node attack's purpose is to reveal covert services. Herein, the relay of the attacker tries to change into a relay linked immediately to the server home to the concealed services. This will instantly make the hidden service's position known. The attacker connects to the server through the malevolent host. Utilizing temporal analytics, the malevolent host decides whether a specific node is an initial node. The Clock Skew attack is another attack that can identify concealed services from a directory of servers. You can find concealed services by contrasting the timestamps of various servers.

4.7 Phishing

An attacker frequently uses phishing tactics (Al-Haija & Badawi, 2021) or assumes a false identity to set up malware or get penetrating data from the user. In this kind of attack, an adversary may send you an email that seems to be from a reliable source. The malware will be downloaded if you open an email containing a malicious link or attachment. According to the literature, there are three types of phishing: spear phishing, whaling phishing, and clone phishing. When a particular organization is a target, spear phishing is used. The purpose of this attack is to steal private information from numerous people. Whaling is aimed at leading or C-level directors of an organization. To deceive the victim, the attackers use specific messaging. Targets are shown a copy or clone of a message they have already seen in clone phishing.

4.8 Reusing Credentials

We must avoid using the same username and password repeatedly. An attacker is aware of an opportunity to be able to log in if they use the same credentials once they can get a compilation of usernames/passwords. Numerous password managers are available, which are beneficial for managing your various login credentials.

5. MALWARE ON THE DARK WEB

A dark web market is where illegal goods can be bought and sold. The Dark Web allows access to various malicious programs and services (Al-Qudah et al., 2023). Users trading these services with malicious intent make sizable profits. Positive Technologies, a security firm, recently released a report emphasizing the booming Dark Web market. A study of over 10,000 advertisements yielded some intriguing findings (Kaur & Randhawa, 2020). Malware is a key component of many cyberattacks. Malware of all different varieties and price points were available for purchase. The most popular products were cryptocurrency miners, according to the advertisements that were found. Below are a few general malware categories that are covered:

Figure 4. Congestion attack

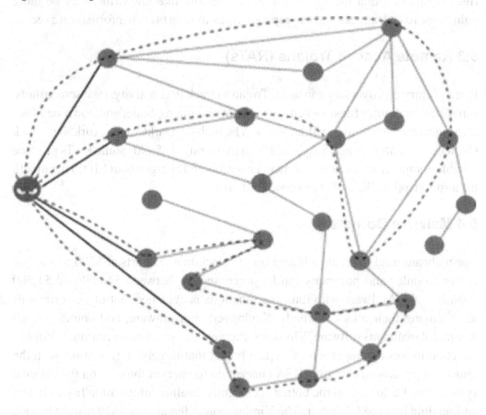

5.1 Trojans for Stealing Data

Passwords can be taken from the clipboard, keystrokes can be recorded, antivirus software can be avoided or disabled, and files can be sent to the attacker's email address. The average price of a thief is $10 (Ibrahim et al., 2022). These thieves' use of stolen data can result in very expensive data.

5.2 Ransomware

Your system or files are encrypted by ransomware (Al-Haija & Alsulami, 2021), which then demands payment to unlock them. The median expense of this malware is $270. A user's computer is hijacked by a malicious attack known as ransomware, which prevents them from using it. There are several ways that attackers can pick which businesses to use ransomware on. Some businesses seem more likely to respond quickly to a ransom demand, which makes them desirable targets. For instance, hospitals and governmental agencies frequently request immediate access to their files. Businesses that handle sensitive information, like law firms, may be more vulnerable to leak attacks and be prepared to pay to keep breach information a secret.

5.3 Remote Access Trojans (RATs)

Remote entry an adversary can use a Trojan to track user activity, take screenshots, run scripts and other files, switch on webcam and microphone, and download files from the Internet. Some well-known RATs include DarkComet, Turkojan, Back Orifice, Cerberus Rat, and Spy-Net. The price is roughly $490. Some RATs produce reliable computer remote management programs and charge about $1000 per month for a subscription (Kaur & Randhawa, 2020).

5.4 Malware Botnets

The malware used to create a botnet on the black market starts at $200. Complete server modules and programs can be purchased for between $1,000 and $1,500 (Dan.Is, n.d.). Malware with multiple functions is evidence that online criminals are changing their attack methods. Keyloggers, ransomware, and botnets are all features of malicious software. Virobot is a botnet ransomware illustration. A Virobot infection makes a computer join a spam botnet that targets more victims with the ransom. The ransomware uses RSA encryption to encrypt the data on the infected system. The keylogger on the botnet is currently stealing information from victims and sending it to the C2 server. The Virobot botnet feature uses Microsoft Outlook on an infected computer to send spam emails to every contact in the user's address book. The malware, which is still being created, was discovered for the first time on September 17 (Chertoff, 2017).

6 INTERNET GOVERNANCE OF DARK WEB

Undoubtedly, the dark web is one of the strongest and most unique networks on the Internet. As the dark web currently has, any robust system could easily spiral out of control. As a result, there needs to be much focus on monitoring and regulating dark web activity. A definition that forbids illegal use while upholding user anonymity should be part of the strategic governance of dark web activities (Satterfeld, 2016). A few dark web regularizations could be implemented with the help of numerous agencies and government authorities, greatly aiding network regulation (N & Zhang, 2020).

A few authorities, including the FBI, have already acted by employing Computer and Internet Protocol Address Verifier (CIPAV), a tool used to monitor anonymous users and proxy servers. To focus on specific suspects, CIPAV can identify TOR traffic among other surface internet traffic. The Defense Advanced Research Projects Agency (DARPA) also employs "Memex," another program that uses pattern recognition to identify illegal behavior.

Several characteristics that help classify the user allow Memex to reveal suspicious behavior (Ferry, 2019). Anonymous user tracking, however, could categorize the entire dark web as an illicit network, giving authorities more control over some people's otherwise harmless actions. The need for such systems may also be more critical in other nations due to the variations in legal systems among various nations. Due to the complex web-like structure of the dark web, reaching conclusions through legislative action has occasionally been both effective and ineffective. As a result, a legal framework that carries out internal and national inspections is necessary (N & Zhang, 2020). The primary focus areas for initiatives to address this issue are discussed in the following paragraphs.

6.1 The Hidden Services Directory Mapping

The remote database, which is kept secret using a Distributed Hashing Table (DHT) technique, is one prominent way the dark web is controlled. Both Tor and I2P use this technique. For DHT to function, a portion of the database must be cooperatively retained and kept in a key-value dictionary structure by the system hosts. With the help of this distributed structure, monitoring nodes can be deployed in response to incoming track requests in the DHT, allowing subsequent hosts to follow the initial hosts and match them for future use.

6.2 Data Monitoring for Customers

Web data of the customers can be examined to trace connections to unauthorized dark web territories. This technique uses destination queries to keep track of top-level domains. The degree of web use differs from person to person, so it could be useless to track each person to find a connection to the dark web. In either case, if done widely across a large network, customer data monitoring could provide insightful information about dark web activities. Since only the request's endpoint is observed during the entire monitoring process, the user's privacy cannot be determined from the data collected. The FBI's "Carnivore," used in inspections going back to 1997, is a well-known model of computer software that observes such information correspondence (Naseem et al., 2016).

6.3 Monitoring Social Sites

Many social media platforms on the dark web, including Pastebin, that function similarly to those on the surface web. The main distinction is that new addresses for new covert services are frequently exchanged instead of personally identifiable information. Monitoring social sites on the dark web entails tracking cryptographic operations (Abu Al-Haija, Al Badawi, & Bojja, 2022; Ahmad et al., 2022). Monitoring social media platforms might make it easier to gain more precise control over the dark web. Therefore, using intelligent methods, it's critical to constantly monitor social media platforms to weed out brand-new illicit dark web domains (Abu Al-Haija & Alsulami, 2022; Abu Al-Haija, Odeh, & Qattous, 2022).

6.4 Monitoring of Hidden Services

One of the few methods for tracing a crime back to its origin is through the hidden service it uses. Due to frequent URL changes made to avoid fees, most hidden services are extremely unstable. Recording a snapshot of any new services or websites as soon as they appear is the best method for keeping track of hidden services. Later analysis of the snapshot can determine whether it vanished or reappeared under a different domain name.

6.5 Semantic Analysis

As with any other problem, gathering the necessary data is a crucial first step in solving it. Similarly, any website or service hidden on the dark web must record its data for post-retrieval and a semantic analysis database. The information about a certain hidden service could then be pulled out of this database as needed to find out who was using it to do bad things and predict what would happen next.

6.6 Market Profiling

Because there are so many dark web marketplaces, it is crucial to concentrate on each and profile its transactions. So, over time, information is gathered about suppliers, buyers, intermediaries, products, and exchanges. This makes it possible to link people who use the dark web to illegal activity.

Many horrible things can be done on the dark web. Because it is decentralized, people frequently use chat forums and covert services to plan and carry out crimes. Several tax refund scammers allegedly shared their deceptive techniques on the dark web (Yang, 2019). This is according to claims. One more prevalent fraudulent practice on the dark web is dealing with malware for obtaining unencrypted financial information during large-scale data breaches. To make money, point-of-sale (POS) systems are remotely used to carry out fraudulent activities by RAM scrapers, another kind of malware bought off the dark web. The precise network traffic and the growth of the different kinds of domains and sites on the dark web are still being determined, though, due to the need for more trustworthy information (Rafiuddin, 2017). However, thanks to the capabilities of the Tor network, it has been possible to approximate the network traffic related to secret service records by about 83 percent. The illicit market activity on the dark web is still too little to determine.

7. PAYMENT ON THE DARK WEB

Payment over the internet is a usual transaction performed by cyberspace users. One of the most common and recent means of payment on the dark web is Bitcoin. Users can transact securely using this decentralized digital currency (Murphy et al., 2015). Bitcoins are typically obtained by "mining" them, exchanging them for fiat currency, or accepting them as payment. Every time a bitcoin is exchanged for cash, the transaction is recorded in the blockchain, a widely accessible ledger. The information kept in the blockchain is the bitcoin addresses of the sender and recipient. An address does not uniquely identify any one bitcoin; it only serves to identify a particular transaction. The addresses of users are stored in and connected to a wallet (Sadeghi et al., 2013). A user's private key is kept in the wallet, a secret code similar to a password that enables them to spend bitcoins from the associated wallet. The transaction address and a cryptographic signature are used to confirm transactions. Because the wallet and private key are hidden from the public ledger, using bitcoin has increased privacy. A hardware device, the internet, desktop or mobile software, or both, can be used to store wallets.

8. THE IMPACTS OF THE DARK WEB

There are many reasons why people might want to protect their online identity. This is the case, for instance, in nations where political censorship is practiced, or the government forbids the free press. In other instances, this is because revealing their identity might put them in danger. Others might use it to lower their risk of becoming victims of crime, such as those who have been cyberstalked or are concerned about online banking security (Gulati et al., 2022). Most traffic on Tor, used for anonymous Internet access, is not generated by Hidden Services. Figure 7.9.1 shows the total number of Tor users in the world.

Figure 5. Tor users around the whole world

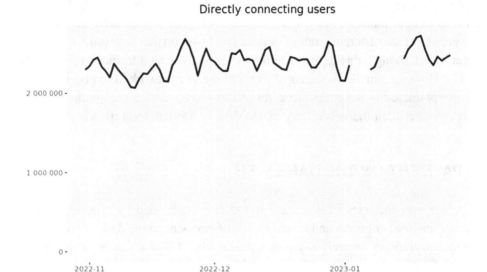

Directly connecting users

The Tor Project - https://metrics.torproject.org/

The anonymity of the user and the website are both protected when using a Hidden Service by Tor. As a result, the website's host, geographic, and content-related information cannot be ascertained, including its IP address. Hidden Services are often called "onion addresses" because the domain name frequently ends in ".onion." This makes it possible for buyers and sellers to stay anonymous from each other when a buyer doesn't want to purchase in public or wants to keep his identity hidden. Some examples of covert services are that those who can't afford them can still access medical assistance and cheaper, illegal lifestyle products (Gulati et al., 2022).

9. CONCLUSION AND REMARKS

In this chapter, we have thoughtfully investigated and discussed the darknet networks, the dark portion of the internet world. The darknet or dark web is a concealed section of our global communication system. Such a section comprises a special address space that is accessible by special communication protocols and authorization techniques. Criminals, attackers, and malware designers extensively use dark address space to perform different types of:

- *Crimes: such as sex trafficking and human trafficking, child pornography, terrorism and the trafficking of arms, drug trafficking, bitcoin and money laundry, proxying, torture, the hidden wiki, gambling, exploit markets, and even murder.*
- *Cyber-Attacks: such as correlation attacks, congestion attacks, attacks based on correlations between traffic and timing, attacks using traffic fingerprinting, attacks caused by a DDoS, attacks using hidden services, phishing, and reusing credentials.*
- *Malware: such as trojans for stealing data, ransomware, remote access trojans (rats), malware botnets, and crypto-loggers.*

Therefore, the development of resilience countermeasures tools and systems to defend users and systems against the exploitations of the dark web is in demand. Several state of art techniques were proposed in the literature to construct defense systems against the darknet. Such systems mainly depended on either signature-based detection or anomaly-based detection.

REFERENCES

Abu Al-Haija, Q., & Al Badawi, A. (2022). High-performance intrusion detection system for networked UAVs via deep learning. *Neural Computing & Applications*, *34*(13), 10885–10900. doi:10.100700521-022-07015-9

Abu Al-Haija, Q., Al Badawi, A., & Bojja, G. R. (2022). Boost-Defence for resilient IoT networks: A head-to-toe approach. *Expert Systems: International Journal of Knowledge Engineering and Neural Networks*, *39*(10), e12934. doi:10.1111/exsy.12934

Abu Al-Haija, Q., & Alsulami, A. A. (2022). Detection of Fake Replay Attack Signals on Remote Keyless Controlled Vehicles Using Pre-Trained Deep Neural Network. *Electronics (Basel)*, *11*(20), 3376. doi:10.3390/electronics11203376

Abu Al-Haija, Q., Krichen, M., & Abu Elhaija, W. (2022). Machine-Learning-Based Darknet Traffic Detection System for IoT Applications. *Electronics (Basel)*, *11*(4), 556. doi:10.3390/electronics11040556

Abu Al-Haija, Q., Odeh, A., & Qattous, H. (2022). PDF Malware Detection Based on Optimizable Decision Trees. *Electronics (Basel)*, *11*(19), 3142. doi:10.3390/electronics11193142

Ahmad, A., AbuHour, Y., Younisse, R., Alslman, Y., Alnagi, E., & Abu Al-Haija, Q. (2022). MID-Crypt: A Cryptographic Algorithm for Advanced Medical Images Protection. *J. Sens. Actuator Netw.*, *11*(2), 24. doi:10.3390/jsan11020024

AL Attack Map. (2020). Distributed Denial of Service Attack. *AL Attack Map.* alattackmap.com/#anim=1&color=0&country=ALL&list=2&time=1643 8&view=map

Al-Fayoumi, M. A., Elayyan, A., Odeh, A., & Abu Al-Haija, Q. (2022). Tor Network Traffic Classification Using Machine Learning Based on Time-Related Feature. *6th Smart Cities Symposium 2022 (6SCS)*. IET. 10.1049/icp.2023.0354

Al-Haija, Q. A., & Alsulami, A. A. (2021). High Performance Classification Model to Identify Ransomware Payments for Heterogeneous Bitcoin Networks. *Electronics (Basel)*, *10*(17), 2113. doi:10.3390/electronics10172113

Al-Haija, Q. A., & Badawi, A. A. (2021). URL-based Phishing Websites Detection via Machine Learning. *2021 International Conference on Data Analytics for Business and Industry (ICDABI)*, Sakheer, Bahrain. 10.1109/ICDABI53623.2021.9655851

Al-Qudah, M., Ashi, Z., Alnabhan, M., & Abu Al-Haija, Q. (2023). Effective One-Class Classifier Model for Memory Dump Malware Detection. *J. Sens. Actuator Netw.*, *12*(1), 5. doi:10.3390/jsan12010005

Androulaki, E., Karame, G. O., Roeschlin, M., Scherer, T., & Capkun, S. (2013). Evaluating User Privacy in Bitcoin. In A. R. Sadeghi (Ed.), Lecture Notes in Computer Science: Vol. 7859. *Financial Cryptography and Data Security, FC 2013* (pp. 34–51). Springer. doi:10.1007/978-3-642-39884-1_4

Andy, G. (2013, November). Meet the 'Assassination Market' Creator Who's Crowdfunding Murder with Bitcoins. *Forbes*, 18.

Butler, S. (2019). The Role of PGP Encryption on the Dark Web. *Tech Nadu*. https://www.technadu.com/pgpencryption-dark-web/57005/

Cambiaso, E., Vaccari, I., Patti, L., & Aiello, M. (2019). Darknet security: A categorization of attacks to the TOR network. In *Italian Conference on Cyber Security*. CEUR-WS.

Charlie, M. (2007). *The Legitimate Vulnerability Market: Inside the Secretive World of 0-day Exploit Sales*. Independent Security Evaluators.

Chertoff, M. (2017). A public policy perspective of the Dark Web. *Journal of Cyber Policy*, *2*(1), 26–38. doi:10.1080/23738871.2017.1298643

Ciancaglini, V., Balduzzi, M., & Goncharov, M. (n.d.). Cibercrime and digital threats. *Trend Micro*. https://www.trendmicro.com/vinfo/pl/security/news/cybercrime -and-digital-threats/deep-web-and-cyber crime-its-not-all-about-tor.

Dan.Is. (n.d.). *TOR Nodes List*. Dan.Is. https://www.dan.me.uk/tornodes

Evers, B., Hols, J., Kula, E., Schouten, J., den Toom, M., van der Laan, R. M., & Pouwelse, J. A. (2015). *Thirteen years of tor attacks*. Github. https://github.com/ Attacks-on-Tor/Attacks-on-Tor.

Ferry, N. (2019). *Methodology of dark web monitoring*. IEEE. doi:10.1109/ ECAI46879.2019.9042072

Gulati, H., Saxena, A., Pawar, N., Tanwar, P., & Sharma, S. (2022, January). Dark Web in Modern World Theoretical Perspective: A survey. In *2022 International Conference on Computer Communication and Informatics (ICCCI)* (pp. 1-10). IEEE. 10.1109/ICCCI54379.2022.9740785

Ibrahim, R. F., Abu Al-Haija, Q., & Ahmad, A. (2022). DDoS Attack Prevention for Internet of Thing Devices Using Ethereum Blockchain Technology. *Sensors (Basel)*, *22*(18), 6806. doi:10.339022186806 PMID:36146163

Kaur, S., & Randhawa, S. (2020). Dark web: A web of crimes. *Wireless Personal Communications*, *112*(4), 2131–2158. doi:10.100711277-020-07143-2

Mirea, M., Wang, V., & Jung, J. (2019). The not so dark side of the darknet: A qualitative study. *Security Journal*, *32*(2), 102–118. doi:10.105741284-018-0150-5

Murphy, E. V., Murphy, M. M., & Seitzinger, M. V. (2015). Bitcoin: Questions, Answers, and Analysis of Legal Issues. *Congressional Research Service*, 1-36. https://fas.org/sgp/crs/misc/R43339.pdf

Naseem, I., Kashyap, A. K., & Mandloi, D. (2016). Exploring unknown depths of invisible web and the digi-underworld. *International Journal of Computer Applications, NCC*, (3), 21–25.

Nasr, M., Bahramali, A., & Houmansadr, A. (2018). DeepCorr: Strong flow correlation attacks on tor using deep learning. In *Proceedings of the 2018 ACM SIGSAC Conference on Computer and Communications Security* (pp. 1962–1976). ACM. 10.1145/3243734.3243824

O'Neill, P. H. (2013). Inside the Bustling, Dicey World of Bitcoin Gambling. *The Daily Dot*.

Rafiuddin, H. (2017). *A Dark Web story in-depth research and study conducted on the dark web based on forensic computing and security in Malaysia*. IEEE. doi:10.1109/ICPCSI.2017.8392286

Rapid 7. (n.d.). *Types of Attacks*. Rapid7. https://www.rapid7.com/funda mentals/types-of-attacks/

Rudesill, D. S., Caverlee, J., & Sui, D. (2015). *The deep web and the darknet: A look inside the internet's massive black box*. Ohio State Public Law Working Paper No. 314.

Satterfeld, J. (2016). FBI Tactic in National Child Porn Sting under Attack. *USA Today*. https://www.usatoday.com/story/news/nation-now/2016/09/05/fb i-tactic-child-pornstingunder-%0Aattack/89892954/

Tor Project. (n.d.a). *Relay Search*. Tor Project. https://metrics.torproject.org/rs.html

Tor Project. (n.d.b). *Types of Relays*. Tor Project. https://community.torproject.org/relay/types-of-relays/

Tor Project. (n.d.c). *Welcome to the Tor Bulk Exit List exporting tool*. Tor Project. https://check.torproject.org/cgi-bin/TorBulkExitList.py

Yang, L. Y. (2019). Dark Web forum correlation analysis research. ITAIC 2019, Chongqing, China. doi:10.1109/ITAIC.2019.8785760

Zane, P. (2014). How to Navigate the Deep Web. *ISSUU*, (3).

Zhang, N. (2020). A generative adversarial learning framework for breaking text-based CAPTCHA in the Dark Web. National Science Foundation.

Zillman, M. P. (2015). *Deep Web Research and Discovery Resources 2015*. LLRX. https:// www.llrx.com/2019/01/deep-web-research-and-discover y-resources-2019/

Chapter 6
Introduction to Ransomware

Qasem Abu Al-Haija

iD https://orcid.org/0000-0003-2422-0297
Princess Sumaya University for Technology, Jordan

Noor A. Jebril
Princess Sumaya University, Jordan

ABSTRACT

Ransomware can lock users' information or resources (such as screens); hence, authorized users are blocked from retrieving their private data/assets. Ransomware enciphers the victim's plaintext data into ciphertext data; subsequently, the victim host can no longer decipher the ciphertext data to original plaintext data. To get back the plaintext data, the user will need the proper decryption key; therefore, the user needs to pay the ransom. In this chapter, the authors shed light on ransomware malware, concepts, elements, structure, and other aspects of ransomware utilization. Specifically, this chapter will extend the elaboration on the ransomware, the state-of-art ransomware, the ransomware lifecycle, the ransomware activation and encryption processes, the ransom request process, the payment and recovery, the ransomware types, recommendation for ransomware detection and prevention, and strategies for ransomware mitigation.

1. INTRODUCTION

In the last years, ransomware was grown very quickly, with many disturbing trends pointing to effective and targeted attacks against either organizations or individuals. These profiteering attackers indiscriminately aim at public and private sector entities for maximum gain (Bajpai & Enbody, 2020).

DOI: 10.4018/978-1-6684-8218-6.ch006

The word ransomware is derived from "ransom" and "malware". Ransom is described as "money paid to release a person who has been captured or abducted" and as "paying to return or required for the free of a person or thing from a vanishment". The word "malware" means malicious software (Ali, 2017) since attackers use it maliciously. Finally, ransomware is a sort of malicious software that applies encipherment as a main weapon (Glassberg, 2016).

O'Gorman and McDonald (2012) gave a fuller description as "a class of malicious software that, when run, disrupts computer functionality in some way. The ransomware shows a message demanding payment to retrieve functionality. This definition illustrates ransomware when it disrupts computer functionality and is later called an "extortion racket." Still, there is no specific information about paying the ransom in exchange for keeping the computer ransom. In other cases, ransomware can do certain harm to computer data files. Although it can access any file on a computer, ransomware often aims at certain types of files (O'Gorman & McDonald, 2012). Luo and Liao (2007) showed that ransomware targets files with the following file name extension: .txt, .doc, .rft, .ppt, .cbm, .cpp, .asm, .db, .db1, .db1, .dbx, .cgi, .dsw, .gzip, .zip, jpeg, .key, .mdb, .pgp, and .pdf.

The Cyber Threat Alliance provided the most integrated meaning for the ransomware. This alliance is a group of cyber security companies formed in 2014 to track cyber threats. In its first report published in 2015, the Cyber Threat Alliance (2015) gave the following definition of ransomware:

Ransomware is a kind of malware that encrypts the victim's files and then orders paying for the key that enables decrypting mentioned files. When ransomware is first installed on a victim's device, it usually aims for critical files such as important financial data, business records, databases, personal files, etc. Personal files, such as photos and home movies, may have sentimental value to the victim.

Since then, the attack has become automated and professional. It is considered very profitable, with past damages estimated at hundreds of millions of dollars annually. For example, the harm from one type of ransomware, CryptoWall3, was estimated to be more than $320 million in 2015 alone (Cyber Threat Alliance, 2015).

Consumers are believed to be the most common victims of ransomware (Symantec, 2017; Kaspersky, 2016). While most attacks are believed to be untargeted, consumers are often less likely to have robust security, which increases the likelihood of dropping victim to an attack (Symantec, 2017). However, regarding the damage that ransomware can cause, relatively little is known about the spread and characteristics of such attacks in the general population. Reliable estimates of ransomware prevalence are essential to understanding the nature of the threat landscape today and for long-term comparison and analysis.

Many government and industry organizations and researchers tried to document the phenomenon but results often needed to be more consistent. This is largely due to the non-representative data on which they are based. Industry reports are usually published by security companies based on the users of their software products. Thus, such samples are inevitably biased toward consumers with adequate security realization and financial resources to buy such products. Thus, their experiences may not mirror those of the common users.

In contrast, government agencies usually report rates based on voluntary victim reports. These assessments are believed to underestimate the true rate (Redmiles et al., 2017) significantly. For example, the US Department of Justice estimates that only 15% of fraud victims in the country report their crimes to law enforcement (United States Department of Justice, 2015); however, it is ambiguous what the correct average of reporting is in the common inhabitance. Apart from the difficulty in characterizing the extent of the problem, little is known about the factors and behavioral patterns that place individuals at risk of such attacks. Devising accurate risk assessment methods to identify the vulnerable population is particularly relevant for ransomware attacks, as infection may impose an especially high cost to consumers. There is often little recourse for victims who want to retrieve their data other than paying the ransom. Just identified, data about vulnerable groups can be used to devise proactive strategies to reduce the effects of ransomware attacks on the most vulnerable individuals.

For example, vulnerable people may be affected through several means, including in-person educational resources and training or discounted offers of services to reduce the effects of contagion (e.g., cloud-based data backup services). If they know they are at risk of infection, consumers may be more motivated to adopt preventative measures to reduce the effects of a possible attack. Ransomware caused a rise the damage to computer resources; Figure 1 shows the annual ransomware damages (Mohammad, 2020). Below are a few more startling statistics regarding ransomware:

- *Ransomware attacks have increased by 97% since 2017 – AttackIQ.*
- *34% of those affected took a week, if not more, to restore full access, up from 29% in 2016 – Kaspersky.*
- *Ransomware generates over $25 million in revenue for hackers each year – Business Insider.*
- *A new business will fall victim to a ransomware attack every 14 seconds in 2019. In 2021, that number will be every 11 seconds – KnowBe4.*

Figure 1. Overview of ransomware damages

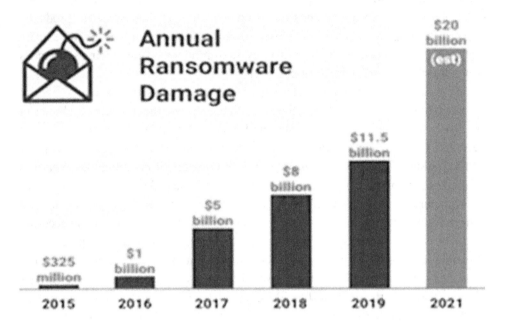

2. RELATED WORK

2.1 Estimates of Ransomware Victimization

There are conflicting versions of the date ransomware began to infect machines, its first form, the amount of money they used to demand in exchange for restoring access to data, and the areas from where it first appeared. The discrepancy in accounts might be primarily attributed to the fact that ransomware initially appeared in a smaller form than what we see now, had a different mechanism for spreading, and required a different method of money collection.

For instance, Glassberg (2016) observed that ransomware has been around for a while but did not offer a timeframe for when this virus first appeared. On the other hand, Hampton and Baig (2016) explained that although extortion of money from users has evolved over the past three decades, these activities remained "unsophisticated" until 2011, when cybercriminals discovered how to entice users to their websites, trick them into downloading malware, encrypt their files, and then demand money from them. According to Kharraz, Robertson, Balzarotti, Bilge, and Kirda (2015), ransomware first appeared in about 2004. Still, the number of assaults did not become substantial until nearly ten years later, in 2013, when ransomware gained considerable media attention.

Salvi and Kerkar (2016) went into much information about ransomware's emergence. They explained that the first virus used to extort money from computers did so in 1989 when it replaced the original "autoexec.bat" file with a new file to infect machines. It would take a while for the new "autoex-ec.bat" to locate the computer and show a message requesting payment. Salvi and Kerkar went on to say that the computer would be locked until a payment was made to a post office box in Panama, at which point they would send a floppy disk with the fix for the locking issue.

The conclusion that can be drawn from the information above is that between 2012 and 2014, ransomware began to take on the form that we see today. Two reports from Cyber Threat Alliance (2015) and Cyber Threat Alliance (2016) provide proof for this. Both reports show that Cryp-to-wall 3 and Cryp-to-wall 4 (two ransomware variants) first appeared in 2014 and 2015, respectively. Different reports make varied notes regarding the amount of money collected due to ransomware attacks. For instance, the American Bankers Association (2016) estimated that the ransom demanded releasing the decryption key might range from a few hundred to thousands of dollars, depending on who was infected with the virus. When cyber extortion initially began, specific data on the amount charged was supplied by Salvi and Kerkar (2016). They pointed out that the first ransom demand was made on victims in 1989 when they were required to mail $189 to PC Cyborg Corp. at a Panama post office box. As online communication improved and new digital currencies began to be used in the exchange of currency, this pricing was altered.

The author discovered through literature research that the price demanded was often equal to one bitcoin (about $500). Similar figures were cited by Salvi and Kerkar (2016), who also pointed out that ransomware attackers typically ask for $300 to $500 in exchange for disclosing the decryption key and restoring functionality. Even Mac users continued to receive the same question. According to Heater (2016), ransomware for Mac began to surface on the market and demanded a $500 payment to decrypt the data. In other instances, various extreme figures were indicated as the amount to be paid for providing access to encrypted data. For instance, Everett (2016) mentioned that a hospital attacked with malware was first demanded to pay a $3.5 million ransom before negotiating it down to $17,000.

The geographical origin of the ransomware and how it propagated to other regions were also subject to conflicting claims. According to Glassberg (2016), ransomware initially only affected Russia and Eastern Europe. According to O'Gorman and McDonald (2012), ransomware originally surfaced in Russia and other countries where the language was spoken in 2009. It subsequently moved to Western Europe, the United States, and Canada around 2010. The information mentioned above all pointed to Russia and Eastern Europe as the original origins of ransomware. A

Table 1. Comparison of top hit countries (Cyber Threat Alliance, 2015)

CryptoWall v3	CryptoWall v4
United States	United States
Canada	India
United Kingdom	Canada
Australia	Mexico
Russia	United Arab Emirates
Germany	France
India	Romania
Italy	Taiwan
France	Jamaica
Netherlands	Bulgaria

separate account of a gang holding captive medical data was available. The group claimed Turkish ancestry (Everett, 2016).

Even if computers in Russia and Eastern Europe may have first been infected by ransomware, that has changed. All G20 nations experience ransomware attacks regularly, with the United States being the most often attacked country (Glassberg, 2016). The two ransomware variants, CryptoWall 3 and CryptoWall 4 were most prevalent in the nations mentioned by the Cyber Threat Alliance (2015). This ranking from the CyberThreat Alliance is shown in Table 1.

In 2016, there were 2,673 reports of ransomware made to the FBI's Internet Crime Complaint Center, which is equivalent to losses of nearly $2.4 million (Federal Bureau of Investigation (FBI), 2016). One thousand seven hundred eighty-three ransomware reports and an estimated $2.3 million in damages were reported in 2017 (Federal Bureau of Investigation (FBI), 2017). However, government figures are known to significantly understate the actual rate as they are based on voluntary self-reports.

Another data source is industry reports that include user feedback on antivirus software. Typically, these claims substitute blocked detections for genuine illnesses. As an illustration, Symantec estimates that between June 2016 and June 2017, it prevented 405,000 consumer ransomware attacks (Symantec, 2017). While security vendor studies benefit from not depending on self-reports, they are biased in other ways. Only the public segment that chooses to buy their security solution is shown in industry reports as having experience.

Because this sample is made up of people who may have a greater security knowledge of online risks, value the product, and have the financial means to purchase protection, it is likely not representative of the internet community. Additionally, blocked detections are erroneous indicators of infection. Modern machine learning techniques suffer from false positives, whereas traditional signature-based solutions

144

can only detect and stop known threats, ignoring newer assaults. Recently, researchers have estimated ransomware outbreaks using data from open Bitcoin transactions. As an illustration, Huang et al. (Symantec, 2017) offer a lower-bound estimate of 19,750 possible victims worldwide who paid a ransom in Bitcoin. They accomplish this by: (1) Scraping ransomware infection reports from public forums and lists of seed ransom addresses from commercial databases that keep track of ransomware victims and the ransom addresses linked to them. (2) extracting ransom addresses by running several ransomware binaries in a controlled environment. Although the available measuring system enables large-scale victim rates assessment, it can only offer insights into one payment type: Bitcoin payments (Al-Haija, 2022). As we shall demonstrate, our data imply that concentrating solely on this payment mechanism could only give a partial view of overall infections.

2.2 Susceptibility to Ransomware (Being Vulnerable to Ransomware)

Traditional malware security strategies have been reactive and victim-agnostic, emphasizing recognizing assaults or attackers (such as phishing emails, malicious websites, and files) (Halawa et al., 2016). For instance, various research (Cabaj & Mazurczyk, 2016; Kharraz et al., 2016; Kharraz, Robertson, Balzarotti, Bilge, & Kirda, 2015; Scaife et al., 2016; Zavarsky & Lindskog, 2016) provide technological, automated techniques to stop ransomware assaults. These designations are frequently given to tiny, unrepresentative sample sizes and include various situations and demographic sub-segments. Because of this, it has been challenging to make conclusions about the overall significance of demographic, environmental, and behavioral variables on victimization risk.

To evaluate the impact of individual and situational characteristics on seven forms of cybercrime victimization, including computer viruses, Ngo and Paternoster (2011) employ the general theory of crime and everyday activities (Choi, 2008). They conducted an online survey of 295 students in the United States who self-report being victims of cybercrime, and they discovered that the risks of getting a computer infection were much greater for non-white students and younger students. Contrary to popular belief, they also discovered that people who frequently opened any unknown attachments or clicked on web links in the emails they received, opened any file or attachment on their instant messengers, and frequently clicked on an intriguing pop-up message, had lower odds (by about 35%) of getting a computer virus.

To research the factors that increase the likelihood of data loss brought on by malware infection, Bossler and Holt (2009) polled 788 college students. The elements investigated include "deviant" behavior (such as downloading pirated media and accessing adult websites), regular activities (such as using social media, programming, and shopping), parental controls (such as using antivirus software and sharing passwords), and computer proficiency. According to the authors, having a job and being a woman made you more likely to become a victim of the malware. Only pirating material raised the chance of malware infection; engaging in deviant conduct was often not a strong predictor of malware infection. Strong computer skills and good password management did not lessen the projected threat of malware victimization, and guardianship had only minor roles in explaining infections.

Milne et al. conducted a nationwide online survey of 449 US internet buyers (Milne et al., 2009). They discover that users' propensity to engage in dangerous online activities, excluding email, is significantly influenced by their gender, age, and number of online hours. They conclude that users who are male, younger, and spend much time online are more vulnerable. Researchers have recently opted for large-scale, data-driven methodologies to anticipate user risk from various cyber dangers. Using DSL data logs of anonymized network traces, Maier et al. (2011) investigate whether the likelihood of producing malicious traffic is connected with security hygiene. They discover that visiting URLs on blocklists more than doubles the risk while maintaining good security hygiene (such as installing operating system software updates) has little to no link with risk. Lévesque et al. (2013, 2014) used instrumented laptops from a clinical trial of an antiviral product to monitor the malware exposure of 50 individuals over four months. They discover that exposure to malware is associated with a high self-reported degree of computer proficiency, more frequent file downloads and app installs, and frequent surfing. The writers come to conflicting conclusions regarding user age and website content categories. Ovelgönne et al. (2017) investigate the association between the number of attempted malware assaults discovered and user profiles using Symantec telemetry for a subset of 1.6 million users over eight months. The authors divide users into four groups (gamers, professionals, software developers, and others). They discover a subpopulation of gamers with particularly destructive behavioral patterns and that software developers are likelier to engage in harmful cyber behaviors.

Strecher and Rosenstock (1997) and Bilge et al. (2017) investigate specific user-level malware exposures in an enterprise environment. Yen et al. leverage user demographics, VPN data, and online proxy logs from a sizable multinational company. The writers look at characteristics of visited website categories, total amounts of web traffic, and links to prohibited or low-reputation websites. They discover that of the three feature categories, user demographics is the highest predictor of risk, followed by VPN behavior, using a logistic regression model to infer the likelihood of hosts encountering malware. Contrary to expectations, only a small amount of the entire model was affected by web activity. The scientists concluded that this was because just 3% of the hosts were exposed to web-based malware.

3. RANSOMWARE LIFECYCLE

The increasing use of Internet of Things (IoT) devices across various industries has created a massive new attack surface for ransomware, comprising billions of endpoints. IoT devices often lack tight security controls to stop multiple cyber-attacks (Ibrahim et al., 2022). Attackers can easily and quickly adapt their malware to industrial sensors, healthcare devices, or self-driving cars. Ransomware is often expanded via phishing emails (or increasingly text messages) that include links or attachments—think of—fake emails from Netflix, your bank, even IRS ads, and fake ads with malicious links. It unleashes malware on the target device. Or—via a drive-by download, infecting users when they visit an infected website. During this visit, malware is automatically downloaded without the victim's knowledge—a big reason why Google warns you when you visit an unsafe website. However, malware enters your network through executable files, and the victim's data/files are encrypted. The hijackers then send a "ransom note" – a request for payment and a deadline, usually in encrypted currency, after which the victim gets the key to decrypt the files. From there, the attackers typically threaten that the files will be destroyed or sold if the ransom is not received by date X.

Figure 2 shows a ransomware lifecycle and a broad description of the functionality of a ransomware attack. It provides a step-by-step life cycle for a typical ransomware attack.

Figure 2. Lifecycle of ransomware

1. **Launch an attack:** Hackers write/modify ransomware programs that are based on the attacking paradigm. After that, they decide on the delivery method - (such as email, SMS, PE files, and others) and obtain preliminary access. This phase can be accomplished using various methods like creating a malicious website, utilizing a vulnerability in a remote computer connection, or initiating an immediate incident on a vulnerability using certain software packages (unverified applications). However, the frequent and usual entry holes are phishing/smishing operations that thrive access to Trojans such as TrickBot, or Bazar tools. The hackers often focus on less-skilled employees since they are likelier to click links or submit contact details. In fact, having more users in the company means a larger attack surface for user-centric attacks. One mistake by a single user is enough for a hacker to execute code and infect the entire system. Various methods are utilized to transmit the malware codes to the victim's host (Anghel & Racautanu, n.d.), including:

    ○ *Drive-by download: this happens whenever the system is configured to automatically downloads malicious codes without the knowledge of the end user.*
    ○ *Phishing emails: may be spam or specifically designed for a company or sector. These emails may provide links to dangerous websites or attachments.*
    ○ *Exploiting security holes in systems accessible via the internet or a network: Without requiring human involvement, hackers can infect susceptible devices by searching networks or the internet for them.*

Ransomware detection is difficult due to the nature of the threat. Despite this, several researchers discuss several methods and strategies that could aid in avoiding and identifying ransomware. We must be aware of the techniques employed by ransomware to infiltrate our devices and organizations to take the greatest precautions for its detection and prevention. Figure 3 displays the majority of ransomware techniques (VPN.com, n.d.). The three most common delivery methods for ransomware are web programs, email attachments, and links in emails, as seen in figure 3.

Table 2 shows various lists of ransomware families, their deployment strategies, the dates they first emerged, the cryptographic methods employed to encipher files, and the C-&-C procedures (VPN.com, n.d.).

2. **Misuse, Develop, Realize:** Whenever the ransomware gains preliminary entrance to the victim's computer, the malicious code establishes a communication path with the attacker by employing some other types of remote access malware such as Metasploit or remote access tools (RATs). This phase turns the victim's

148

Figure 3. Ransomware distribution methods

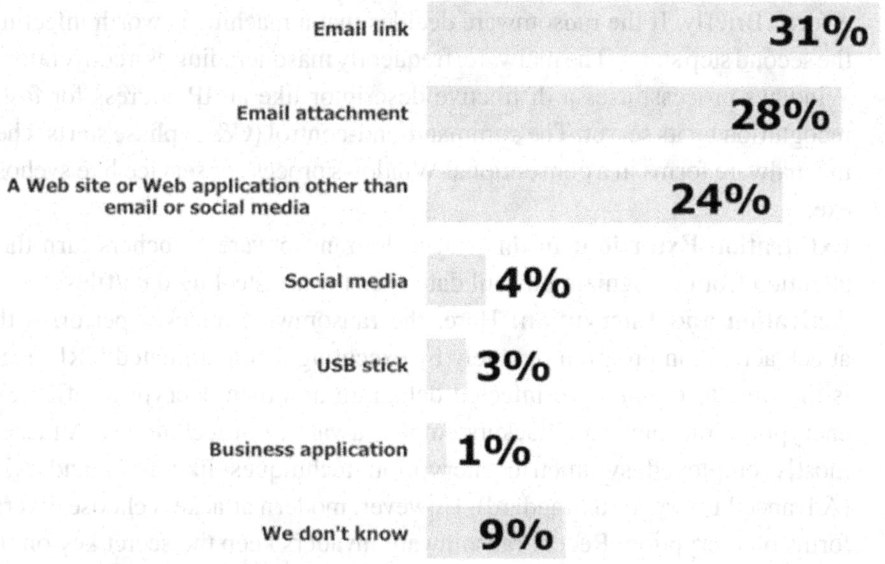

Email link	**31%**
Email attachment	**28%**
A Web site or Web application other than email or social media	**24%**
Social media	**4%**
USB stick	**3%**
Business application	**1%**
We don't know	**9%**

Table 2. Different ransomware families

Families	Deployment	Date Appeared	Cryptographic Techniques	Command and Control
Reveton	Drive-by downloads	2012	RSA and DES	MoneyPak
GpCode	Email Attachments	2013	660-bit RSA and AES	Tor Network
CryptoLocker	Compromised websites and Email Attachments	2013	2048-bit RSA	Tor Network
CryptoWall	Compromised websites and Email Attachments	2013	2048-bit RSA	Tor Network
FileCrypto	Compromised websites and Email Attachments	2013	2048-bit RSA	Tor Network
TeslaCrypt	Compromised websites and Email Attachments	2013	2048-bit RSA	Tor Network
CTB-Locker	Email Attachments	2014	ECC	Onion Network
Shade	Spam Email	2015	RSA-3072 and AES-256	Fixed Server
Jigsaw	Word Document with Javascript	2016	RSA and AES	Onion Network
WannaCry	Samba Vulnerability	2017	RSA and AES	Onion Network

computer to be under the control of the C&C server. It's a reconnaissance mission by cyber attackers to figure out how to get the most ransom from their efforts. Briefly, If the ransomware decides that a machine is worth infecting, the second step starts. The malware, frequently masquerading as a conventional Windows process, uses a distinctive descriptor like an IP address for faster recognition by its server. The command-and-control (C&C) phase starts when the malware forms in a conventional Windows process or service like svchost. exe.

3. **Exfiltration-Extortion:** In this stage, the ransomware launchers turn their attention from recognizing useful data/files toward stealing data/files.

4. **Activation and Encryption:** Here, the ransomware attacker performs the attack activation program remotely by executing a contaminated URL. Here is the time to execute the infected datagram and then encrypt the file. For encryption, ransomware attackers employ a variety of techniques. Attackers mostly employed symmetric encryption techniques like RC4 and AES (Advanced Encryption Standard). However, modern attackers choose diverse forms of encryption. Recent ransomware invaders keep the secret key on the C&C's servers, which only transfer the public key to the infected system. The first phase of encryption uses a symmetric key, and the next phase utilizes an RSA cryptosystem. This indicates that symmetric encryption is carried out initially, followed by asymmetric encryption in the second stage. Files cannot readily be decrypted as a result (Adamov & Carlsson, 2017; Celiktaş, 2018; Lee, 2019; Liska & Gallo, 2016).

The quality of the utilized encryption algorithm influences the ransomware's strength. Ransomware encryption uses hybrid techniques that are carried out in three phases (Adamov & Carlsson, 2017; Celiktaş, 2018; Kotov & Rajpal, 2014; Lee, 2019; Liska & Gallo, 2016). The ransomware attacker creates an asymmetric pair of keys and inserts them inside the ransomware in step 1. After the ransomware has been loaded or activated, encryption using symmetric keys is executed on the victim workstation in phase 2. Moreover, the symmetric key used for encryption in step 2 is likewise encrypted using the public key supplied by the C&C server. Phase 2 also involves encryption and the deletion of the symmetric key, making it harder to recover the original data. Phase 2 concludes with a pop-up message containing instructions for making a payment. In phase 3, assuming money has been received, the attacker will use the victim's private key to decrypt the asymmetric ciphertext before sending them the symmetric key; Figure 4 shows the lifespan of a typical crypto-ransomware key (Luo & Liao, 2007).

Figure 4. General crypto-ransomware key lifecycle

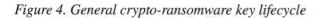
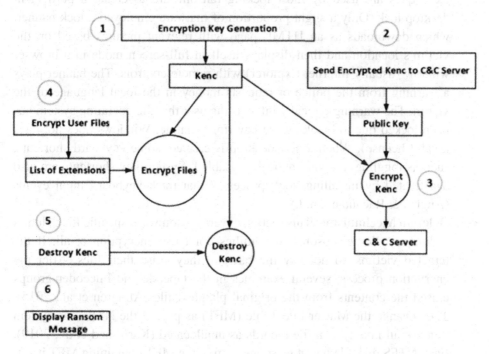

Significantly a host is controlled by a server run by the illicit attacker, as the malware program is currently active on the victim's host. The host will use a predetermined handshake protocol to verify that it is interacting with the right server. The generation and exchange of keys come next after the client and server successfully authenticate one another. While some ransomware uses simple ciphers, others, like 4.096-bit RSA, use complicated ones. Public key algorithms (Asytmetrics) only send the public key to the host; the attacker's server keeps the private key. This is when the ransomware significantly alters the target computer by locking it or encrypting the contents (Figure 4). In this event, after the command stage, all the files (which could be papers, photos, sound files, etc.) identified for encryption are now being encrypted. It can be challenging for the victim to specify how far the assault has evolved or what is been encrypted because some malware encrypts the files and filenames.

- **Locking Procedure:** The attack's success in the case of locking ransomware depends on the system's locking. Usually, this is accomplished by setting up a new persistent desktop. Simple changes to the Desktop environment and the removal of pointless programs are all that ransomware samples do. The instance that takes input from the victim is the new desktop. Similar techniques are used by most locking ransomware to create a permanent desktop lock. Only a slight proportion of malware employed a lock banner, which downloads as an HTML page with relevant pictures based on the victim's location and then displays itself in full-screen mode in a browser window (usually Internet Explorer) with remote controls. The banner plays a warning from the police or requests money in the local language of the victim. The warning typically informs the user that the operating system has been locked due to breaking the law (for example, Windows does not have a valid license). When a new desktop is created, some keyboard shortcuts, such as toggling, are automatically disabled. System key shortcuts are also deactivated by installing hook processes that track keyboard input events (Anghel & Racautanu, n.d.).

- **Deletion Mechanisms:** This section covers ransomware-specific file deletion techniques. Some ransomware families don't use encryption at all. If the targeted victims do not pay the ransom, they erase their files. After the encryption process, several examples in the Gpcode and Filecoder groups erased the contents from the original plaintext file (Almgren et al., 2015). They change the Master File Table (MFT) as part of the NTFS file system to mark all non-system file records as unallocated (Karresand et al., 2019). Each NTFS disk file has at least one item in the MFT containing MFT itself. A file's metadata, including its size, time and date stamps, permissions, and data content, are either kept in MFT entries or in external storage spaces that MFT entries specify. The size of the MFT remains the same, though the disk space set aside for these entries needs to be reassigned.

- **Changing Master Boot Records (MBR):** MBR is the first sector where the data of a hard or detachable drive is first located. For the computer's random-access memory (RAM) to be loaded, it needs to be aware of the mechanism and location of the operating system (OS) in the RAM. Such a program (boot loader) responsible for the OS loading into RAM is integrated into the MBR. Therefore, MBR is a rich area to be targeted by several ransomware groups such as Seftad or Petya (Ghania, 2016), which house the partition table and the executable boot code.

5. **Ransom Request:** This is the phase when the victim receives a message describing how things should work - ransom demand, amount, timeframe, wallet/transfer method used to send money, and consequences for non-acquiescence. This is also affected by the currency type and the funds' distribution. It is worth noting that cryptocurrencies are essentially imperceptible due to the decentralized environment of the blockchain. There are several ways to make the extortion payment. Some ransomware variations permit the victim to decrypt one file for free to demonstrate

that a key to the system exists. Escalating payments is another kind where the cost rises over time.

6. **Payment (Or Not...) and Recovery:** Things can proceed differently based on the victim's choice. Choosing to pay the ransom, the victim's files are/should be unencrypted, informing law enforcement of the intrusion, revising cybersecurity statements, and learning as of the practice. Table 3 shows the major ransomware families' payment methods (Almgren et al., 2015).

Although it is uncommon, recovering money is feasible. Here, law enforcement there has been some success, especially in the case of the attack on the Colonial Pipeline. In one instance, the Bitcoin price fell soon after the attack, allowing law enforcement to recoup 63.7 of the 75 BTC, or $2.3M. So, yes, even if everything goes as planned, you risk suffering significant losses due to the inherent volatility of cryptocurrencies.

Table 3. The major ransomware families' methods of payment

Families	Type of Charge			
	Premium Number	Untraceable Payments	Online Shopping	Bitcoin Transactions
Reveton		✓	✓	
Cryptolocker		✓		✓
CryptoWall				✓
Tobfy		✓		
Seftad	✓			
Winlock				
Loktrom	✓			
Calelk	✓			
Urausy		✓	✓	
Krotten		✓		
BlueScreen		✓		
kovter		✓	✓	
Filecoder		✓		
GPcode		✓		
Weelsof		✓		
WannaCry				✓

Choosing not to pay the ransom, robbed information is lost, and significant intellectual property and user information may be sold or leaked into the world. Your business's reputation suffers significantly. This probably leads to losing clients and people's confidence, which makes it much more difficult to get back lost clients. Then there are the issues of regulatory penalties, legal actions, and damages. However, remember that neither paying the ransom nor restoring data from a backup assures your safety. With payment, resale is still a possibility. There is only sometimes unequivocal evidence that the hacker has no copy of the ciphered data. Additionally, the danger of resale increases with the value of the information. Please remove any malicious files or ransomware code that may still be present in your system and close any residual security holes or vulnerabilities before continuing.

However, the assault itself might offer you a general sense of the ransomware that affected your machine, making it simpler to find and get rid of. However, it's a good idea to clean up your system in a secure sandbox environment to guard against the possibility of reactivation.

7.   **Final Thoughts:** Even with the finest security measures, ransomware attacks can still access your system. It's crucial to comprehend how ransomware functions, from penetration through activation, encryption, and beyond, to stop assaults from occurring in the earliest spot.

Properly securing them knows how to assist in mapping important endpoints, susceptibilities, users, and credentials. More importantly, it enables you to make security a key component of company culture and adopt solutions that simplify for everyone to adopt good habits, ensuring that security is integrated into the daily workflow. Velosio's Microsoft professionals work with clients to find and implement safety solutions that keep their operations secure in the cloud, actively benefit their clients, and outperform the competition.

4. RANSOMWARE TYPES

There are various varieties of ransomware. According to some analysts, ransomware has several variations and develops 100 new forms and patterns yearly (Caivano et al., 2017; Finkle, 2016; Wyke & Ajjan, 2015). Other studies discuss two basic categories of ransomware, while others discuss three or more. In this study, the author will exclusively concentrate on crypto and locker ransomware, which are thought to be the only two major types of ransomware by many experts and corporations. Figure 5 depicts the anticipated harm-related growth caused by ransomware attacks (Runciman, 2020).

4.1 Crypto Ransomware

A particularly hazardous form of ransomware, crypto-ransomware encrypts victims' files (common examples include WannaCry, Crypto-Wall, and Crypto-Locker). After that, the attackers demand payment within a predetermined window to unlock the files. There are various ways to get paid. Bitcoins are the most often used form of payment nowadays. Several sites claim that each attack costs about $300 in bitcoin (Christensen & Beuschau, 2017). The prevalence of crypto-ransomware is increasing, as is the number of attacks of this nature.

Several techniques exist for encrypting files, including getting the file, browsing its content, and enciphering it (and changing the file directory path is a further choice). Additional approaches entail read-overwrite-make new-delete original files (Held, 2018).

Figure 5. Growth of ransomware damage

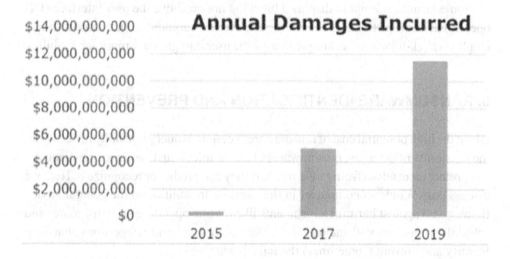

4.2 Locker Ransomware

Users cannot access their operating systems or computer resources due to this sort of locker ransomware (Runciman, 2020). Infecting a victim's computer and encrypting user files is known as locker ransomware. Users can access their files or data on the PC once the ransom is paid because of this lock.

4.3 Leakware (Doxware)

This type of ransomware differs from others in that it does not restrict the victim from accessing the operating system or other system resources; rather, it covertly collects personal data from the compromised computer and employs it to blackmail the victim. The acquired data is maintained on the attacker's server, and then the attacker threatens to disseminate the private data in case the victim decides not to pay (Anghel & Racautanu, n.d.).

4.4 Mobile Ransomware

Such type of ransomware targets portable devices such as smartphones and tablets. It is a simple blocker that stores and restores data to/from the cloud or any other backup solution. Therefore, in contrast to typical locker viruses, they depend on the value of the mobile device itself to pay the ransom rather than the data that has been saved on it.

Some viruses attempt to display a blocking notice above the user interface (UI) once it has been installed or started on the victim's computer. In contrast, others employ a click-hijacking technique to trick the user into giving it more accessibility.

5. RANSOMWARE IDENTIFICATION AND PREVENTION

Given the high potential cost of ransomware, a common query is usually raised about the possibility of detecting ransomware before the infection. Users and organizations can protect themselves from ransomware if they can predict or recognize it. Here we discuss only Windows computers in this section. In addition to the general advice to safeguard against harmful attachments like routine operating system updates and schedule scans, we will include up-to-date comments and suggestions that help identify and prevent (sometimes) the ransomware.

5.1 Observing File Activity and Incident Tracking

The observation of the file system activities can be effectively used to find ransomware. The System Service Descriptor Table (SSDT) contains features on the OS's services to distribute system calls. It could spot malicious requests by filtering out the process name and identity. Note that stopping ransomware spread is achievable if a log of SSDT calls is made. Shutting down every associated process accomplishes these (Christensen & Beuschau, 2017; Held, 2018; Lee, 2019). The CyberPoint research team conducted a study in 2017 (Christensen & Beuschau, 2017) and reported that

Windows operating systems' event tracing could be used to identify ransomware. Their strategy was focused on scrutinizing file events, including read, write, and size changes. To do this objective, an algorithm was created. This algorithm's high rate of false positives is a significant flaw (Mohammad, 2020).

5.2 Honeypots

A deceptive network system that is used to interest hackers and then grabs them. It uploads files to the network to catch the attacker off guard. If the attacker accesses honeypot files, the system will respond and become aware of the intrusion. Organizations benefit more from this form of detection than individuals do. Although ransomware detection using a honeypot may seem odd to some individuals, it is a legitimate security tool (Albulayhi & Al-Haija, 2022).

5.3 Educate Users

Most ransomware attacks, among other cyber security attacks, negligent target personnel. Most employees use simple and predictable passwords, while others may share theirs with friends and family or write it down in their office. The author thinks teaching and training users on security policy can prevent many cyber-attacks. Users also play a crucial part in cybersecurity. Utilizing security policies is crucial in every firm. Every firm must adhere to a clear security strategy to defend itself from dangerous malware. For instance, creating a culture that prioritizes security, training new hires, implementing a security strategy, and assessing how well it works (Albulayhi & Al-Haija, 2022).

5.4 Using Antiviruses

The most popular method for defending against harmful malware is antivirus software. Various companies created numerous antivirus programs. Antivirus software uses various methods, including anomaly-based and signature-based detection methods. Each antivirus program has a separate signature database. If the file is looked at, it is evaluated, and its signature is checked against a database of signatures. Some antivirus software uses the heuristic module to evaluate the code directly.

Unfortunately, antivirus software does not provide a complete solution to the difficulties caused by malicious software and ransomware. Antivirus programs use ransomware behavior analysis to detect the threat. Most antivirus software can identify ransomware, but it cannot be stopped once it has taken over your laptop. The answer to the question "Can antivirus software stop ransomware?" is yes and no. While antivirus software can stop several forms of ransomware before they ever start, they are powerless to stop it after it takes hold.

5.5 Machine Learning Approaches

Various applications, such as anomaly recognition, malware categorization, and spam identification, use machine learning techniques (Abu Al-Haija & Zein-Sabatto, 2020; Zitar & Mohammad, 2011). Doubtless, ransomware detection using machine learning techniques is possible. The traffic inside the network can be categorized into connection-based, encryption-based, and certificate-based (Modi, 2019). However, it is challenging to detect ransomware, so tools and a strategy for keeping an eye on network and file activities are required. A method to anticipate ransomware will likely be developed by examining the typical behaviors of malware. A methodology for ransomware detection using reverse engineering has been developed by Subash Poudyal (Poudyal et al., 2018). The foundation of this approach is a working multi-stage investigation model, which includes row binaries, assembly codes, libraries, and function calls which can a detection ranged with accuracy ranging from 76%-97% based on the machine learning technique method.

6. STRATEGIES FOR MITIGATION

Given the efficiency of phishing emails and drive-by downloads from otherwise trustworthy websites, ransomware outbreaks are only partially prevented (Anghel & Racautanu, n.d.). However, users can significantly lower this risk by employing network security techniques and enhancing threat consciousness and protection procedures. A thorough data backup mechanism is the most efficient way to reduce the likelihood of data damage from a fortunate ransomware incident. The following provides comprehensive suggestions that can lessen the hazard caused by ransomware:

1. Safeguard of Data:
   ◦ *Backups made offsite and regularly checked for consistency*
   ◦ *Backup of files online*
2. System Management
   ◦ *up-to-date antivirus software*
   ◦ *The SO, software, plugins, and browsers receive automatic upgrades*
   ◦ *Reliable user access management*
   ◦ *Turn off unnecessary wifi connections by using the application allow listing.*
   ◦ *Turn off macros in Microsoft Office programs.*

- ○ *Use ad-blocking browser add-ons to ward off "drive-by" infections*
- ○ *Deactivate Windows PowerShell and ScriptHost.*
- ○ *Deactivate remote connection protocols (Telnet, SSH).*
- ○ *Configure the OS using the Group Policy Editor (or using CryptoPrevent: free tool) to restrict PE files (such as .exe,.rar,.pdf) from operating in places like appdata, localappdata, temp, and the recycle bin.*
- ○ *Use access policy control to prevent harmful websites, check incoming/ outgoing emails, downloads, and attachments, and prevent emails with doubtful file extensions (such as .exe, .vbs, and. Scr) from being dispatched.*
- ○ *Think about using an anti-ransomware program that is either free or sold by reputable computer security companies.*

3. Network Administration
    - ○ *Verify that the firewall is activated and set up correctly.*
    - ○ *Lock and pay attention to unused ports*
    - ○ *Block Tor IP addresses that are known to be harmful.*

4. Device Management for Mobile
    - ○ *Only download material from the official stores, turn on two-factor authentication for Apple iOS devices and avoid "jailbreaking" the device.*

Figure 6. Quantity of ransomware types included in each survey paper

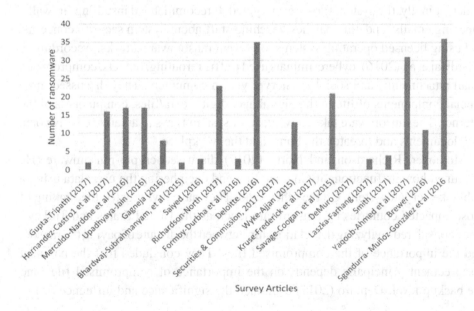

　○　*Prevent "rooting" an Android device by unchecking the "unknown sources" option.*

Finally, Figure 6 provides other relevant surveys in ransomware research (Brewer, 2016) that can provide more insights into the ransomware problems and solutions.

Hernandez-Castro et al. (2017) examined the financial evaluation of several significant ransomware (such as Crypto-locker, Tesla-Crypt, etc.). This drives up the price for the offender, yet consistent data backup will deter them from their illegal behavior. Because ransomware and other malware can easily infect Android devices, Mercaldo et al. (2016) and Albert et al. (2016) focused on analyzing the Android environment. They proposed an automated model-checking technique that can inevitably scrutinize malware instances by manually inspecting their behavior and employing predefined logic regulations to recognize ransomware instances.

According to Upadhyaya and Jain (2016), the executable files, such as task manager and command prompt, are often blocked by ransomware, which makes the prospective system inoperable. The research narrows its emphasis on CTB Locker, analyzes its attack method, and describes its method to accomplish a Bitcoin wallet meant for each victim and a disbursement method employing the dark web (i.e. Tor) gateway. Some physicists suggested creating quantum encryption systems free of mirage-like flaws. In contrast, others suggest that the best course of action is to regularly back up all digital assets and to safeguard them beforehand against assault. Additionally, Gagneja (2017) highlights several methods that ransomware uses system security flaws to spread infestations by executing out-of-date software on victims' computers. Then, to hinder system restoration, it deletes the backup files and folders. Finally, the system files are encrypted. It recommended installing firewalls, checking emails, updating patches, teaching staff about system security concerns, and using licensed operating systems to prevent ransomware attacks. According to Bhardwaj et al. (2016), cybercriminals used traffic rerouting, botnets, compromised email attachments, and social to encrypt victim computers. They discussed three crucial components of life in the digital age: digital data/files, computers, and the internet. The ransomware takes advantage of system flaws to assault sensitive data and documents and threaten the home and the workplace.

Moreover, Richardson and North (2017) discussed crypto-ransomware. He examined how its functionality in operating and enciphering the user data using a public key cryptosystem. On the other hand, They have investigated the ranking of most impacted countries by ransomware (such as the United States and Turkey). They considered analyzing the relationship between the dispute of paying the ransom and the importance of the compromised files. They concluded that the payment disagreement principally depends on the importance of compromised files and the backup level. Demuro (2017) examined the significance and influence of law in the negotiation context. They recommended several options/methods, such as burdensome ransom payments, insisting on harsher cyber examination constraints, mandating examinations by specialists in the field of cyberspace, and employing effective intelligent methods for attacks detection with high sensitivity (Al-Qudah et al., 2023), which can be useful legitimate tools to address the ransomware issue.

Like Formby et al. (2017), they focused their investigation on the recent ransomware assaults on prominent hospitals. It was shown that failing to upgrade their network systems regularly and perceiving a long-term lack of danger to the industrial control system (ICS) were both factors in exposing sensitive data to hackers. Based on this, a revolutionary strategy of defense against ransomware that targets hospital PLCs was developed, along with strict regulations that would deter further attacks.

In their analysis of the history of ransomware, Deloitte (2016) divided ransomware into two types: (1) locker, which locks down computers after compromising, and (2) crypto, which encrypts user data and files and requests a ransom to be paid in exchange for the decryption key. It included widespread ransomware vectors and kinds and suggested methods for detecting, removing, and recovering from them. Securities and Commission (2017) also investigated the WannaCry ransomware attack. They demonstrated its infiltration mechanism that uses Remote Desktop Protocol (RDP) to manipulate some crucial Windows Server Message Blocks to perform its evil deeds.

It occasionally used phishing and pharming methods on harmful emails and websites. Reviewing the advisory released by the Computer Emergency Readiness division of the US Department of Homeland Security is necessary for protection against the WannaCry ransomware. Additionally, for Microsoft Windows (XP/8/Ser2003) OSs, providing efficient security requires reviewing relevant Microsoft updates. Wyke and Ajjan (Poudyal et al., 2018) thoroughly investigate the four most common ransomware variants—CryptoWall, TorrentLocker, CTB Locker, and TeslaCrypt—whose origins can be traced in the first days of FakeAV and Locker with an emphasis on providing insight into the current status of ransomware. It covered many facets of their functioning, infection modes, and each version's worldwide geographical prevalence. It also looked at Sophos HIPS Technology's proactive blocking of crypto-ransomware attacks. Three distinct searches were used by Kruse et al. (Strecher & Rosenstock, 1997) to perform their systematic review: CINAHL, PubMed (MEDLINE), and the Nursing and Allied Health Source through ProQuest databases. The outcome demonstrates that the healthcare sector is lagging in addressing the core information security and cyber threat challenges. It has been noticed that governmental policy actions and technical developments have significantly increased the healthcare sector's exposure to cyberspace. The usage of VLANs, DE authentication, cloud computing, and a defined method for software upgrades and addressing data breaches were all suggested as mitigating measures to be done to reduce the danger of a cyber assault on the healthcare industry. From a technological and psychological perspective, Otawi et al. (1982) .'s ransomware analysis demonstrates ransomware's progression from less appealing forms to direct income generating through deceptive programs to a further destructive degree employing the computer performance application programs.

Note that 11 of the top 12 countries affected by ransomware are G20 members, representing industrialized and developing economies that account for roughly 85% of the global gross domestic product (GDP). This is because the cybercriminals behind ransomware affect more wealthy or crowded nations in the hopes of discovering precious selections (Lee & Lee, 2017). It was claimed that limiting the effects of ransomware attacks on enterprises might be accomplished by frequent employee training on new trends and ransomware behavior, as well as system and application updates.

The first game-theoretic model of the ransomware ecosystem was created by Laszka et al. (2017) and depicts a multi-stage scenario with hospitals and universities confronting a sophisticated ransomware attack. The model concentrated on crucial elements of the adversarial interaction between organizations and ransomware attackers and deduced the circumstances under which a victim organization will pay the demanded ransom after assessing the effort the organization put into ransomware mitigation techniques, policies, security features, and level of backup will help in preventing the threat of ransomware attack. Last but not least, Coccaro (2017) examined the danger of ransomware to law enforcement organizations in the hope that their good name would deter any possible attackers on its infrastructure and units. In 2016, the research produced several significant findings, including the number of ransomware attacks, the most frequently targeted organizations, the number of securities breaches, and the inclusion of law enforcement agencies among the organizations targeted due to a lack of incident response capabilities. Understanding the organizational structure, creating an incident response plan, establishing a dual-purpose digital forensic unit to respond to the announcement of the internal security program agenda, combining information technology experts with other similar internal security units, and other measures, among others, will protect, prevent, and mitigate against any cyber security breaches especially the use of darknet part of the internet. Al-Haija et al. (2022) proposed a precise dynamic learning machine designed to support antivirus. It includes feature selection and classification phases (Albulayhi et al., 2022) that make analyzing and categorizing new and old variants of ransomware attacks easier while producing results quickly. Despite the benefits, the internet of things (IoT) has for human activities.

Brewer (2016) examined ransomware attacks on commercial organizations. He suggested that to build a defense or lessen their impact, it is important to understand the five phases of ransomware attacks: exploitation and infection, delivery and execution, backup spoliation, image encryption (Ahmad et al., 2022), and user notification. Also, Abu Al-Haija et al. (2021) discussed ransomware attacks and security issues over bitcoin. The study reveals the potential threats and vulnerabilities to IoT devices and suggests several countermeasures. Data integrity, minimal security measures, better software security, features for upgrades and patches, physical safety for trillions of devices, privacy, and trust are only a few of the suggested actions.

The earlier survey pieces included various suggestions for preventing the spread of ransomware. The articles do, however, also include certain flaws that may be related to prioritization. The most common gaps in the survey articles are shown below (Maigida et al., 2019):

- *In some papers, the victims' willingness or unwillingness to pay the ransom is the main subject. It did not consider people in a difficult situation and could choose an incremental payment arrangement over an incremental delivery of encrypted content. If they cannot decrypt the files for usage, they risk losing both the money and the vital papers.*
- *Some overview publications conduct preventative experiments using a manual examination of a small number of ransomware samples, which cannot be applied as a preventive strategy.*
- *While other ransomware analyses could identify and stop an attack, they could not ensure that the affected files would be restored.*
- *There aren't many alternative analyses that can combat ransomware that relies on signatures or whose signatures have been included in intrusion detection systems or detecting network device indicators.*
- *According to some, using a cloud-based sandbox for malware detection might be risky for the entire network's systems if the sandbox were to get infected by the malware; the infections could then readily spread to other parts of the network.*
- *While others are restricted to geographically focusing on the American healthcare system.*

7. CONCLUSION

Ransomware assaults have grown in popularity among online criminals over the past few years. This malware extorts money from its victims via online payment systems or cryptocurrencies in exchange for functionality, private information, and files, or data to which the user's access has been restricted. This chapter covers the basic types of ransomware, how they behave, and how to install them on a computer host. We've also compared some of the most well-known ransomware families and provided information on how to mitigate or respond to a ransomware assault.

REFERENCES

Abu Al-Haija, Q., & Al-Dala'ien, M. (2022). ELBA-IoT: An Ensemble Learning Model for Botnet Attack Detection in IoT Networks. *J. Sens. Actuator Netw.*, *11*(1), 18. doi:10.3390/jsan11010018

Abu Al-Haija, Q., Krichen, M., & Abu Elhaija, W. (2022). Machine-Learning-Based Darknet Traffic Detection System for IoT Applications. *Electronics (Basel)*, *11*(4), 556. doi:10.3390/electronics11040556

Abu Al-Haija, Q., & Zein-Sabatto, S. (2020). An Efficient Deep-Learning-Based Detection and Classification System for Cyber-Attacks in IoT Communication Networks. *Electronics (Basel)*, *9*(12), 2152. doi:10.3390/electronics9122152

Adamov, A., & Carlsson, A. (2017). The state of ransomware. Trends and mitigation techniques. *2017 IEEE East-West Design & Test Symposium (EWDTS)*. IEEE. 10.1109/EWDTS.2017.8110056

Ahmad, A., AbuHour, Y., Younisse, R., Alslman, Y., Alnagi, E., & Abu Al-Haija, Q. (2022). MID-Crypt: A Cryptographic Algorithm for Advanced Medical Images Protection. *J. Sens. Actuator Netw.*, *11*(2), 24. doi:10.3390/jsan11020024

Al-Haija, Q. A. (2022). Time-Series Analysis of Cryptocurrency Price: Bitcoin as a Case Study. *2022 International Conference on Electrical Engineering, Computer and Information Technology (ICEECIT)*, Jember, Indonesia.10.1109/ICEECIT55908.2022.10030536

Al-Haija, Q. A., & Alsulami, A. A. (2021). High Performance Classification Model to Identify Ransomware Payments for Heterogeneous Bitcoin Networks. *Electronics (Basel)*, *10*(17), 2113. doi:10.3390/electronics10172113

Al-Qudah, M., Ashi, Z., Alnabhan, M., & Abu Al-Haija, Q. (2023). Effective One-Class Classifier Model for Memory Dump Malware Detection. *J. Sens. Actuator Netw.*, *12*(1), 5. doi:10.3390/jsan12010005

Albulayhi, K., Abu Al-Haija, Q., Alsuhibany, S. A., Jillepalli, A. A., Ashrafuzzaman, M., & Sheldon, F. T. (2022). IoT Intrusion Detection Using Machine Learning with a Novel High Performing Feature Selection Method. *Applied Sciences (Basel, Switzerland)*, *12*(10), 5015. doi:10.3390/app12105015

Albulayhi K., & Al-Haija Q.A. (2022). Security and Privacy Challenges in Blockchain Application. In *The Data-Driven Blockchain Ecosystem.* CRC Press.

Ali, A. (2017). Ransomware: A research and a personal case study of dealing with this nasty malware. *Issues in Informing Science and Information Technology.*

Anghel, M., & Racautanu, A. (n.d.). *A note on different types of ransomware attacks.* Computer Science Faculty, Al. I. Cuza University.

Bajpai, P. & Enbody, R. (2020). Attacking Key Management in Ransomware. *IT Professional, 22*(2), 21-27. . doi:10.1109/MITP.2020.2977285

Bhardwaj, A., Avasthi, V., Sastry, H., & Subrahmanyam, G. V. (2016, April). Ransomware digital extortion: A rising new age threat. *Indian Journal of Science and Technology, 9*(14), 1–5. doi:10.17485/ijst/2016/v9i14/82936

Bilge, L., Han, Y., & Dell'Amico, M. (2017). Riskteller: Predicting the risk of cyber incidents. In *Proceedings of the 2017 ACM SIGSAC Conference on Computer and Communications Security,* (pp. 1299–1311). ACM. 10.1145/3133956.3134022

Bossler, A. M., & Holt, T. J. (2009). Online activities, guardianship, and malware infection: An examination of routine activities theory. *International Journal of Cyber Criminology, 3*(1).

Brewer, R. (2016, September 1). Ransomware attacks: Detection, prevention and cure. *Network Security, Elsevier, 2016*(9), 5–9. doi:10.1016/S1353-4858(16)30086-1

Cabaj, K., & Mazurczyk, K. (2016). Using software-defined networking for ransomware mitigation: The case of cryptowall. *IEEE Network, 30*(6), 14–20. doi:10.1109/MNET.2016.1600110NM

Caivano, D., Canfora, G., Cocomazzi, A., Pirozzi, A., & Visaggio, C. A. (2017). Ransomware at X-Rays. *2017 IEEE International Conference on Internet of Things (iThings) and IEEE Green Computing and Communications (GreenCom) and IEEE Cyber, Physical and Social Computing (CPSCom) and IEEE Smart Data (SmartData).* IEEE. 10.1109/iThings-GreenCom-CPSCom-SmartData.2017.58

Celiktaş, B. (2018). *The ransomware detection and prevention tool design by using signature and anomaly-based detection methods* [Master Thesis]. Istanbul Technical University.

Choi, K.-S. (2008). Computer crime victimization and integrated theory: An empirical assessment. *International Journal of Cyber Criminology, 2*(1).

Christensen, J. B., & Beuschau, N. (2017). *Ransomware detection and mitigation tool.* Tech. Univ. Denmark, Lyngby.

Coccaro, R. (2017). *Evaluation of Weaknesses in US Cybersecurity and Recommendations for Improvement* [Doctoral dissertation]. Utica College.

Cyber Threat Alliance. (2015). *Lucrative ransomware attacks: Analysis of the cryptowall version 3 threat.* Cyber Threat Alliance. https://www.cyberthreatalliance.org/resources/lucrative-ran somware-attacksanalysis-cryptowall-version-3-threat/

Deloitte. (2016) *Ransomware holding your data.* Deloitte Threat Intelligence and Analytics. https://www2.deloitte.com/content/dam/Deloitte/us/Documents/risk/us-aers-ransomware.pdf

Demuro, P. R. (2017). Keeping internet pirates at bay: ransomware negotiation in the healthcare industry keeping internet pirates at bay: ransomware negotiation in the healthcare industry. *Nova Law Review, 41*(3), 5.

Everett, C. (2016). Ransomware: To pay or not to pay? *Computer Fraud & Security, 2016*(4), 8–12. doi:10.1016/S1361-3723(16)30036-7 PMID:27382895

Federal Bureau of Investigation (FBI). (2016). Internet Crime Complaint Center (IC3). *IC3 Annual Report, 2016 State Reports.* FBI. https://www.ic3.gov/Home/AnnualReports?redirect=true

Federal Bureau of Investigation (FBI). (2017). Internet Crime Complaint Center (IC3). *IC3 Annual Report, 2017 State Reports.* FBI. https://www.ic3.gov/Home/AnnualReports?redirect=true

Finkle, J. (2016). *Ransomware: Extortionist hackers borrow customer service tactics.* Reuters. https://www.reuters.com/article/us-usa-cyber-ransomware-idUS KCN0X917X

Formby D, Durbha S, & Beyah R. (2017). *Out of control: Ransomware for industrial control systems.* InRSA.

Gagneja, K. K. (2017). Knowing the ransomware and building defense against it - specific to healthcare institutes. *2017 Third International Conference on Mobile and Secure Services (MobiSecServ)*, Miami Beach, FL, USA. 10.1109/MOBISECSERV.2017.7886569

Ghania, A. S. (2016). Analyzing Master Boot Record for Forensic Investigations. *International Journal of Applied Information Systems, 10*, 2249–0868.

Glassberg, J. (2016). Defending against the ransomware threat. *POWERGRID International, 21*(8), 22–24.

Halawa, H., Beznosov, K., Boshmaf, Y., Coskun, B., Ripeanu, M., & Santos-Neto, E. (2016). Harvesting the low-hanging fruits: defending against automated large-scale cyber-intrusions by focusing on the vulnerable population. In *Proceedings of the 2016 New Security Paradigms Workshop*, (pp. 11–22). ACM. 10.1145/3011883.3011885

Heater, B. (2016, May). How ransomware conquered the world. *PC Magazine Digital Edition*, 109-118.

Held, M. (2018). *Detecting Ransomware* [Thesis]. University Konstanz.

Hernandez-Castro, J., Cartwright, E., & Stepanova, A. (2017). Economic analysis of ransomware. arXiv preprint arXiv:1703.06660.

Ibrahim, R. F., Abu Al-Haija, Q., & Ahmad, A. (2022). DDoS Attack Prevention for Internet of Thing Devices Using Ethereum Blockchain Technology. *Sensors (Basel), 22*(18), 6806. doi:10.339022186806 PMID:36146163

Kaspersky Lab. (2016). *Ransomware in 2014-2016 Technical report.* Kaspersky Lab. https://media.kasperskycontenthub.com/wp-content/uploads/sit es/43/2018/03/ 07190822/KSN_Report_Ransomware_2014-2016_fina l_ENG.pdf

Karresand, M., Axelsson, S., & Dyrkolbotn, G. O. (2019, July 1). Using NTFS cluster allocation behavior to find the location of user data. *Digital Investigation, 29*, S51–S60. doi:10.1016/j.diin.2019.04.018

Kharraz, A., Arshad, S., Mulliner, C., Robertson, W. K., & Kirda, E. (2016). Unveil: A largescale,automated approach to detecting ransomware. In *USENIX Security Symposium*, (pp. 757–772). USENIX.

Kharraz, A., Robertson, W., Balzarotti, D., Bilge, L., & Kirda, E. (2015). Cutting the gordian knot: A look under the hood of ransomware attacks. In *Proceedings of the Detection of Intrusions and Malware, and Vulnerability Assessment Conference* (pp. 3-24). Springer International Publishing. 10.1007/978-3-319-20550-2_1

Kotov, V. & Rajpal, M. S. (2014). In-Depth Analysis of the Most Popular Malware Families. *Understanding Crypto-Ransomware Report*. Bromium.

Laszka, A., Farhang, S., & Grossklags, J. (2017). On the economics of ransomware. In *Decision and Game Theory for Security: 8th International Conference, GameSec 2017*, (pp. 397-417). Springer International Publishing.

Lee, B. (2019). *What is Ransomware? The Major Cybersecurity Threat Explained.* Spin Backup. https://spinbackup.com/blog/what-is-ransomware-the-major-cyb ersecurity-threat-explained/.

Lee, J., & Lee, K. (2017, November 15). Spillover effect of ransomware: Economic analysis of web vulnerability market. *Research Briefs on Information and Communication Technology Evolution., 3,* 193–203. doi:10.56801/rebicte.v3i.59

Lévesque, F. L., Fernandez, J. M., & Somayaji, A. (2014). Risk prediction of malware victimization based on user behavior. In *Malicious and Unwanted Software: The Americas (MALWARE), 9th International Conference* (pp. 28–134). IEEE. 10.1109/MALWARE.2014.6999412

Levesque, F. L., Nsiempba, J., Fernandez, J. M., Chiasson, S., & Somayaji, A. (2013). A clinical study of risk factors related to malware infections. In *Proceedings of the ACM SIGSAC conference on Computer & communications security* (pp. 97–108). ACM. 10.1145/2508859.2516747

Liska, A., & Gallo, T. (2016). *Ransomware: Defending Against Digital Extortion.* O'Reilly Media, Inc.

Luo, X., & Liao, Q. (2007). Awareness education is the key to ransomware prevention. *Information Systems Security, 16*(4), 195–202. doi:10.1080/10658980701576412

Luo, X., & Liao, Q. (2007, September 4). Awareness education as the key to ransomware prevention. *Information Systems Security., 16*(4), 195–202. doi:10.1080/10658980701576412

Maier, G., Feldmann, A., Paxson, V., Sommer, R., & Vallentin, M. (2011). An assessment of overt malicious activity manifest in residential networks. In *International Conference on Detection of Intrusions and Malware, and Vulnerability Assessment,* (pp. 144–163). Springer. 10.1007/978-3-642-22424-9_9

Maigida, A. M., Abdulhamid, S. M., Olalere, M., Alhassan, J. K., Chiroma, H., & Dada, E. G. (2019). Systematic literature review and metadata analysis of ransomware attacks and detection mechanisms. *Journal of Reliable Intelligent Environments, 5*(2), 67–89. doi:10.100740860-019-00080-3

Mercaldo, F., Nardone, V., & Santone, A. (2016). *Ransomware Inside Out.* 2016 11th International Conference on Availability, Reliability and Security (ARES), Salzburg, Austria. 10.1109/ARES.2016.35

Mercaldo, F., Nardone, V., Santone, A., & Visaggio, C. A. (2016). Ransomware Steals Your Phone. Formal Methods Rescue It. In E. Albert & I. Lanese (Eds.), Lecture Notes in Computer Science: Vol. 9688. *Formal Techniques for Distributed Objects, Components, and Systems. FORTE 2016.* Springer. doi:10.1007/978-3-319-39570-8_14

Milne, G. R., Labrecque, L. I., & Cromer, C. (2009). Toward an understanding of the online consumer's risky behavior and protection practices. *The Journal of Consumer Affairs, 43*(3), 449–473. doi:10.1111/j.1745-6606.2009.01148.x

Modi, J. (2019). *Detecting ransomware in encrypted network traffic using machine learning* [PhD dissertation]. UVIC.

Mohammad, A. H. (2020, February). Ransomware evolution, growth, and recommendation for detection. *Modern Applied Science, 14*(3), 68. doi:10.5539/mas.v14n3p68

Ngo, F. T., & Paternoster, R. (2011). Cybercrime victimization: An examination of individual and situational level factors. *International Journal of Cyber Criminology, 5*(1).

O'Gorman, G., & McDonald, G. (2012). *Ransomware: A growing menace.* Symantec Corporation.

Otway, H. J., & Von Winterfeldt, D. (1982). Beyond acceptable risk: On the social acceptability of technologies. *Policy Sciences, 14*(3), 247–256. doi:10.1007/BF00136399

Ovelgönne, M., Dumitra, T., Prakash, B. A., Subrahmanian, V. S., & Wang, B. (2017). Understanding the relationship between human behavior and susceptibility to cyber-attacks: A data-driven approach. *ACM Transactions on Intelligent Systems and Technology, 8*(4), 51. doi:10.1145/2890509

Poudyal, S., Subedi, K. P., & Dasgupta, D. (2018). A Framework for Analyzing Ransomware using Machine Learning. *2018 IEEE Symposium Series on Computational Intelligence (SSCI)*, Bangalore, India. 10.1109/SSCI.2018.8628743

Redmiles, E., Kross, S., Pradhan, A., & Mazurek, M. (2017). *How well do my results generalize? comparing security and privacy survey results from mturk and web panels to us.* Technical report.

Richardson, R., & North, M. M. (2017). *Ransomware: Evolution, Mitigation and Prevention.* Faculty Publications. 4276. https://digitalcommons.kennesaw.edu/facpubs/4276

Runciman, B. (2020). Cybersecurity Report 2020. *ITNOW, 62*(4), 28–29. doi:10.1093/itnow/bwaa103

Salvi, M. H. U., & Kerkar, M. R. V. (2016). Ransomware: A cyber extortion. Asian Journal of Convergence in Technology, 2(3), Solander, A. C., Forman, A. S., & Glasser, N. M. (2016). Ransomware—Give me back my files! *Employee Relations Law Journal, 42*(2), 53–55.

Scaife, N., Carter, H., Traynor, P., & Butler, K. (2016). Cryptolock (and drop it): stopping ransomware attacks on user data. In *Distributed Computing Systems (ICDCS) 36th International Conference*. IEEE.

Securities and Commission. (2017). Cybersecurity: Ransomware alert. *Natl Exam Progr Risk Alert, 5*(4), 15–16.

Strecher, V. J., & Rosenstock, I. M. (1997). The health belief model. Cambridge handbook of psychology, health and medicine, 113–117.

Symantec. (2017a). *Internet security threat report, 22*. Symantec. https://www.symantec.com.

Symantec. (2017b). *2017 internet security threat report, 22*. Symantec.

United States Department of Justice. (2015). Financial crime fraud victims. *Technical report*. The United States Attorney's Office, Western District of Washington. https://www.justice.gov/usao-wdwa/victim-witness/victim-info/financial-fraud

Upadhyaya, R., & Jain, A. (2016). Cyber ethics and cyber crime: A deep dwelved study into legality, ransomware, underground web and bitcoin wallet. *2016 International Conference on Computing, Communication and Automation (ICCCA)*, Greater Noida, India. 10.1109/CCAA.2016.7813706

VPN.com. (n.d.). *Malware, Adware, Spyware, and Ransomware: What Do These Terms Mean*. VPN. https://www.vpn.com/guides/malware-adware-spyware-and-ransomware-what-do-all-these-terms-mean

Wyke, J., & Ajjan, A. (2015). *The current state of ransomware*. Sophos Labs.

Zavarsky, P., & Lindskog, D. (2016). Experimental analysis of ransomware on windows and android platforms: Evolution and characterization. *Procedia Computer Science, 94*, 465–472. doi:10.1016/j.procs.2016.08.072

Zitar, R. A., & Mohammad, A. H. (2011). Spam detection using genetic assisted artificial immune system. *International Journal of Pattern Recognition and Artificial Intelligence, 25*(8), 1275–1295. doi:10.1142/S0218001411009123

Chapter 7
Demystifying Ransomware: Classification, Mechanism and Anatomy

Aaeen Naushadahmad Alchi

ⓘD https://orcid.org/0000-0002-0802-5363
Gujarat University, India

Kiranbhai R. Dodiya

ⓘD https://orcid.org/0009-0001-9409-7303
Gujarat University, India

ABSTRACT

Malware, classified as ransomware, encrypts data on a computer, preventing individuals from accessing it. The intruder then demands a ransom from the user for the password that unlocks the files. Recent cyberattacks against prominent corporate targets have increased the extensive media attention on ransomware. The primary reason for computer intrusions is financial gain. Ransomware targets individual owners of information, keeping their file systems captive until a ransom is paid, compared to malware, which permits criminals to steal valuable data and then use it throughout the digital marketplace. Ransomware's terrifying complexity level heralds a paradigm shift in the cybercrime ecosystem. Ransomware has become more mysterious, with some latest forms working without ever connecting to the Internet. In this chapter, the authors will discuss the overview of ransomware, the history and development of ransomware, some of the famous cases, the anatomy of ransomware attacks, types of ransomware attack vectors, and the prevention of such kinds of attacks in cyberspace.

DOI: 10.4018/978-1-6684-8218-6.ch007

1. LET US KNOW ABOUT RANSOMWARE

Ransomware is wicked software that encrypts a victim's documents, making them unreachable, and demands a ransom payment in exchange for the decryption key. The price of a cryptocurrency like Bitcoin and the ransom amount are often relatively high. Ransomware can be delivered through various means, such as malicious email attachments or software vulnerabilities, and can significantly impact individuals and organizations. It is a cyber-attack and can cause serious business interruption and data loss. (FinCEN, 2021)

2. HISTORY AND DEVELOPMENT OF RANSOMWARE

Ransomware has been around in various forms since the late 1980s, with the first known instance being the "AIDS Trojan", distributed on floppy disks in 1989. However, it was not until the mid-2000s that ransomware began to gain widespread attention as a serious cyber threat. Early versions of ransomware typically just locked the victim's screen and displayed a message demanding a ransom payment, but over time the malware has evolved to include encryption of files, making them inaccessible until paid the ransom.

In the 2010s, ransomware began to be distributed on a large scale via email phishing campaigns and exploit kits. The use of cryptocurrency as a means of payment also became more common, providing a way for attackers to receive the ransom payment while remaining anonymous. The malware also began targeting individuals, businesses, healthcare organizations, and government agencies(CryptoDeFix, n.d.).

In recent years, ransomware has become even more sophisticated, with some variants using double extortion techniques, not only encrypting the files but also exfiltrating sensitive data and threatening to release it if the ransom is unpaid. In addition, some ransomware can spread laterally across a network, encrypting multiple machines and causing widespread disruption.

Overall, ransomware has evolved from a nuisance to a severe cyber threat that can cause significant damage to organizations and individuals.

2.1 History of Ransomware

One of the first examples of this type of malware was the AIDS Trojan, discovered in 1989. The malware encrypted the victim's files and demanded payment for the decryption key.

In the late 1990s and early 2000s, ransomware began to distribute via email as a form of "spam." This early ransomware was typically not very sophisticated, and antivirus software could easily remove it.

In 2005, a new form of ransomware called Gpcode was discovered, which used a more advanced encryption algorithm; this ransomware instance was challenging to remove.

In 2008-2009, ransomware became more prevalent and more sophisticated with the emergence of Trojans like CryptoLocker and Cryptowall. This ransomware used more robust encryption and was more challenging to remove, making it much more difficult for victims to recover their files without paying the ransom.

Overall, the history of ransomware till 2009 was relatively simple and not much sophisticated, but as the technology improved, it became more complex and harder to remove.

In 2011, the "Police Trojan" began to appear, which locked a victim's computer and displayed a message purporting to be from a law enforcement agency, claiming that the laptop had viewed illegal content.

In 2013, Crypto Locker appeared, and it was the first ransomware to use strong encryption to lock victims' files. It quickly became a significant threat and was spread via spam emails.

By 2016, ransomware had become a significant problem, and high-profile attacks such as the WannaCry attack afflicted more than 200,000 victims in 150 countries. (BBC News, n.d.)

In 2017, a new ransomware named "Not Petya" was found, designed to spread rapidly and cause damage to the infected systems. It was a devastating attack and caused hundreds of millions of dollars in damage.

Since then, ransomware attacks have continued to evolve, becoming more sophisticated and targeted, with some ransomware families specifically designed to target specific industries such as healthcare and manufacturing.

In 2020, ransomware attacks increased significantly due to the COVID-19 pandemic, as many organizations were forced to shift to remote work and were less prepared to defend against cyber-attacks.

As of 2022, ransomware remains a significant threat, with attacks increasing in frequency and sophistication. Furthermore, many experts believe that ransomware will continue to be a significant problem in the years to come.

Figure 1. Classification of ransomware

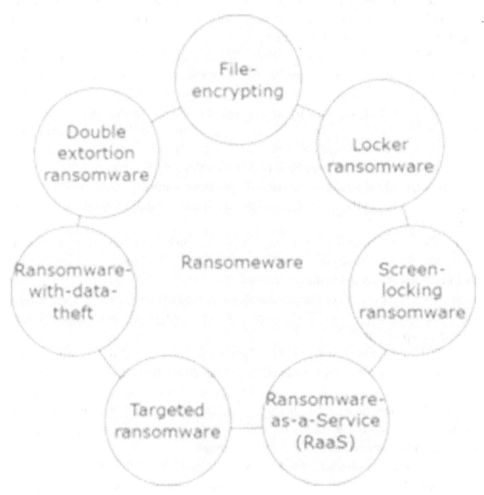

3. CLASSIFICATION OF RANSOMWARE

The classification of ransomware into several categories is based on various characteristics, such as the type of encryption used, the method of delivery, and the intended target. Some standard classifications of ransomware include:

File-encrypting ransomware: This ransomware encrypts the files on a victim's computer or network, making them inaccessible until paid the ransom. It is the most common type of ransomware.

Locker ransomware: This type of ransomware locks the victim's entire computer or network, preventing them from accessing any files or applications until they pay the ransom.

Screen-locking ransomware: This type of ransomware locks the victim's screen, preventing them from accessing their computer until they pay the ransom.

Ransomware-as-a-Service (RaaS): This ransomware is a service to other attackers, who can customize and distribute the malware.

Targeted ransomware: This ransomware targets specific organizations, such as healthcare providers or government agencies.

Ransomware-with-data-theft: This type of ransomware not only encrypts the files but also steals sensitive data.

Double extortion ransomware: This ransomware not only encrypts the files but also threatens to leak sensitive data to the public.

It is important to note that ransomware can also combine these types, and new variants and variations continue to emerge.

4. ATTACK MECHANISMS OF RANSOMWARE

Ransomware typically uses one or more of the following attack mechanisms to infect a victim's system and encrypt their files:

There are several types of ransomware attack vectors that attackers use to deliver the malware and infect a victim's system; some of them are:

Phishing emails: Emails which contain malicious attachments or links are phishing ransomware. And when someone clicks the mail, the malware is installed on the victim's system when the extension or the link is open.

Exploit kits: Ransomware is delivered through exploit kits, which allow attackers to exploit software vulnerabilities to infect a victim's system.

Malicious websites: Ransomware sent through an attacker who has compromised the website. The malware is automatically downloaded and installed when the victim visits the website.

Remote access tools: Ransomware is sent through remote access tools, such as Remote Desktop Protocol (RDP), that allow an attacker to access a victim's system.

Malvertising: Ransomware is sent through malicious advertising, also known as "malvertising," which is the use of online advertising to deliver malware.

Supply Chain Attack: Ransomware can be delivered through an attack on the software supply chain; this type of attack usually occurs when the malware is embedded into a software update or application.

Figure 2. Attack mechanisms of ransomware

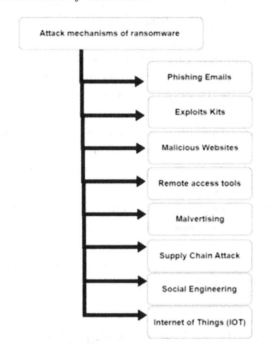

Social Engineering: Ransomware can be delivered through social engineering tactics where attackers trick the victims into running the malware.

Internet of things (IoT) devices: Ransomware can also be delivered through vulnerable Internet of things (IoT) devices, like routers, cameras, and other intelligent devices.

These are the most common ransomware attack vectors, but new ones may appear as cybercriminals evolve their tactics. Once the installation is complete, malware encrypts the victim's files using a robust encryption algorithm. The encryption process is designed to make it very difficult to recover the files without the decryption key, which is only provided to the victim after they pay the ransom.

5. ANATOMY OF A RANSOMWARE ATTACK

The anatomy of a ransomware attack typically consists of several stages, including:

Ransomware is malevolent software that encrypts a target's files and hassles payment in interchange for the decryption key. The anatomy of a typical ransomware attack includes several components.(*Poolz.Finance*, n.d.)

Figure 3. Anatomy of ransomware attack

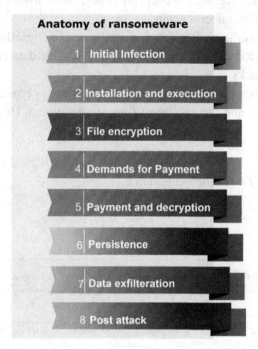

Initial infection: The malware is typically delivered through a phishing email or a malicious website that tricks the victim into downloading and running the malware. (*Understanding Ransomware: Types, Prevention Techniques and Containment Measures for Best Cybersecurity*, n.d.)

Installation and execution: Once the malware is delivered, it is installed and executed on the victim's system. It typically involves the malware creating a persistence mechanism to ensure it remains on the system even after a reboot

File Encryption: Once the malware is executed, it encrypts the victim's files using a robust encryption algorithm. The encryption process may also include creating a unique encryption key for each victim.

Demands for payment: The malware then displays a ransom note on the victim's screen, demanding payment in exchange for the decryption key. The note typically includes instructions for paying the ransom and a deadline for payment.

Payment and decryption: If the victim decides to pay the ransom, they will typically be instructed to send payment in cryptocurrency to a specified address. Once payment is received, the attacker will provide the victim with the decryption key, which can be used to decrypt the encrypted files.

Persistence: Some Ransomware malware is designed to persist on the system even after the encryption process is complete. This means that even if the victim pays the ransom and receives the decryption key, the malware will remain on the system and potentially infect it again.

Data exfiltration: Some Ransomware malware is designed to extract sensitive data from the infected system before encrypting it. This data can then be sold or used for further attacks.

Post-attack: After the ransom is paid, the malware is typically removed from the victim's system, and the encrypted files are decrypted. However, it is crucial

Figure 4. Process of a ransomware attack

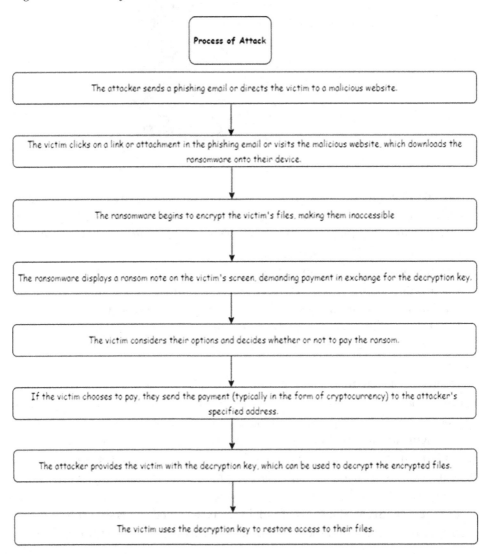

to note that the malware may have already stolen sensitive data or caused other damage to the system.

It's imperative to note that paying the ransom does not surety that the victim will regain access to their files and that ransomware is a continuously evolving threat with new variants and techniques being developed. (*Morgan Office*, n.d.)

6. DETECTION AND ANALYSIS OF RANSOMWARE

6.1 Detection of Ransomware

A malware that encrypts a victim's files, making them inaccessible, and demands a ransom payment in exchange for the decryption key. Detection of ransomware can be accomplished through a variety of methods, including(Wayne, n.d.-a)

Signature-based detection: This method involves identifying known ransomware by comparing the malware's code to a database of known ransomware signatures. This approach is practical for detecting known ransomware but may not be able to see new or unknown variants.

Behavioural-based detection: This method involves identifying ransomware by its behaviour, such as creating many new files or encrypting files in a specific location. This approach can detect new or unknown ransomware variants but may have a higher rate of false positives.

Heuristic-based detection: This method involves identifying ransomware by its characteristics, such as the presence of a ransom note or encryption. This approach can detect new or unknown variants of ransomware but may also have a higher rate of false positives

Sandboxing: This method involves running a suspicious program in an isolated environment and monitoring its behaviour to determine if it is malware. This approach can detect new or unknown variants of malware, including ransomware.

Network-based detection: This method involves monitoring network traffic to detect signs of ransomware, such as large amounts of data being transferred or connections to known command and control servers.

File analysis: This method involves analyzing the ransomware executable to understand its capabilities and behaviour and using various tools, such as disassemblers, debuggers, and reverse-engineering tools.

Backup and recovery: This method involves having a backup of all critical data and being able to restore it in case of a ransomware attack. Having a backup can prevent paying the ransom and also allow restoring the system to a previous state before the attack.

Endpoint detection and response (EDR) software: EDR software monitors all activity on a device and can detect, alert, and respond to malware infections, including ransomware.

There is no single method that can provide complete protection against ransomware. Multiple ways are typically used to provide the best possible defence. Additionally, regular backups of important files and software updates can help mitigate a ransomware attack's impact. (*The Latest Ransomware Targeting Businesses and Individuals in Nepal*, n.d.-a)

6.2 Analysis of Ransomware Using Different Tools and Technique

There are a variety of tools and techniques to analyze and detect ransomware. Some popular tools and techniques include:(BetterBI, n.d.)

Antivirus and anti-malware software: These programs use signature-based and behavioural-based detection to identify and remove known and unknown ransomware variants.

Sandboxing: As mentioned earlier, Sandboxing is a technique that allows you to run a suspicious program in an isolated environment and monitor its behaviour to determine if it is malware.

Network traffic analysis: Network-based detection tools such as Wireshark and Network Miner capture and analyze network traffic for signs of ransomware, such as large amounts of data is transferred or connections to known command and control servers.

Memory analysis: Memory tools such as Volatility and Rekall extract information from a system's memory, such as running processes and open files, to detect and analyze ransomware.

File analysis: File analysis tools, such as Virus Total and Hybrid-Analysis, analyze files for signs of malware, including ransomware.

Endpoint detection and response (EDR) software: EDR software detects, alert, and respond to malware infections, including ransomware. This software typically provides real-time monitoring of system activity and can provide detailed information about the malware, including its behaviour, origin, and extent of the infection.

Reverse Engineering: Disassembling and analyzing the malware code, identifying the encryption and decryption technique used and the location of the ransom note.

It's worth noting that none of these tools and techniques can provide 100% protection against a ransomware attack. However, a combination of multiple devices and methods provides the best possible defence, and regular backups of essential files, software updates, and user education can aid to mitigate the impact of a ransomware attack (Check Point Software, n.d.).

7. RANSOMWARE ATTACK ON DIFFERENT PLATFORM

Ransomware attacks can occur on a variety of platforms, including:

Windows: Windows is the most common platform targeted by ransomware attacks, primarily because it is the most widely used operating system. Ransomware on Windows can encrypt files on the local system and connected network shares.

MacOS: MacOS is considered more secure than Windows, but it is not immune to ransomware attacks. Ransomware on MacOS can encrypt files on the local system and connected network shares.

Linux is considered more secure than Windows or MacOS, but it is still vulnerable to ransomware attacks. Ransomware on Linux can encrypt files on the local system and connected network shares.

Mobile Devices: Ransomware attacks on mobile devices, such as smartphones and tablets, are becoming increasingly common. Ransomware on mobile devices can encrypt files and lock the device's screen.

Cloud: Ransomware can also target cloud services, such as Microsoft Office 365, Google Drive, and Dropbox. Ransomware on cloud services can encrypt files stored and spread to other connected devices and systems.

Industrial Control Systems: Ransomware can also target industrial control systems (ICS), such as those used in manufacturing, transportation, and critical infrastructure. Ransomware on ICS can encrypt files and disrupt the control system, leading to production downtime and possible physical damage.

In all cases, the ransomware encrypts the victims' files and commands a ransom payment in exchange for the decryption key. The attacker can deliver the malware via email attachments, malicious links, exploited vulnerabilities, or a software supply chain attack. Additionally, the attackers use different social engineering tactics to trick users into installing the malware.

It is important to note that a ransomware attack can have severe consequences and can cause significant financial losses. Therefore, regular backups, software updates, and user education are the best defence against ransomware attacks. Additionally, conglomerates should have incident response plans to quickly contain and recover from an attack and minimize the attack's impact.

8. FAMOUS CASE STUDY OF RANSOMWARE ATTACK

There have been many high-profile ransomware attacks in recent years, but a few notable examples include the following:

1. **Colonial Pipeline**: In May 2021, Colonial Pipeline, one of the largest fuel pipelines in the US, was forced to shut down its operations after a ransomware attack. The attack encrypted the company's data and demanded a ransom of $5 million in Bitcoin. The incident caused widespread fuel shortages and panic buying across the US East Coast. The Colonial Pipeline ransomware attack occurred in May 2021 and was one of the most significant ransomware attacks to date. The attack targeted Colonial Pipeline, a major fuel pipeline operator that supplies gasoline and other fuel products to the United States East Coast. The attackers used ransomware known as Dark Side, a ransomware-as-a-service (RaaS) that encrypts the target's files and presumes ransom payment in exchange for the decryption key. The attackers demanded a ransom of $5 million in Bitcoin to restore access to the company's data. The attack resulted in the shutdown of Colonial Pipeline's entire pipeline system, which supplies gasoline and other fuel products to the United States East Coast. It caused widespread fuel shortages and panic buying across the region, leading to long lines at gas stations and higher prices at the pump. Colonial Pipeline shut down its entire pipeline system to contain the attack and engaged cybersecurity experts to investigate the incident. The company also contacted the FBI for assistance in the investigation. The company paid the ransom of $4.4 million in bitcoin on May 7 2021, which enabled them to get the decryption key to recover their data. The company also reported that it could restore some of its operations using a combination of its backups and manual processes. However, the attack caused significant disruptions to the company's operations and the broader fuel supply chain. The episode highlights the considerable impact ransomware attacks can have on critical infrastructure and the importance of robust cybersecurity measures to protect against them. The company already had an incident response plan, but the episode shows that the incident response plan should be regularly up to date. Additionally, regular backups and software updates can help to mitigate the influence of a ransomware attack, but it's important to note that no single security measure can provide complete protection against ransomware.

2. **JBS**: In June 2021, the world's largest meat processor, JBS, fell victim to a ransomware attack. The attack resulted in the shutdown of the company's plants in the US and Australia and caused meat shortages across the globe. JBS ransomware attack occurred in June 2021 and was one of the most significant ransomware attacks. JBS is the world's biggest meat processing company, operating in over 15 countries. The attackers used ransomware known as REvil (also known as Sodinokibi), which rizeencrypts the target's files and demands a ransom settlement in exchange for the decryption key. The attackers demanded a ransom of $11 million in Bitcoin to restore access

to the company's data. The attack resulted in the shutdown of JBS's meat processing plants in the US and Australia, causing meat shortages across the globe and significant disruptions to the company's supply chain. To contain the attack, JBS shut down its affected plants and engaged cybersecurity experts to investigate the incident. The company also contacted the FBI for assistance in the investigation. The company did not disclose if it paid the ransom but could restore some of its operations using a combination of its backups and manual processes. However, the attack caused significant disruptions to the company's operations and the broader meat supply chain, leading to meat shortages and increased prices in some areas. The episode highlights the considerable impact ransomware attacks can have on critical infrastructure and the importance of robust cybersecurity measures to protect against them. The company already had an incident response plan, but the episode shows that the incident response plan should be up to date. Additionally, regular backups and software updates can help to mitigate the inf of a ransomware attack, but it's important to note that no single security measure can provide complete protection against ransomware.

3. **Garmin:** In July 2021, the fitness and navigation technology company Garmin got hit by a ransomware attack. The attack led to the company's website and app being down for several days and caused significant disruption to the company's operations. The Garmin ransomware attack occurred in July 2021 and was a powerful cyber attack on the fitness and navigation technology company. The attackers used ransomware known as WastedLocker, which encrypts the ta's files and demands a ransom settlement in exchange for the decryption key. The attackers demanded a ransom sum in exchange for the decryption key and to not disclose stolen data. The attack resulted in the shutdown of the company's website and app, causing significant disruptions to the company's operations, including the production and supply chain and the company's customer service. The company also reported data stolen during the attack. Garmin engaged cybersecurity experts to investigate the incident and contacted the FBI to contain the attack. As a result, the company paid the assailants the ransom and could restore its operations and services using a combination of its backups and manual processes. The episode highlights the significant impact ransomware attacks can have on a company's operations and the importance of robust cybersecurity measures to protect against them. The company already had an incident response plan, but the attack shows that the incident response plan should be up to date. Additionally, regular backups and software updates can help to mitigate the influence of a ransomware attack, but it's important to note that no single security measure can provide complete protection against

ransomware. It's also essential for companies to have an incident response plan in place and to be prepared to respond to an attack quickly and effectively.

4.   **Acer**: The Acer ransomware attack occurred in August 2021 and was a significant cyber attack on the computer manufacturer. The attackers used ransomware known as Ragnar Locker, which encrypts the target's files and demands a ransom settlement in exchange for the decryption key. The attackers demanded a ransom settlement in exchange for the decryption key and not disclosing stolen data. The attack resulted in the company's production and supply chain shutdown, causing significant disruptions to the company's operations. To contain the attack, Acer engaged cybersecurity experts to investigate the incident and contacted the FBI for assistance. As a result, the company paid the attackers the ransom and could restore its operations and services using a combination of its backups and manual processes. The attack highlights the significant impact ransomware attacks can have on a company's operations and the importance of robust cybersecurity measures to protect against them. Additionally, regular backups and software updates can help to mitigate the influence of a ransomware attack, but it's important to note that no single security measure can provide complete protection against ransomware. It's also important for companies to have an incident response strategy and be prepared to respond to an attack quickly and effectively.

5.   **Iron Mountain:** In September 2021, Iron Mountain, a leading data management and storage company, fell victim to a ransomware attack. The attack encrypted the company's data and demanded a ransom payment in Bitcoin. The incident caused significant disruptions to the company's operations and customer service. The Iron Mountain ransomware attack occurred in September 2021 and was a powerful cyber attack on the leading data management and storage company. The attackers used ransomware known as Ragnar Locker, which encrypts the target's files and demands a ransom settlement in exchange for the decryption key. The attackers demanded a ransom settlement in exchange for the decryption key and not disclosing stolen data. The attack resulted in the shutdown of the company's data centres and caused significant disruptions to the company's operations and customer service. Iron Mountain engaged cybersecurity experts to investigate the incident to contain the attack and contacted the FBI for assistance. As a result, the company paid the attackers the ransom and could restore its operations and services using a combination of its backups and manual processes. The attack highlights the significant impact ransomware attacks can have on a company's operations and the importance of robust cybersecurity measures to protect against them. The company already had an incident response plan, but the attack shows that the incident response plan should be up to date. Additionally, regular backups and software updates

can help to mitigate the influence of a ransomware attack, but it's important to note that no single security measure can provide complete protection against ransomware. It's also crucial for companies to have an incident response plan in place and to be prepared to respond to an attack quickly and effectively.

6. **CNA Financial**: In October 2021, the attackers used ransomware known as Ryuk, which encrypts the target's files and demands a ransom payment in exchange for the decryption key. The attackers demanded a ransom deposit in exchange for the decryption key and to not disclose stolen data. The attack resulted in the shutdown of the company's operations and caused significant disruptions to the company's customer service. CNA Financial engaged cybersecurity experts to investigate the incident to contain the attack and contacted the FBI for assistance. As a result, the company paid the attackers the ransom and could restore its operations and services using a combination of its backups and manual processes. The attack highlights the significant impact ransomware attacks can have on a company's operations and the importance of robust cybersecurity measures to protect against them. Additionally, regular backups and software updates can help to mitigate the influence of a ransomware attack, but it's important to note that no single security measure can provide complete protection against ransomware. It's also crucial for companies to have an incident response plan in place and to be prepared to respond to an attack quickly and effectively.

7. **The City of Newark**: In November 2021, the City of Newark, New Jersey, got hit by a ransomware attack. The attack encrypted the city's data and demanded a ransom sum in Bitcoin. The incident caused significant disruptions to the city's operations, including emergency and online services. The City of Newark ransomware attack occurred in November 2021 and was a substantial cyber-attack on the city. The attackers used ransomware known as Ryuk, which encrypts the target's files and demands a ransom payment in exchange for the decryption key. The attackers demanded a ransom payment in Bitcoin in exchange for the decryption key and not disclosing stolen data. The attack resulted in shutdown of the city's operations and services, including emergency and online services, and caused significant disruptions to the city's operations. Although the town reported that some data was stolen during the attack to contain the attack, the city of Newark engaged cybersecurity experts to investigate the incident and contacted the FBI for assistance. The city paid the attackers the ransom and could restore its operations and services using a combination of its backups and manual processes. The attack highlights the significant impact ransomware attacks can have on a city's operations and the importance of robust cybersecurity measures to protect against them. The city already had an incident response plan, but the attack shows that the incident response plan should be

up to date regularly. Additionally, regular backups and software updates can help to mitigate the influence of a ransomware attack, but it's important to note that no single security measure can provide complete protection against ransomware. It's also crucial for cities to have an incident response strategy and be prepared to respond to an attack quickly and effectively. (*Why Go for TISAX If You Have Already Reached the ISO27001?* n.d.)

8. **Dixons Carphone**: In December 2021, Dixons Carphone, a leading consumer electronics retailer, was hit by a ransomware attack. The attack encrypted the company's data and demanded a ransom sum in Bitcoin. The incident caused significant disruptions to the company's operations and supply chain. The Dixons Carphone ransomware attack occurred in December 2021 and was a substantial cyber-attack on the leading consumer electronics retailer. The attackers used ransomware known as Conti, which encrypts the target's files and demands a ransom sum in exchange for the decryption key. The attack resulted in the company's production and supply chain shutdown, causing significant disruptions to the company's operations and customer service. The company also reported stolen data during the attack. Dixons Carphone engaged cybersecurity experts to investigate the incident and contacted the FBI for assistance in containing the attack. As a result, the company paid the attackers the ransom and could restore its operations and services using a combination of its backups and manual processes. The attack highlights the significant impact ransomware attacks can have on a company's operations and the importance of robust cybersecurity measures to protect against them. Additionally, regular backups and software updates can help to mitigate the influence of a ransomware attack, but it's important to note that no single security measure can provide complete protection against ransomware.

9. **Magellan Health:** Health got hit by a ransomware attack. The attack encrypted the company's data and demanded a ransom payment in Bitcoin. The incident caused significant disruptions to the company's operations, including patient care and billing. The Magellan Health ransomware attack occurred in January 2022 and was a substantial cyber-attack on the healthcare company. The attackers used ransomware known as Ryuk, which encrypts the target's files and demands a ransom sum in exchange for the decryption key. The attackers demanded a ransom payment in Bitcoin in exchange for the decryption key and not disclosing stolen data, which resulted in the shutdown of the company's operations and services, including patient care and billing. The company also reported stolen data during the attack. To contain the attack, Magellan Health engaged cybersecurity experts to investigate the incident and contacted the

FBI for assistance. As a result, the company paid the attackers the ransom and could restore its operations and services using a combination of its backups and manual processes. The attack highlights the significant impact that ransomware attacks can have on a healthcare company's operations and the importance of robust cybersecurity measures to protect against them. It's also essential for healthcare companies to have an incident response strategy and be prepared to respond to an attack quickly and effectively.

10. **The City of Baltimore**: In February 2022, the City of Baltimore got hit by a ransomware attack. The attack encrypted the city's data and demanded a ransom payment in Bitcoin. The incident caused significant disruptions to the city's operations, including emergency and online services. The City of Baltimore ransomware attack occurred in February 2022 and was a significant cyber-attack on the city. The attackers used ransomware known as Robbin Hood, which encrypts the target's files and demands a ransom settlement in exchange for the decryption key. The attackers demanded a ransom payment in Bitcoin in exchange for the decryption key and not disclosing stolen data. The attack resulted in a shutdown of the city's operations and services, including emergency and online services, and caused significant disruptions to the city's operations. The town also reported the data to be stolen during the attack. To contain the attack, the city of Baltimore engaged cybersecurity experts to investigate the incident and contacted the FBI for assistance. The city initially refused to pay the ransom and attempted to restore its operations and services using a combination of its backups and manual processes. However, the recovery process was slow, and the city had to pay a ransom to decrypt the files and prevent further damage. The attack highlights the significant impact ransomware attacks can have on a city's operations and the importance of robust cybersecurity measures to protect against them. The city already had an incident response plan, but the attack shows that the incident response plan should be up to date. Additionally, regular backups and software updates can help to mitigate the influence of a ransomware attack, but it's important to note that no single security measure can provide complete protection against ransomware. It's also vital for cities to have an incident response plan, be prepared to respond to an attack quickly and effectively, and weigh the costs and risks of paying the ransom. These case studies demonstrate the severity and impact of ransomware attacks and the importance of implementing robust security measures to protect against them. It's also essential to have an incident response strategy to quickly contain and recover from an attack and minimize its impact.

9. PREVENTIVE MEASURES FOR A RANSOMWARE ATTACK

There are several preventive measures that organizations can take to protect against ransomware attacks:

Regularly back up essential data: Regularly backing up important data and storing it in a separate, secure location can help organizations recover from a ransomware attack. This way, if the files are encrypted, they can restore them from the backup. (Idoko, n.d.)

Keep software and operating systems up-to-date: Regularly updating software and operating systems can help protect against known vulnerabilities that attackers may exploit.

Implement endpoint protection: Solutions such as antivirus and anti-malware software can help detect and block ransomware before it can encrypt files.

Limit user privileges: Limiting user privileges to the minimum necessary can help prevent malware from spreading throughout an organization.

Train employees: Employee education and training can help prevent phishing and other social engineering ransomware attacks.

Implement multi-factor authentication: multi-factor authentication can help prevent unauthorized access to systems and data, even if an attacker manages to steal login credentials.

Keep regular backups and test them: Regular backups are essential to restore the data in case of a ransomware attack. In addition, regularly testing backups is critical to ensure that they are working and can be used to restore the data.

Have an incident response plan in place: An incident response plan can help organizations respond to a ransomware attack quickly and effectively, minimizing the damage and downtime caused.

There is no single security measure that can provide complete protection against ransomware. Organizations should implement a layered defence and regularly assess their security posture to guard against the latest threats. (*The Latest Ransomware Targeting Businesses and Individuals in Nepal*, n.d.-b)

10. CONCLUSION

Ransomware is a severe cyber threat that can significantly impact organizations of all types and sizes. A variety of malware encrypts the target's files and demands a ransom settlement in exchange for the decryption key. Ransomware attacks can cause disruptions to operations, loss of data and revenue, and can also lead to regulatory fines and reputational damage(*Wayne*, n.d.-b).

Preventative measures such as regular backups, keeping the software and operating systems up-to-date, implementing endpoint protection, and training employees can help protect against ransomware attacks. Additionally, having an incident response plan and regularly assessing the organization's security posture can help minimize the damage and downtime caused by a ransomware attack.

Ransomware is categorized based on characteristics such as the type of encryption used, the method of delivery, and the intended target. The most common type is file-encrypting ransomware, but other styles, such as locker ransomware, screen-locking ransomware, Ransomware-as-a-Service, and targeted ransomware, also exist.

It's essential for organizations to be aware of the potential risks posed by ransomware and to take proactive steps to protect against attacks. These include implementing robust cybersecurity measures, regularly backing up important data, and planning to respond to an attack. With suitable preventive measures and an incident response plan, organizations can minimize the impact of a ransomware attack and quickly restore normal operations.

11. FUTURE TRENDS AND DEVELOPMENTS IN RANSOMWARE ATTACKS

Future directions and effects in Ransomware attacks are expected to become more sophisticated and targeted, making it harder for organizations to protect themselves. Some of the key trends to take care of includes the following:

The surge of Artificial Intelligence and Machine Learning: Ransomware attackers will use AI and machine learning to improve their targeting capabilities, making it harder for organizations to identify and block attacks. This technology will also help attackers to evade detection and bypass security measures.

Ransomware-as-a-Service (RaaS) is becoming more prevalent: RaaS is a business model where attackers offer their malware and support services to other cybercriminals for a fee. This trend will continue, making it easier for less skilled attackers to launch successful ransomware attacks.

More targeted and sophisticated attacks on critical infrastructure and government organizations: Ransomware attackers will focus on high-value targets, such as critical infrastructure and government organizations, to cause widespread disruption and financial losses.

The growth of double extortion attacks involves encrypting data, stealing it, and threatening to disclose it publicly unless a ransom. These attacks will become more common as attackers seek new ways to monetize their efforts.

Increase in attacks on IoT devices and cloud-based systems: The increasing number of IoT devices and the growing adoption of cloud-based systems provide new attack vectors for ransomware attackers. In addition, these devices and systems are often less secure than traditional systems, making them easier targets.

The increasing acceptance of mobile devices and the fact that many users don't take the same security precautions as they do on their PCs makes mobile devices an attractive target for ransomware attackers. More attackers using multi-language ransom notes: To reach a wider audience, attackers will start using ransom notes in multiple languages, making it harder for organizations to identify and respond to attacks. Use of cryptocurrency as a form of deposit for ransoms: Cryptocurrency provides a way for attackers to receive payment anonymously, making it harder to trace and shut down their operations. Cryptocurrency as a payment form will become more common in future ransomware attacks.

Overall, these trends suggest that ransomware will continue to be a significant threat to organizations of all sizes and industries and that the attackers will continue to evolve their tactics to evade detection and maximize their profits. Organizations must stay vigilant and take proactive measures to protect themselves from these increasingly sophisticated attacks.

REFERENCES

Wayne, R. (2022). Best Practices for Protecting Businesses Against Ransomware Attacks. *Medium*. https://medium.com/@rick.wayne.2022/best-practices-for-protecting-businesses-against-ransomware-attacks-740221ebaf5d

Diagnostic analytics. (n.d.). BetterBI. https://www.betterbi.dk/diagnostic-analytics/

FinCEN's Financial Trend Analysis. (2021). Fincen. https://www.fincen.gov/sites/default/files/2021-10/Financial%20Trend%20Analysis_Ransomware%20

Check Point Software. (n.d.). *How To Prevent Ransomware Attacks*. Check Point Software. https://www.checkpoint.com/cyber-hub/threat-prevention/ransomware/how-to-prevent-ransomware/

Idoko, N. (n.d.). *How to Protect Your Data From Cyber Attacks*. Nicholas Idoko Blog. https://nicholasidoko.com/blog/2023/01/12/how-to-protect-your-data-from-cyber-attacks/

CryptoDeFix. (2021). *Indian authorities will ban the purchase of goods with cryptocurrency*. CryptoDeFix. https://cryptodefix.com/articles/cryptocurrency-in-india-will-be-regulated-as-asset

Poolz.Finance. (n.d.). *Ransomware: - Blog*. Poolz Finance. https://blog.poolz.finance/glossary/ransomware

BBC News. (n.d.). *Ransomware cyber-attack threat escalating.* BBC. https://www.bbc.com/news/technology-39913630

The Latest Ransomware Targeting Businesses and Individuals in Nepal. (n.d.-a). LinkedIn. https://www.linkedin.com/pulse/latest-ransomware-targeting-businesses-individuals-nepal-bashyal

Morgan Office. (n.d.). *What is Ransomware?* Morgan Office. https://www.morganoffice.co.uk/help-advice/what-is-ransomware

Why go for TISAX if you have already reached ISO27001? (n.d.). LinkedIn. https://www.linkedin.com/pulse/why-go-tisax-you-have-already-reached-iso27001-guido-b%C3%BCcker

Chapter 8
IoT in Real-Life:
Applications, Security, and Hacking

Pawan Whig

iD https://orcid.org/0000-0003-1863-1591
Vivekananda Institute of Professional Studies, India

Kritika Puruhit
Jodhpur Institute of Engineering and Technology, India

Piyush Kumar Gupta
Jamia Hamdard, India

Pavika Sharma
Bhagwan Parshuram Institute of Technology, India

Rahul Reddy Nadikattu
University of Cumbersome, USA

Ashima Bhatnagar Bhatia
Vivekananda Institute of Professional Studies, India

ABSTRACT

Now, computers and smartphones have become more powerful than internet connectivity. Every 'smart' device in our environment now aspires to use digital interventions to solve real-world problems. The buzz around IoT is, of course, huge. This disturbing technology penetrates various industries, develops new IoT applications, and connects all internet-enabled devices around us. One survey shows that 61 billion connected devices are expected to be available by 2025. But some of them shine more than others, in the mad rush of "newer" and "better" IoT applications. The chapter aims to present a summary of the challenges in the applications that the internet of things must face in their research and development, which are to be explored in this book chapter.

DOI: 10.4018/978-1-6684-8218-6.ch008

INTRODUCTION

Intelligent technologies have been extended from humans to the stuff affecting humans with the advancement of information technology(Whig, Velu, & Naddikatu, 2022). We are surrounded by many smart devices that make our lives simple and comfortably smart, linked to an IoT network, from smart TV to smart farming. The convergence of sensor nets, artificial intelligence, machine learning and others are built with network and information technology, items that allow IoT as a smart system from activity tracking to autonomous vehicle(Alkali et al., 2022a). The Internet of Things describes the whole process from gathering, sorting, taking an action that correlates to the nature of this knowledge to storing it in the cloud. Sensors are used to gather and transmit information to a computing unit that performs comfortable behaviour. The data would then be stored locally and subsequently sent to the cloud service via the internet. IoT systems produce tremendous data where the sensors used in different manufacturing and healthcare applications are interfaced with the devices(Whig, Velu, & Sharma, 2022).

IoT touches each and every industry sector including Agriculture, Healthcare, Business and finance, retail and manufacturing. IoT applications are focusing on societal needs, The Internet of Things will help boost the quality of life of people and provide employees with new and improved careers, business opportunities and market development(Whig, Kouser, Velu, et al., 2022). The Internet of Things would allow objects to acquire knowledge about their location in the universe, to communicate with other objects, and to access data obtained in their proximity for comparative information. IoT may have many applications in various areas like home automation, smart agriculture, autonomous vehicle, smart solar panel, smart energy meter, smart lock, smart village, smart city and much more. In the next decade we would be surrounded by technology and in that technology there would be billions of smart devices connected over the internet for millions of applications. As IoT devices are sharing internet network so there can be an intense concern about the security and privacy of the information that should be a serious issue and challenge must be faced by our Internet security provider from the hacker over the internet(Whig, Velu, & Ready, 2022).

Fundamental Components of IoT

Devices: Sensors and devices are the main components of the device connectivity layer. IoT Devices must be enabled with sensors and actuators, control source, radio communication borders, operating system, user border, device organization, execution setting(Whig, Velu, & Bhatia, 2022).

Figure 1. Components of internet of things

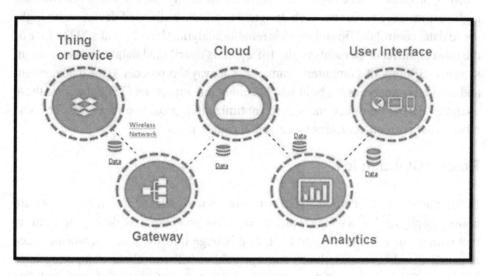

Devices can also be classified as a basic devices or advance devices. Basic devices are those which can perform simple operations like alarm, smart thermostats and communication of devices is not possible without gateways. Advanced plans are execution request at complex level reason and provision message procedures. The majour component of IoT is shown in Figure 1

Gateway: The bidirectional data traffic between various networks and protocols is handled by the IoT Gateway. Another role of the gateway is to convert various network protocols and ensure that the connected devices and sensors are interoperable.

Until transferring it to the next level, gateways can be programmed to pre-process the collected data from thousands of sensors locally. It will be appropriate in some situations because of TCP/IP protocol compatibility. IoT gates may derived in terms, including device setups, software updates, and error handling, as well as perform data gathering, visualisation, storage, and other tasks. With higher order encryption methods, the IoT gateway provides some degree of protection for the network and transmitted data. It serves as a middle layer between computers and the cloud to avoid malware attacks and unintended access to the system(Whig, Velu, & Nadikattu, 2022). IoT Gateways combine many networking protocols and interfaces, performing translations, converter, and connecting operations across various network levels and apis. ALG is present in IoT gateways.

Cloud: The Internet of Things provides vast data that needs to be handled in an appropriate manner from computers, apps and consumers. The IoT cloud provides tools for the real-time compilation, retrieval, monitoring and storage of massive volumes of data. Industries and utilities can conveniently view such data remotely and, where appropriate, make important decisions. Basically, the IoT cloud is a complex high-performance server network designed to handle billions of devices with high-speed data, control traffic and provide reliable analytics(Jupalle et al., 2022). One of the most significant elements of the IoT cloud is distributed database management systems. Billions of computers, cameras, gateways, protocols, data management and predictive analytics are built into the cloud infrastructure. Companies use these analytics data for product and service optimization, proactive measures for some moves, and correctly construct a current business model.

Role of Cloud in IoT

In IoT there can be hundreds of sensors connected with the internet. Sensors are collecting physical data which is processed by the processor into digital information that data needs massive storage to store this huge information, that storing place is called cloud. The underlying concept behind IoT and cloud computing is to improve efficiency without disrupting the consistency of processed or shared data in daily activities. Although the partnership is widespread, both of them effectively compliment each other. The IoT becomes the source of the data, while the Cloud

Figure 2. Role of cloud in IoT

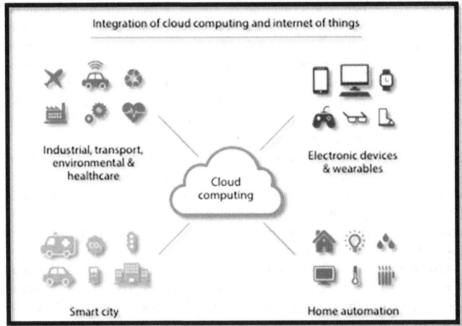

becomes the final destination for storing it(Tomar et al., 2021). As there is no need for on-site equipment for store, analysis, and research, cloud computing offers a cost-effective alternative. Role of cloud in IoT is shown in Figure 2.

Google Cloud platform and more are the main mist vendors in the US. They have massive farms of servers that, as part of their cloud services, they sell to enterprises(Whig, Nadikattu, & Velu, 2022). If we speak of the users who use Google cloud to save their data using Google cloud services on Google drive.

Various Cloud Services

There are three major types of cloud computing, generally known as Infrastructure as a Service (IaaS), Application as a Service (PaaS), and Software as a Service (SaaS). Selecting the correct form of cloud calculation contains of considering the desires to reach an optimum degree of control while thinking over needless tasks. These forms are described by Microsoft as follows and shown in Figure 3:

Infrastructure-as-a-Service (IaaS)

The most simple cloud computing services group. With IaaS, you rent IT resources from a cloud service on a pay-as-you-go basis - servers and virtual machines (VMs), storage, networks, operating systems.

Platform as a Service (PaaS)

Platform-as-a-service (PaaS) refers to cloud infrastructure platforms offering an on-demand environment for software application creation, monitoring, distribution and management. PaaS is intended to brand it easy for creators to easily build web or smartphone applications without thinking about scenery up or maintaining the server, storage, network and database underlying infrastructure required for development.

Software as a Service (SaaS)

Software-as-a-service (SaaS) is a system used on demand and usually on a contract basis to distribute software applications over the Internet. With SaaS, the web framework and underlying technology are hosted and managed by cloud vendors and grip any updates, such as package updates and safety patching. Operators, typically with a web page on their computer, tablet or PC, connect to the programme over the Internet.

Figure 3. Various cloud services

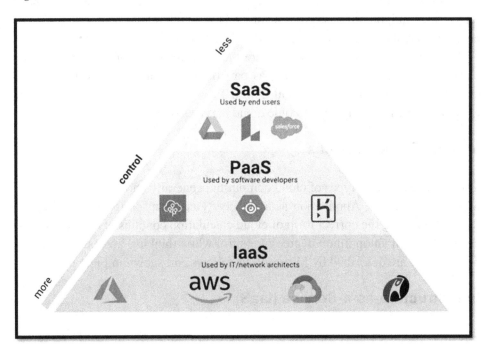

Analysis

Analytics is the method of translating analogue knowledge into practical observations from billions of smart devices and sensors that can be analysed and used for comprehensive research. Smart analytics solutions for the monitoring and enhancement of the whole system are unavoidable for the IoT system(Anand et al., 2022).

Interface With the Consumer

The noticeable, tangible component of the IoT framework which can be reached by users is the user interfaces(Alkali et al., 2022b). For less user effort, programmers would need to maintain a well-designed user experience and promote further experiences.

Basic Concept Behind the technology

Internet of Things as it is clear from its definition the feasibility of complete IoT is depends on mainly two technologies.

i.   Web Technology
ii.   Sensor Technology

Web Technology

Web technology is very essential components of IoT and web applications are very useful for real time desktop applications "addresses." The initial level IoT, IP addresses are seems to be a big hurdle in the feasibility of IoT because we have Limited IP addresses in the IPV4(Chopra & Whig, 2022; Madhu & WHIG, 2022). For putting life in everything we need unlimited IP addresses so that they can communicate with each other. But with the Progress in web technology and with the Development of IpV6 with a provision of unlimited IP addresses added advantage and it is one of the great help in the direction of feasibility of IoT. Through studying the following table, this can easily be appreciated. IPv4 provides roughly 4.3 billion unique Internet Protocol (IP) addresses and is now widely secondhand Internet speaking scheme. However, as operating systems and IoT applications have expanded the number of devices connected to the Internet, all available addresses in the IPv4 scheme have been allocated. Any Net operators are moving to IPv6 of the Internet Protocol, which offers around 340 trillion trillion (3.4x1038) unique IP addresses.

Sensors and Actuators

A sensor is a physical device which is used to detect and measure physical environment. Sensor technology relies on the production of more sensitive sensors in terms of speed cost and power consumption. Some sensors are shown below in Figure 4. The details of these sensors and their use will be covered in the Chapter

Business Processes in IoT

The IoT is transforming the way we live our lives and that's something that's only going to expand and grow, and it's probably something that corporations need to adapt to. A specific ordering of tasks and events through time and place to fulfil a business purpose is defined by business processes. Management of business processes is a well-established discipline that deals with the exploration, recognition, review, design or overhaul, execution, deployment, control, and evolution of business processes. IoT Market Size with time is increases and it can be easily visualize from Figure 5.

Figure 4. Types of sensors

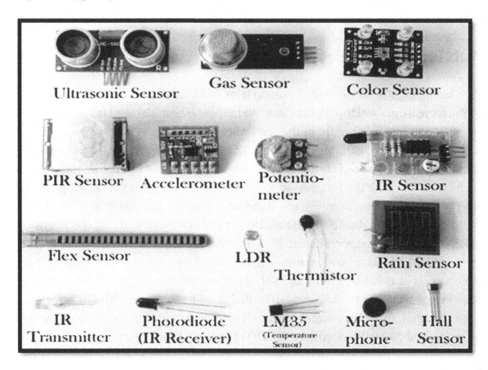

Figure 5. IoT Market Size vs. Years

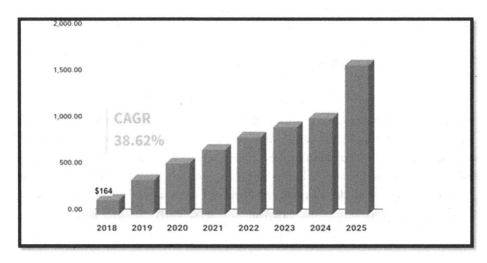

Figure 6. Security issues and challenges with IoT

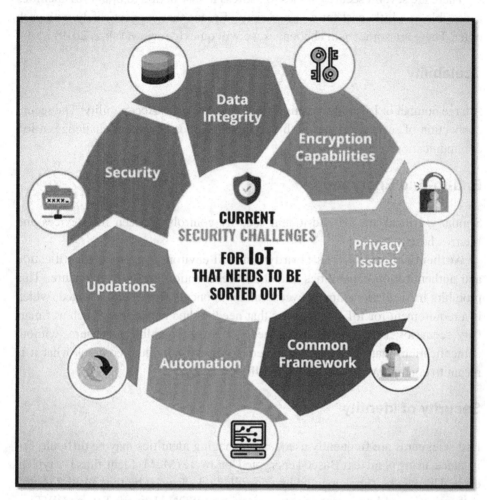

Security Issues and Challenges

Single of the main tests in upcoming of IoT is 'Hacking'. A new report on IoT warns that devices are at risk from computer worms and allow hackers to attack on thousands of devices from one single attack (Khera et al., 2021; Mamza, 2021). Growing IoT Devices applications with previously security challenges cryptography, secure communication, privacy assurances also created new security challenges like regulatory issues, legal issues, economy and development issues and Right issues. Some latest Security Issues and Challenges with IoT is shown in Figure 6.

There are several security issues to address in instruction to brand IoT facilities accessible at a little cost for a vast variety of plans that connect safely with each other. There are some main challenges we will quickly review(Whig, 2019).

Scalability

A large number of IoT nodes requires flexible safeguards. Accessibility: The secure connection of several devices with varying capacities is another challenge in IoT communications.

End-to-End Encryption

Similarly critical are the end-to-end security controls between IoT devices and Internet hosts.

Authentication and Trust: In a dynamic IoT environment, proper identification and authentication capabilities and their orchestration are not yet mature. This prohibits trust relationships between IoT components from being created, which is a requirement for IoT applications that need ad-hoc networking, such as Smart City scenarios, between IoT components. It is not possible to ensure, without authentication, that the data flow generated by an individual contains what it is meant to contain(Whig & Ahmad, 2014b).

Security of Identity

Bad safeguards are frequently used, and managing identities may be difficult. For instance, using plain text/Base64 encrypted hardware (M2M) Identifiers is a typical error. This can be supplemented by controlled tokens used by the OAuth/OAuth2 authentication and authorization system, such as JSON Network Tokens (JWT).

IoT Applications

There is a clear growth in the development of connected devices in this age of technology that emphasises on device interoperability. In order to allow new business models, IoT technology retains more focus on incorporating sensors, devices and information systems through industry verticals and organizations(Whig & Ahmad, 2014a). For more sophisticated networking mechanisms and middleware that enable the convergence of device networks and apps, IoT requires more complicated solutions. The following groups can be classified by the IoT applications as shown in Figure 7.

Figure 7. IoT applications

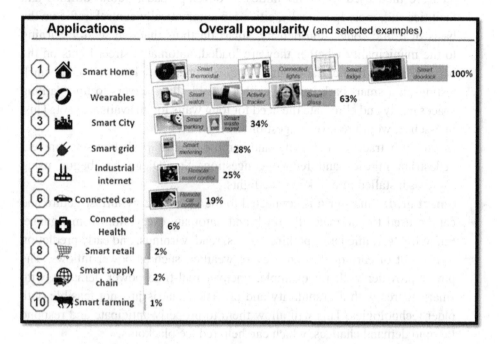

Applications	Overall popularity (and selected examples)
(1) Smart Home	Smart thermostat, Connected lights, Smart fridge, Smart doorlock — 100%
(2) Wearables	Smart watch, Activity tracker, Smart glass — 63%
(3) Smart City	Smart parking, Smart waste mgmt — 34%
(4) Smart grid	Smart metering — 28%
(5) Industrial internet	Remote asset control — 25%
(6) Connected car	Remote car control — 19%
(7) Connected Health	6%
(8) Smart retail	2%
(9) Smart supply chain	2%
(10) Smart farming	1%

- **Smart Home or Home Automation:** IoT can Helps to design smart home with various applications like user can control various home appliances smart LED Bulb, Air conditioner, smart TV, smart Kitchen, smart door lock and many more. For example using smart Led light user can control the intensity of light, user can ON/OFF lights without Human Intervention, smart bathroom user can control the temperature of water just with the help of audio command. In smart kitchen the devices can send information on your phone about expiry date of products available in the kitchen.

- **Wearable Devices:** The smart gadgets that can be worn as external attachments incorporated in the human body, clothing, are wearable devices. Smart belt, smart watch, smart clothes, fitness tracker, smart sunglasses, smart jewellery, smart accessories, tattoos, stickers, shades, and even more can be smart. Following Bluetooth headphones, smartwatches, and web-enabled glasses, health activity trackers became the first major wave of wearable technology on the market. More wearables were added by the gaming industry. In wearable devices, however, the main life-altering uses are mainly used in fitness management and medical use cases.

- **Smart cities:** Smart cities use devices such as wired cameras, lamps, and metres from the Internet of Things (IoT) to gather and interpret data. This data are then used by communities to develop roads, public utilities and facilities and, most specifically, cities can automatically control traffic on the basis of traffic congestion density, smart dustbins that can send a warning to the municipality whether they are loaded, automated street lights on the basis of light or traffic. Earlier this year, London revealed that it would launch testing on a smart parking project that would allow drivers to find parking spaces easily and eliminate the need for long waits for an available spot. This, in essence, would relieve congestion in urban traffic.

- In order to track foot activity and reroute cars at peak hours to prevent pedestrian injuries and reduce congestion, San Diego has begun using cameras installed into linked streetlights.

- **Smart grid:** Smart grid is connected to the smart energy infrastructure and can be used to automatically track and automate everything from signage, traffic lights, traffic jams, parking spaces, road warnings, and early prediction as a result of earthquakes and severe weather, such as power inflows. The power provider will, for example, uncover real-time power demands with smart metres with a granularity and precision that is just not feasible with older technologies. This will allow them to properly anticipate and respond to rapid demand changes, which can help reduce blackouts.

- **Connected car:** Any vehicle which is equipped with internet connectivity can be called a connected car. IoT can helps vehicle to vehicle communication, in remote parking, identifying the location of car using Geo-fencing connected with the internet, multimedia applications.

- **Connected health:** Connected health technologies can be used remotely to link physicians and patients, IoT healthcare applications can help minimise medical expenses, increase quality, and make patients' healthcare more personal, accessible, and affordable.

- **Smart retail:** Retailers may use IoT technologies such as RFID tags and GPS sensors to provide a detailed picture of the flow of items from output to where they are put in a shop to where a buyer buys them. There are many retail uses, such as control of foot traffic, repair of facilities, warehousing of demand warnings, etc. IoT has transformed the retail sector's future and produced better supply chain performance, expanded marketing conversions, and more. As a result, it is clear that IoT-enabled solutions have the potential to boost both consumer service and brand loyalty.

- **Smart supply chain:** supply chain management can help to increase the efficiency of supply chain organization procedures to balance the corporation's marketplace frameworks.

- **Smart farming:** smart agricultural is an request that uses modern and tall knowledge, to GPS technology, water, light, humidity and temperature management, automatic irrigation, precision fertilization, data management and the use of technology in many different areas. Many devices are helping in smart farming such as smart drone for agriculture can help watering, identifying health and diseased plants, for monitoring live stocks etc.

IoT Safety Threats

Internet linked devices deliver a variety of safety risks. However, while the IoT has provided connection to new gadgets.

Authentication Flaws

PINs are unity of the primary appearances of security against attempted pony-trekking. However, if your key is insufficiently robust, your expedient is insecure. End users can also choose a password that is simple to recall. However, if it's simple to recall, it's generally simple to enter.

Most Iot systems require little identification. Smooth if no critical data is kept on the expedient, a susceptible IoT expedient can serve as a entry to an whole net or be combined hooked on a bot, anywhere hacks can utilise its calculation control to spread malware and conduct distribution DDoS assaults.

Poor identification is a major source of worry for Iot. Suppliers may assist to improve authentication security by difficult many steps, utilising strong evasion PINs, and configuring locations that consequence in harmless operator made PINs.

Bequest Possessions

If a programme was not initially developed for fog connectivity, it is likely to be vulnerable to contemporary cyber threats. Creation obsolete apps Internet complete deprived of making major modifications is dangerous, but this isn't always doable with historical assets. They've been cobbled composed (perhaps decades), making safety enhancements seem tremendous.

Access to a Communal Net

In order to keep a security breach on the device isolated, each IoT request must utilise a different net consume a safety exists. This is only one benefit of cellular IoT. A VPN can help shield your devices from assaults from the outside world, but if your application connects to other devices via a shared connection, it may still be at risk of attack if those machines are compromised.

Varying Security Requirements

Once it originates to safety values, the IoT is a bit of a free. Because there isn't a single, industry-wide standard, different businesses and market segments must create their own procedures and rules. IoT device security is made more difficult by a lack of standards, which also makes it more difficult

Incomplete Firmware

If devices are released into the field with a defect that introduces vulnerabilities, that is another of the largest IoT security dangers. Manufacturers must have the capacity to release firmware updates but that isn't always possible. You may have problems if a network has slow data transmission rates or insufficient messaging capabilities.

IoT Security Breaches Examples

There have been several instances recently, especially, of how even harmless IoT devices may be misused and utilised maliciously. Some of the more well-known instances have only served as illustrations of what is conceivable, while other instances have featured real assaults.

LISTED BELOW ARE A FEW INSTANCES OF IOT SECURITY BREACHES.

Botnet Mirai

One of the most well-known IoT security breaches is the Mirai botnet assault from 2016, which is notable not just since it was the biggest bout seen to date but too since the source cypher is motionless available. The code was published online by the original hacker, allowing imitators to use it.

Essentially, a botnet is a dispersed network of computers. Although they are not necessarily malicious, a botnet is an army of devices that may bring down systems when it falls into the wrong hands. The botnet in Mirai's scenario had 134,605 audiovisual plotters. The hacker, a school scholar, used the botnet to consume over a terabyte of data per second to execute an unprecedented attack on the French web hosting company OVH.

IoT devices that were not secured made the botnet feasible. Surprisingly unimportant, the initial attack's goal was to bring down certain Minecraft servers. But roughly a month later, the Dyn service provider was the victim of the Mirai botnet. Then, Mirai shut down significant portions of the Internet, including The Guardian, Twitter, Reddit, Netflix.

Target's Hacking of Credit Cards

Target's network was successfully infiltrated by hackers in 2013, and the data from millions of transactions was taken. How did they do it? An HVAC provider that was assisting Target in tracking their energy usage and improving the efficiency of their systems had their login information stolen.

Although an IoT device may not be at blame in this instance, anybody developing IoT applications, especially apps like smart metres, should be aware of the ramifications. To gain access to a client's network, someone can try to hack your network. Even though it has nothing to do with your application, your devices could always be just a few steps away from incredibly important (and confidential) data.

Hackable Automobiles

In 2015, two cybersecurity professionals set out to compromise the multimedia system of a brand-new Jeep Grand Cherokee. They were prosperous. Additionally, they showed how they could link to another piece of software inside the car using the entertainment system, reprogram it, and then manage the gearbox. This example serves as a chilling reminder that isolating connected devices is a key element of IoT security in the era of self-driving automobiles.

IoT Security Measures

It is crucial to be secure. When implementing M2M devices, companies and hardware providers must take a number of additional security concerns into account brought on by the introduction of new technologies and the rise in worldwide deployments.

Physical Protection

Physical security is vital for avoiding unwanted access to a device because IoT applications are frequently remote in nature. Use of robust components and specialised gear that makes accessing your data more challenging is beneficial in this situation.

For instance, a lot of crucial data is held on the SIM card in cellular IoT devices. This data is particularly susceptible since detachable SIM form factors are increasingly common. On the circuit board, however, an eSIM is soldered.

Personal Networks

Security risks exist even while sending and receiving communications using remotely installed devices. Using public-access networks, like WiFi, to connect devices and enable communication exposes such communications to interception.

While message encryption is a positive step, more care must be taken while sending sensitive data over public networks. In order to assure that no data ever traverses the public Internet, we advise creating private networks on top of the current security measures.

Detection of Anomalies

You must be aware of any efforts to compromise your device or unusual network behaviour. In order for you to determine if there was an employee error or a major danger, we transfer the pertinent connection information to your operational dashboards using the cloud communication platform from EMnify.

Locked IMEI

The distinctive ID number present on the majority of mobile devices is called an International Mobile Station Equipment Identity (IMEI). To prevent a SIM card from being removed and used in another device, you may configure a SIM's functionality to a single IMEI in real time using an IMEI lock. You must encrypt network data transfers in order to safely move data to and from your devices. However, even if your network and application may be safe, there is a loophole in which your data might be captured.

Instead of using public IPs, EMnify also enables you to create a secure VPN for your whole deployment using intra-cloud connect.

Internet-Based Firewall

Small M2M devices typically have constrained computing power. They are unable to set up firewalls as a result. However, a network-based firewall safeguards your data as soon as it connects to the network. This prevents harmful data from ever being delivered being able to access the by removing the time-consuming task of packet filtering from the device.

Businesses can monitor and prohibit traffic outside of your VPN using network-based firewalls, or they can simply block certain conversations. Additionally, it may spot hacking attempts or incursions that do not follow set standards.

Limited Level of Connection

Your IoT application was created with a specific objective in mind. The more you can separate the network connectivity of your gadget from its essential operations.

CONCLUSION

At each tier, the IoT framework is subject to assaults. As a result, there are several security concerns and needs that must be spoke. Present state of IoT research is primarily absorbed on verification and admission switch procedures, but with the fast progression of skill, it is critical to integrate new schmoosing procedures such as IPv6 and 5G in order to accomplish the gradual mash up of IoT topology. The main goal of this chapter was to highlight important IoT security challenges, namely security threats and responses. Many IoT devices become soft targets due to a lack of security mechanisms, and the victim is uninformed. This chapter deliberates safety needs such as privacy, integrity, and verification, among others. Various IOT applications are addressed in this study. We expect that this article will be valuable to security researchers by assisting in the identification of important challenges in IoT security and providing a better knowledge of the dangers and its attributes originating from numerous attackers such as companies and spy activities.

REFERENCES

Alkali, Y., Routray, I., & Whig, P. (2022a). Strategy for Reliable, Efficient and Secure IoT Using Artificial Intelligence. *IUP Journal of Computer Sciences, 16*(2).

AlkaliY.RoutrayI.WhigP. (2022b). Study of various methods for reliable, efficient and Secured IoT using Artificial Intelligence. *Available at* SSRN 4020364. doi:10.2139/ssrn.4020364

Anand, M., Velu, A., & Whig, P. (2022). Prediction of Loan Behaviour with Machine Learning Models for Secure Banking. [JCSE]. *Journal of Computing Science and Engineering : JCSE, 3*(1), 1–13.

Chopra, G., & Whig, P. (2022). Energy Efficient Scheduling for Internet of Vehicles. *International Journal of Sustainable Development in Computing Science, 4*(1).

Cui, A. & Stolfo, S. (2022). *A Quantitative Analysis of the Insecurity of Embedded Network Devices: Results of a Wide-Area Scan Proceedings of the 26th Annual Computer Security Applications*. NYU.

Baker, F. (1995). *Requirements for IP version 4 routers.*

Da Costa, F. (2013). *Rethinking the Internet of Things: A scalable approach to connecting everything.*

Fu, X. & Qian, K. (2008). SAFELI-SQL Injection Scanner Using Symbolic Execution. *Proceedings of the workshop on Testing*, (pp. 34 – 39). IEEE.

Iqbal, A., Saleem, R., & Suryani, M. (2020). Internet of Things (IOT): ongoing Security Challenges and Risks. *International Journal of Computer Science and Information Security, 14.*

Sen, J. (2009). A Survey on Wireless Sensor Network Security. *International Journal of Communication Networks and Information Security (IJCNIS), 1*(2).

Jupalle, H., Kouser, S., Bhatia, A. B., Alam, N., Nadikattu, R. R., & Whig, P. (2022). Automation of human behaviors and its prediction using machine learning. *Microsystem Technologies, 28*(8), 1–9. doi:10.100700542-022-05326-4

Khera, Y., Whig, P., & Velu, A. (2021). efficient effective and secured electronic billing system using AI. *Vivekananda Journal of Research, 10*, 53–60.

Li, Q., Feng, X., Wang, H., & Li, Z. L SunTowards Fine-grained Fingerprinting of Firmware in Online Embedded Devices IEEE INFOCOM 2018-IEEE Conference on Computer Communications, p. 2537 – 2545 Conference (ACSAC '10), p. 97 – 106 Posted: 2010

Madhu, M., & WHIG, P. (2022). A survey of machine learning and its applications. *International Journal of Machine Learning for Sustainable Development, 4*(1), 11–20.

Mamza, E. S. (2021). Use of AIOT in Health System. *International Journal of Sustainable Development in Computing Science, 3*(4), 21–30.

Senie, D. (1998). Network ingress filtering: defeating denial of service attacks which employ IP source address spoofing. *Network.*

Tomar, U., Chakroborty, N., Sharma, H., & Whig, P. (2021). AI based Smart Agricuture System. *Transactions on Latest Trends in Artificial Intelligence, 2*(2).

Wen Huang, F. Y., Yu, C., Hung Hang, C., Tsai, D. T., Lee, S., & Yenkuo, A. (2022). Testing Framework for Web Application Security Assessment. *Journal of Computer Networks, 5*, 739 – 761.

Whig, P. (2019). Exploration of Viral Diseases mortality risk using machine learning. *International Journal of Machine Learning for Sustainable Development, 1*(1), 11–20.

Whig, P., & Ahmad, S. N. (2014a). Development of economical ASIC for PCS for water quality monitoring. *Journal of Circuits, Systems, and Computers, 23*(06), 1450079. doi:10.1142/S0218126614500790

Whig, P., & Ahmad, S. N. (2014b). Simulation of linear dynamic macro model of photo catalytic sensor in SPICE. *COMPEL: The International Journal for Computation and Mathematics in Electrical and Electronic Engineering.*

Whig, P., Kouser, S., Velu, A., & Nadikattu, R. R. (2022). Fog-IoT-Assisted-Based Smart Agriculture Application. In *Demystifying Federated Learning for Blockchain and Industrial Internet of Things* (pp. 74–93). IGI Global. doi:10.4018/978-1-6684-3733-9.ch005

Whig, P., Nadikattu, R. R., & Velu, A. (2022). COVID-19 pandemic analysis using application of AI. *Healthcare Monitoring and Data Analysis Using IoT: Technologies and Applications*, 1.

Whig, P., Velu, A., & Bhatia, A. B. (2022). Protect Nature and Reduce the Carbon Footprint With an Application of Blockchain for IIoT. In *Demystifying Federated Learning for Blockchain and Industrial Internet of Things* (pp. 123–142). IGI Global. doi:10.4018/978-1-6684-3733-9.ch007

Whig, P., Velu, A., & Naddikatu, R. R. (2022). The Economic Impact of AI-Enabled Blockchain in 6G-Based Industry. In *AI and Blockchain Technology in 6G Wireless Network* (pp. 205–224). Springer. doi:10.1007/978-981-19-2868-0_10

Whig, P., Velu, A., & Nadikattu, R. R. (2022). Blockchain Platform to Resolve Security Issues in IoT and Smart Networks. In *AI-Enabled Agile Internet of Things for Sustainable FinTech Ecosystems* (pp. 46–65). IGI Global. doi:10.4018/978-1-6684-4176-3.ch003

Whig, P., Velu, A., & Ready, R. (2022). Demystifying Federated Learning in Artificial Intelligence With Human-Computer Interaction. In *Demystifying Federated Learning for Blockchain and Industrial Internet of Things* (pp. 94–122). IGI Global. doi:10.4018/978-1-6684-3733-9.ch006

Whig, P., Velu, A., & Sharma, P. (2022). Demystifying Federated Learning for Blockchain: A Case Study. In Demystifying Federated Learning for Blockchain and Industrial Internet of Things (pp. 143–165). IGI Global. doi:10.4018/978-1-6684-3733-9.ch008

Yang, Y., Wu, L., Yin, G., Lifie, L., & Hongbin, Z. (2017). A Survey on Security and Privacy Issues. Internet-of-Things IEEE Internet of Things Journal.

Chapter 9
Maintaining Cybersecurity Awareness in Large-Scale Organizations:
A Pilot Study in a Public Institution

Muhammed Aslan
Cankaya University, Turkey

Tolga Pusatli
ⓘD https://orcid.org/0000-0002-2303-8023
Cankaya University, Turkey

ABSTRACT

Research was conducted to increase the awareness of employees with regard to cyber security to fill the gap in the literature where few studies on how effective the measures implemented in organizations were reported. This research uses the outcome of the phishing drills that a public institution applied to its personnel, participation of said personnel in awareness training, and the reading statistics of regularly published information security bulletins. This has been beneficial in determining the methods to increase the cyber security awareness of personnel in organizations with 1,000 or more personnel; users were considered as a whole, and not individually evaluated. Findings report that organizations can increase users' cybersecurity awareness by systematically conducting phishing exercises, providing awareness training, and regularly publishing information security bulletins. The awareness of reading bulletins rapidly increased after phishing exercises and training and decreased in the following months; however, an increase was observed in the long term.

DOI: 10.4018/978-1-6684-8218-6.ch009

INTRODUCTION

Ensuring information security in corporate organizations is of vital importance. In both public institutions and private sector companies, many precautions are taken to protect information assets like all other assets and not to share them with third parties without permission. Humans, as the weakest link in the security chain, must be investigated for the protection of information.

Cyber security is commonly ensured by implementing hardware and software solutions in information systems infrastructures. However, even though the necessary infrastructure investments have been made, it cannot be said that information assets are completely secure. To achieve real cyber maturity, increasing the cyber security awareness of personnel working in corporate organizations is necessary. Despite all advanced security products, the user cannot be prevented from sharing his/her password with someone else or from opening an e-mail containing fake links. It is obvious that the investments to be made to increase the awareness of users regarding information security are at least as important as infrastructure investments such as hardware and software assets.

From this perspective, we have opted in to study how to strengthen the human dimension. We reviewed the studies on this subject in the literature and we have observed that similar methods are applied in Turkey, from where the research data is taken, and abroad. To prepare users against a possible threat by conducting phishing drills, to educate them on basic issues by providing information security awareness training, to regularly publish information security bulletins to inform personnel about current cyber events and cyberattacks, to inform users via e-mail and to attract the attention of the personnel through digital screens can all be listed as featured methods.

Although it is observed that some studies have been carried out to increase information security awareness both in public institutions and in the private sector, we found few studies examining how these activities benefited in terms of increasing user awareness.

We aim to reveal how much phishing exercises, awareness training, and the publishing of bulletin activities in public institutions and private sector businesses with 1,000 or more users work in increasing personnel awareness. Moreover, we aim to determine which of these three methods is most effective and to show how long this effect lasts. As a result of the determinations, we aim to guide public institutions and large-scale enterprises in increasing personnel awareness and ensuring information security.

Within the scope of our study, we investigated the contribution of these three measures to garner information security awareness in an institution with roughly 1,000 personnel. In this chapter, which focuses on the human factor, technical cyber security measures taken through hardware and software products are excluded or minimized.

The main purpose is to show how phishing exercises, the provision of information security awareness training, and issuing monthly information security bulletins have an impact on personnel within one year.

The research question is:

Can conducting phishing drills, doing awareness training, and preparing information security bulletins increase the cyber security awareness of personnel in the long term?

Information Security and Cyber Security

Information is an asset that can be an opportunity when one possesses it. However, it can become a threat if it is lost and can be changed without notice if it is stolen. For this reason, the security of an information asset is of great importance to the owner of the information. As reported in (Telecommunication Standardization Sector of ITU, 2009), cybersecurity is the collection of security policies, risk management approaches, training, tools and techniques, and best practices with regard to technologies that can be used to protect the cyber environment and assets of organizations and users. On the other hand, the assets are the sum of the information infrastructure, applications, services, and information transmitted and/or stored in the cyber environment. Cyber security endeavors to ensure the security of these assets in a sustainable manner against security risks in the cyber environment. Information assets that must be kept secure within the scope of our work include personal information such as identity numbers and telephone numbers, credentials such as computer user names and passwords, and commercially confidential information such as price offers.

In the literature, the user is often referred to as the weakest link; local examples are found in (Arslan, 2021) and (A.Naci Unal & Ergen, 2018). Bulletins, drills, and training sessions are among the popular approaches to increase awareness. Before going into the details, we introduce these approaches for the purpose of familiarization.

Information Security Bulletin

In general, content that is shared online through web pages, in which recent developments in the field of cyber security are compiled, are called information security bulletins.

An information security bulletin is regularly published every month in the workplace from where we obtained our dataset. There are generally ten pieces of content in the bulletin, which are sent to all personnel via e-mail. The bulletin is about corporate cyber security activities and news about global cyber security that may be of interest to the end user.

The bulletin also includes detailed information on what users should do to protect information security while using digital and social media in both their business and private lives. For example, applications that use a user's biometric data for entertainment purposes may violate personal data security. These applications carry the risk of fraudulent access to bank accounts by using the user's biometric data. There is news that the information we provide on our social media accounts may be in the hands of hackers. There are examples in a bulletin from cyber incidents around the world to understand the level reached by the attackers in cyberattacks and hacking. Bulletins also include some information about the legal aspects of information security personnel data protection law.

The bulletin includes popular incidents to attract the reader's attention. Warnings about voice imitation, which may lead to personal data breaches and be used for fraudulent purposes, the dangers that await users as a result of the change in the user agreement of the WhatsApp messaging application, and warnings and important information about the major cyber threats that will stand out in the world in the close future can be given as examples of a number of topics in the bulletin. The aim of publishing a newsletter every month is to keep users aware of information security.

On the other hand, in our study, a phishing drill was carried out with scenarios compatible with the news content in the bulletin. There are reports that personal information and password information can be obtained by redirecting to fake pages.

It is aimed to increase the interest of users in the bulletin by sharing news, including the scenario to be used before a phishing drill in the bulletin.

Phishing Drills and the ISMS

Since there is no international authority that controls the content of web pages, it is possible to create fake web pages on the Internet. The phishing attacker designs a fake interface similar to, for instance, a real bitcoin account interface. Then, trap content is prepared for the user to access the link in the e-mail sent to the user and redirect to this fake interface. When the user tries to enter their account information on the fake web interface, this information is captured by hackers. During the coronavirus pandemic, which affected the whole world from 2020, many cyberattacks were experienced. Phishing drills should be carried out within the organization to increase users' awareness of phishing attacks.

Institutions and organizations must create ISMS (Information Security Management System) policies within the scope of ISO 27001. In line with these policies, corporate information security processes operate. Although cyber security products endeavor to provide security in information infrastructures, it is vital to prepare the weakest link, the human, against possible threats. Phishing drills are accepted as an effective method in both measuring user awareness and increasing awareness (Arslan, 2021).

Before phishing attacks are experienced against an organization, phishing drills that are similar to phishing attack scenarios are performed to determine the level of readiness of human resources.

It is thought that the most effective method to prevent information security violations that may occur as a result of personnel mistakes in the institution where we performed our study is to train and inform personnel. Phishing drills are implemented to cover all personnel for this purpose. Practice scenarios similar to current phishing attacks are determined by the IT department. The details of the phishing exercises conducted in our study are discussed in section "Phishing Drills".

Training should be organized to inform users after phishing exercises and awareness is expected be increased against cyber threats such as phishing attacks.

Information Security Awareness Training

The leading risk that threatens information security is the lack of awareness of employees with regard to security. When information security violation incidents experienced by world-famous IT companies in recent years are examined in detail, it is revealed that the problem is mostly caused by the lack of information security awareness of the employees.

A typical information security awareness training is expected to equip employees with the information they need to protect an organization's information assets from loss and damage.

In the workplace where we carried out our work, online training was held at certain intervals within the scope of information security activities. When the personnel history was examined, it was observed that there were employees from different disciplines who may have little experience in the field of information technologies. For this reason, many subjects in the training were covered at the basic level.

Examples that all users can easily understand and encounter in daily life were included in the training. In addition, the training content was prepared interactively to make the training more efficient. Immediately after providing information about a certain subject in the training, questions about the subject were asked to the user, and unless the correct answer was received, it was not possible to move on to the next subject.

The access address was sent to the e-mail of the personnel via the training platform. In this way, users were able to complete their training at a time convenient for them. Information on personnel who attended or did not participate in the training were reported. In addition, reminder e-mails were sent periodically to those who did not attend the sessions.

The aim of the awareness training was to increase the information security awareness level of the personnel and keep their information up-to-date on information security issues.

Training contains the definition and details of secure information that is explained in the context of Confidentiality, Integrity, and Accessibility. Details on issues such as disruption of business continuity are also included. Internal and external threats that may endanger information security are explained in detail. Details of technical measures may be insufficient if the user violates security. There is detailed information about prevention, storing the information for which the user is responsible, how mobile devices should be carried, leaving a computer session locked, and not leaving documents in meeting rooms. In addition, how to diagnose a social engineering attack and clues about the manner of defense if an attack is noticed are also examined with sample scenarios. Risks that may arise in wireless networks in open areas are detailed.

A number of basic habits are expected from users at the end of the training. Some of these habits include determining passwords that are sufficiently complex and with a large number of characters, refraining from clicking on connection addresses that do not have SSL certificates, being careful when sharing personal information such as identity numbers and phone numbers, the avoidance of using corporate e-mail information while joining e commerce sites, contacting the technical support unit in cases of receiving suspicious e-mails, not connecting to wireless networks serving in open areas, and not leaving confidential information on office desks.

So far, we discussed the fact that information is a valuable asset that needs to be protected and we touched on the training of human resources, which is the weakest link in the context of information protection. In this context, we discussed bulletins, phishing drills, and awareness training. Next, we examine the studies in the literature on these issues.

LITERATURE REVIEW

A literature search through national and international sources was conducted on the studies conducted in the field of cyber security. We took studies that fall within the scope of measuring cyber security awareness after training, drills, and publishing bulletins.

A study that was carried out in South Korea recently (Inho Hwang et al., 2021) included factors that have an impact on information security awareness. These factors included awareness training, participation of management in security, and physical security. In addition to the impact of the titles on cyber security awareness, there are evaluations of the interaction between each other. In the study, based on the foresight that corporate training will increase information security awareness, the hypothesis of "Information Security Education is positively related to Information Security Awareness" was tested. As the research instrument, a questionnaire was sent to 3,000 people from three different institutions operating in the public sector in South Korea. As a result of the study, it was understood that the participation of employees in awareness training had a positive effect on their focus on information security processes and procedures. When the results of the survey were examined, the hypothesis that says there is a positive relationship between the level of cyber security awareness and awareness training was confirmed.

A study (Surachai Chatchalermpun et al., 2020) was conducted to measure the cyber security awareness of the employees of a financial institution operating in Thailand. In the financial sector, where monetary transactions occur instantly, it is important to avoid experiencing any problems regarding the services received or to provide a quick solution to the problem. As a result, there is a SLA with its stakeholders. SLA values being applied for both internal and external customers are important in the finance sector. Employees are required to respond quickly and effectively to requests from customers. This study was carried out by following the e-mail phishing drill method with Bank of Thailand employees. There were 20,500 employees and 700 managers within the scope of the study. A cyber scenario was determined for redirecting to a fake address with the URL in the e-mail text content and typing the password information on the page that opens; details of the study are interesting. According to the results of the study, it was observed that 72.9% of the managers rejected the e-mail. 3% of the administrators, that is 21 people, only opened the e-mail. 85 people, corresponding to 12.3%, opened the e-mail and clicked on the link. 81 people, that is 11.72%, opened the e-mail, clicked the link, and completed the password information. 76.77% of other employees did not open the e-mail. 1.32% of them opened the e-mail but did not perform any clicks. 6.96% of them opened the e-mail and clicked the link. 3,063 people, 14.95% of the employees, opened the e-mail, clicked the link, and entered the password on the fake web page. As a result of this study, a prediction was made about the cyber security awareness levels of the company employees. It was determined that some studies should be performed to reduce the risk. This research aimed to measure and increase cyber security awareness by performing a phishing exercise. However, there was no test study on whether the exercise increased the awareness of the employees thereafter.

In a recent local study (Arslan, 2021), to measure cyber security awareness, a phishing exercise was conducted in a public institution with 33,000 employees and 400 provincial units. The provincial organization and the central organization and the IT personnel of the institution also participated in the study. Fake web pages and fake e-mails were used. This study was carried out in two different phases and for three different personnel types with seven scenarios. The address of the fake website was chosen to be very similar to the name of the institution, just like attacks seen in real life. Users can access the site by sending an e-mail over the domain name used for the exercise. The fake website was designed similarly to the interface of the corporate remote access service. It can be seen in the browser that the site is not safe. On the landing page, there is a username and password field where user information can be entered. As a result of this comprehensive study, it was concluded that cyber security awareness is quite low. However, there is no study conducted to increase awareness after this exercise, and benefits were not reported.

Another local study (Ayse Ozdemir & Uluyol, 2021) aimed to determine the information security awareness levels of people working in public institutions based on various criteria. IT department employees and employees from other departments from many different institutions participated in the study. Most of the 501 personnel were undergraduates. One of the criteria in the survey was education. In the statistical evaluation made according to education level, it was observed that university graduate employees have higher information security awareness than participants with high school or below education levels. In the study, there was no data on the examination of the change after giving awareness training to the same people. However, it was revealed that there is a link between the increase in education level and the increase in information security awareness in general. However, it was argued that one time would not be sufficient to increase the awareness of information security and that it should be a constantly renewed process.

In another enterprise-wide study (Ileri, 2018), employees in a local hospital were taken as a case. Password habits of physicians using the Hospital Information Management System (HIMS) were examined. Since access to institutional information resources is provided through HIMS, physicians must have a high level of information security awareness and use strong passwords. A questionnaire form was sent to 420 physicians via e-mail. The evaluation was made based on the answers of 203 physicians who responded to the questionnaire. Password meter software developed by local authorities, TÜBİTAK BİLGEM (Informatics and Information Security Research Center of TÜBİTAK) was used to measure password security levels. Password security levels are collected in five categories. When the results of the study were examined, it was observed that none of the physicians participating in the study used it at the "strong" or "very strong" level. On the other hand, only 9% used "good/medium" passwords. The password security levels of the remaining 91% were determined as "weak" or "very weak." As a result of the study, it was recommended to increase the inspections for the end users in health institutions and to disseminate information security awareness training as a necessity. On the other hand, although it was suggested to provide awareness training to increase awareness of information security, it is observed that no research thereafter has been conducted to measure the contribution of this training to awareness.

Another local study was carried out on the information security awareness of the employees of a public health directorate (Durmuş Ozdemir & Aslay, 2016). For employees' awareness of information security, an evaluation was made based on factors such as age, gender, education level, title, and experience of using information and communication technologies. Information security awareness training was given and a situation analysis was made before and after the training. The training included information about social engineering attacks, malware types and precautions, password security, and legal responsibilities. A total of 53 personnel, including a doctor, midwife, nurse, dietitian, public servant, technician and manager, participated in the study. A significant difference was found between the information security awareness of the public health personnel participating in the study before and after the training. It was concluded that the majority of the employees did not have enough information about information security before the training and they increased their level of information security after the training. It was concluded that the training given in the research was beneficial and created awareness against cyber and social engineering attacks.

In the literature, another study of the experiences and results obtained during the establishment and development of the information security management system of a Medical Faculty Hospital are discussed (Ileri, 2017). In this 1,200 bed hospital, there is an automation system. All medical procedures and administrative work in the hospital are carried out with this automation system with 2,000 users being able to operate on this system at the same time. During the three-year-long ISMS installation process, six specialist personnel provided regular training to the hospital staff every year. An annual average of 1,217 personnel received an average of 23 hours of training. After the training, regular e-mails and system messages were sent to the employees in order to keep the awareness of the employees high and to keep the issue on the agenda. In addition to these factors, information security warnings were placed at certain points within the institution, and reminders were provided. As a result of the study, thirteen cyber threats at the major level in the previous year before the ISMS integration within the institution interrupted the information systems at an annual rate of 6%. It was determined that 72% of the threats created before the establishment of the ISMS system were caused by the human factor. After three years of training, the total number of threats to information resources decreased by 95%. The rate of human error decreased to 40%. Within two years after the system was installed, the system outage rate decreased to 1%. At the end of the third year, no major threats to information resources were detected. With these results, it was understood that the training given to the employees gave results and that progress was made in the awareness of information security.

In another study, data from the personnel working in the IT departments, which are responsible for providing cyber security, of universities operating in Turkey were used (Gunduzalp, 2021). Within the scope of the study, a personal information form, a data security awareness scale, and a personal cyber security provision scale were sent to the e-mail addresses of 1,440 personnel working in 174 of 206 universities. The evaluation was carried out with the information collected from 410 people who responded voluntarily. It was determined that the awareness of the working personnel of providing digital data security and personal cyber security is high, and there is a positive relationship between these two types of awareness.

In recent research on the example of a city' bar association (Damla Mursul & Kaya, 2019), it is stated that several practices occurred in order to transfer the services offered by the association to the digital environment on a daily basis so as to minimize the risks that may arise due to this and to increase the awareness of cyber security. With developing technology, various security risks arise in the Internet environment at the individual or public level, and accordingly, public institutions tend to take various measures. The legislative information system, which envisages regulations on information security and cyber security in order to create a security culture in public institutions, constitutes the source of the institutional documents that the study deals with. Although the aim is to raise awareness through training, announcements, briefings and bulletins, there is no evaluation of the extent to which these applications contribute to cyber security awareness.

In another study (Sena Nezgitli & Gokcearslan, 2022), a survey model was used to measure the level of information security awareness for a working group composed of public institutions and private sectors. 138 people working in the informatics from different organizations in Ankara were included in the study. According to the results of the study, there was no significant difference in gender, type of institution, age, position and where information security is learned and the frequency of courses; however, a significant difference was observed when information security awareness training is given in the workplace where there exists information security policy. Still, they do not report any evaluation of the effect of repetitive trainings.

Within the scope of another study, the questionnaire applied to the gendarmerie and police officers in the Provincial Gendarmerie Command and the Provincial Security Directorate in a city center was evaluated (Emre Taner & Kilic, 2019). 404 personnel, including 207 gendarmerie personnel and 197 police personnel, participated in the study. In the first part of the questionnaire, questions about socio-demographic characteristics were included, and in the second part, a 34 item information security awareness scale set of questions was included. When the results of the study were evaluated, the information security awareness levels were found to be low in general. The reason for this was thought to be insufficient in-house training. Based on the results of the study, it was recommended to provide information security and awareness training and exams together with basic computer training during the orientation training before starting the profession to at least maximize the information security awareness level of the personnel.

Another study, (Ihsan Tugal et al., 2021) is on cyber security awareness in universities. Within the scope of the study, the importance of cyber security of university infrastructures is emphasized. The awareness levels of university staff on this issue are examined. It was stated that all employees should trained; a guidance list of training is also given. However, there is no pilot study to test awareness after the training given - to our knowledge.

In (Eroglu, 2023) the relationship between the size of the organization, human factor, cyber awareness, education and cybercrime victimization was examined in the light of the data collected from 1419 businesses with the survey method. As a result, it was determined that the human factor increased the risk of phishing crime. It has been evaluated that providing trainings on cyber security to the employees of the enterprises will reduce the victimization of cybercrime. However, there is no examination of the increase in awareness in the case of training.

In another study (Aslay, 2017), the literature observed that if enough attention is not paid to the issue of cyber security, how much damage can be caused to both public organizations, private sector organizations, individuals, and states, and that the issue of cyber security cannot be only with technological devices, it was pointed out that there are many factors in the cyber security management model. In the study, the theoretical aspects of the cyber security management model, which can be used to cope with the cyber security issue of any organization and ensure the security of their critical infrastructures, are examined for organizations to survive cyberattacks with the least damage with a good policy. Regardless of being in the public or private sector, it was stated that the necessary infrastructure should be created to ensure the cyber resistance of corporate organizations, the awareness of personnel should be increased, and management should make decisions first. Within the scope of the said study, there is no evaluation of the contribution of the activities to be made although the measures recommended within the scope of cyber security management are included.

Within the scope of another study (Yildirim, 2018), a questionnaire was administered to the information technologies managers of 150 small and medium-sized enterprises selected randomly among 5,000 enterprises registered in a city's chamber of commerce and industry operating in the city. After the evaluation of the questionnaires were made in line with the information security criteria, it was determined that 65% of the employees in the enterprises do not have sufficient knowledge about cyber security. It has been suggested that awareness studies on cyber security should be carried out in order to increase the awareness and knowledge levels of employees with regard to cyber risks. On the other hand, although the problem point has been determined, concrete suggestions for the solution of the problem and measurements regarding the benefits of the suggestions were not included.

The results of the study conducted in an airline company operating in Turkey in the recent period were examined (Erdogan, 2020). Several activities were carried out within the scope of the ISMS installation at the said airline company. After the documentation, studies started to increase the awareness of information security. Increasing the number of personnel who received information security awareness training above 60% was determined as a target. The change in the number of personnel receiving training over the years was examined. It is stated that 1,000 of the existing 5,000 personnel received training in 2018. After the ISMS studies, 4,000 of the 5,500 personnel received this training in 2019. The participation rate, which was 20%, increased to 72% within one year. Although the rate of receiving awareness training in this business had increased thanks to ISMS studies, there was no analysis of the level of information security awareness in real terms. There was also no information about the change in the level of personnel awareness before and after the training.

In a recent study (Akal, 2022), information security awareness levels of IT companies operating in Ankara were examined. An analysis was carried out based on the survey data made to a manager from 253 different companies. As a result of the survey, answers to five different questions were sought. One of the questions, namely "Does the information security awareness of companies change according to the education level of company managers?" was in the form. The educational status of 253 participants was grouped as high school, associate degree, undergraduate, and graduate. When the findings were evaluated, it was observed that as the education level of the company manager increases, the level of awareness also increases. Within the scope of the study, the relationship between education status and awareness levels was examined in general; however, there is no data on which graduate level employees should receive information security awareness training. Therefore, no connection was established between mindfulness training and awareness level.

Another study (Gulhan, 2021) aimed to measure the information security awareness of the personnel working at a university, where 111 academic staff and 58 administrative staff participated. An information security awareness scale was sent to 169 employees who were asked to answer 34 questions. The feedback received from the employees was analyzed according to variables such as demographic characteristics, title, working time, and gender. In line with the findings of the study, it was concluded that training should be organized and visual materials should be used in order to increase the information security awareness of the employees. Within the scope of the research, the level of information security awareness was examined, but there was no evaluation of how much it would contribute to levels of awareness if the training recommended to be organized to increase information security awareness were implemented.

In a recent study conducted in Ankara (Altiner, 2021), the cyber security awareness levels of secondary school teachers were examined. Personal information forms and some other questions were used in the research. Responses from 455 teachers were analyzed according to demographic characteristics, branch, etc. and evaluated according to the criteria. It was observed that the level of providing personal cyber security to teachers in the branch of information technologies is significantly higher than for teachers in other branches, regardless of the branches, courses, etc. for information and communication technologies. It was observed that the level of providing cyber security to trainees was higher. According to the results of the research, it was observed that the awareness of teachers who received training in information and communication technologies and cyber security was higher. From this perspective, it was evaluated that training should be given to all personnel in the school, especially teachers, in order to increase cyber security awareness.

The literature reviewed so far shows a considerable amount of work on preventing phishing attacks. When we go into more detail, we observe that these tests were conducted in large organizations, similar to our case.

However, the literature did not concentrate on measuring the preventions; hence, solid results on how the tests were beneficial for the organizations are missing from our knowledge. Most mention that the test was beneficial However, such claims would remain vague and cloudy because they missed rigorous measurements to highlight positive changes.

As we have seen in the literature review, the effect of information security awareness training on increasing the cyber security awareness of people in samples has been examined in many studies. In the studies we have examined within the scope of the literature review, it is clear that the solution predictions for increasing the cyber security awareness of the users in corporate organizations remain as suggestions and incentives.

Studies were carried out in the literature on the methods and benefits to be applied to increase cyber security awareness. In the articles examined, there is information that the training to be given to the employees in order to ensure corporate cyber security would be beneficial. Awareness training is diversified with developing technology.

A few examples of awareness training methods include:

- collective training in the form of video conferences;
- organizing surveys;
- sending digital bulletins to all personnel;
- sharing on information screens in the workplace;
- organizing in-class training;

- conducting phishing drills; and
- providing interactive training by preparing computer games.

On the other hand, although the importance of awareness training in the field of information security is emphasized in the studies, it can be observed that the studies on measuring the level of benefits of the training in question are limited.

From this point of view, in this study, the benefits of some of the methods applied to increase cyber security awareness are discussed. The changes in the awareness level were attempted to be measured as a result of the awareness-raising activities carried out regularly and the phishing exercises carried out afterward.

Considering our discussion of the works we have studied, we have opted to concentrate on changes in the awareness of employees after the drills and tests.

"Method" section reports our approach to understanding what and how such tests can change awareness in an organization.

METHOD

When the studies in the literature are examined, it is emphasized that the level of cyber security awareness of employees in organizational structures should be increased. It is foreseen that information security awareness training will contribute to this goal in general.

A number of methods may be applied to increase cyber security awareness. Some of these include:

- Phishing drills,
- Awareness training,
- Digital bulletins,
- Information screens,
- Survey studies.

Within the scope of our study, the contribution of phishing exercises, information security awareness training, and digital bulletins in increasing the awareness level of employees was examined.

In this study, data obtained from repetitive and regular studies were used to measure and increase their awareness levels with regard to the cyber security awareness of the employees in an organization with roughly 1,000 employees. The organization requested that their names be anonymous in the study; we explain how the institution had collected the data.

Our measurement method for increasing the cyber security awareness of the employees was in accordance with the following criteria:

1- Increase in participation rate in awareness trainings to be implemented after the phishing exercises

2-Increase in the rate of reading bulletins after the phishing exercise and awareness training practices

3- Decrease in the number of personnel clicking malicious links in fake e-mail content in the phishing exercises after the trainings and bulletins

4- The effects of activities in changing awareness and bulletin reading rates in the long term.

Phishing exercises were carried out with fake e-mails sent to the employees. Information security bulletins were sent regularly every month and an information security awareness training was provided. Afterwards, a re-phishing exercise was carried out and the awareness levels of the users were measured.

The first phishing exercise was that of a discount agreement having been made with a fuel company. Employees were expected not to click on the button that redirects to an unsafe address in the e-mail.

After the phishing exercise was terminated and announced to all the employees, the online training video was shared with all the employees to raise awareness of information security. With regular reminder notifications, more employees became able to complete this training.

A monthly digital cyber security bulletin was sent to every employee via e-mail. The newsletter contained informative content on current cyber events, phishing attacks, and information security. In addition to the awareness training, users were expected to be vigilant against phishing attacks, thanks to the digital bulletins.

After a certain period, the phishing practice was renewed. The scenario of the second exercise was that of a vaccination campaign that was started against the COVID 19 epidemic with the thought of being a current issue. With the "Vaccination Appointment" button in the fake e-mail, users were directed to a fake "e government" page and they were asked to enter their citizenship Identity Number and password information.

After the second phishing exercise, information security awareness training was sent to users. Users were reminded for approximately two months and they were informed to complete the training.

Along with awareness training, digital bulletins continued to be published. Details of the phishing exercise carried out were as follows:

- Issues of what users should be aware;
- How to detect whether a redirect to a fake page has been made;
- Security differences between http and https protocols; and
- Some statistical information.

These criteria were shared with users in the monthly bulletin.

After the phishing drills and other awareness-raising activities, a third phishing exercise was conducted. The scenario of this exercise was determined as a discount agreement with a chain of stores selling electronic products and a draw among the participants. In the e-mail that was sent, a button to be clicked to benefit from the discount and participate in the lottery campaign was presented. It was analyzed whether the users noticed that the relevant button was redirecting to a fake web page.

RESULTS

Phishing Drills

Three phishing exercises are carried out; these are contracted gas station campaign, vaccination appointment and electronics store discount and lottery campaign scenarios.

The data regarding the phishing exercise carried out with the contracted fuel station campaign scenario are given in Table 1. E-mails were sent to 1,010 personnel. The number of personnel reading the e-mail was 334. The number of personnel who clicked on the link that led to downloading the malicious file contained in the e-mail to the computer was 94. In addition, 47 of the 94 personnel activated the application file in the ".exe", executable format which they had downloaded to the computer. While the rate of those who clicked on the malicious link among the personnel who read the e-mail was 28.14%, the rate of the personnel who ran the malicious file among those who clicked on the link was 50%.

The data regarding the phishing exercise with the scenario of starting a vaccination campaign due to the COVID 19 epidemic are given in Table 2.

An e-mail was sent to 1,019 personnel for phishing purposes. 822 personnel read the e-mail. 237 personnel clicked on the link in the e-mail. The number of personnel who entered the information requested for the appointment on the page that opened was 132. Among those who read the e-mail, the rate of those who clicked on the harmful link was 28.83%. The rate of personnel entering information among those who clicked on the link is 55.70%. The rate of personnel who entered the information among those who read the e-mail was 16.06%.

Table 1. Results of first phishing drill

Number of personnel who were sent e-mail	1,010
Number of personnel reading the e-mail	334
Number of personnel clicking on the link	94
Number of personnel running the malicious file	47
Percentage of personnel clicking on the link among readers (%)	28.14
Percentage of personnel who run the malicious file among those who clicked on the link (%)	50

Table 2. Results of second phishing drill

Number of personnel who were sent e-mail	1,019
Number of personnel reading the e-mail	822
Number of personnel clicking on the link	237
Number of personnel entering information	132
Percentage of personnel clicking on the link among readers (%)	28.83
Percentage of personnel entering information among those clicking on the link (%)	55.70
Percentage of personnel who entered information among those who read the e-mail (%)	16.06

Table 3. Results of third phishing drill

Number of personnel who were sent e-mail	1,034
Number of personnel reading the e-mail	639
Number of personnel clicking on the link	55
Percentage of personnel clicking on the link among readers (%)	8.61

The data regarding the discount and lottery agreement campaign scenario with the chain of stores that sell electronic products and the phishing exercise are given in Table 3.

The number of personnel to whom the e-mail was sent is 1,034. The number of personnel who read the e-mail was 639 and the number of personnel who clicked on the malicious link was 55. The rate of those who clicked on the harmful link among the personnel who read the e-mail was 8.61%.

Table 4. Results of awareness trainings

	First Training	Second Training
Number of personnel who were sent e-mail	1,015	1,024
Number of personnel reading the e-mail	921	925
Number of personnel completing training	525	650
Percentage of personnel who completed education (%)	57.00	70.27

Cyber Security Awareness Training

The data regarding the information security awareness training held shortly after the Contracted Fuel Station Campaign Phishing exercise are listed in Table 4 under the first training column.

The number of personnel sent e-mails is 1,015. The number of personnel reading the e-mail is 921. The number of personnel who completed the awareness training reached 525. The rate of those who completed the training among those who read the e-mail is 57%.

The data regarding the information security awareness training carried out after the phishing exercise with the vaccination appointment scenario are given under the second training column in Table 4.

The number of personnel sent e-mail is 1,024. The number of personnel reading the e-mail is 925. The number of personnel who completed the training is 650. The rate of those who completed the training sessions among those who read the e-mail is 70.27%.

Reading Levels of Digital Bulletins

Prior to the phishing exercises and awareness training, information security bulletins containing news about current cyber threats were shared with the personnel. Statistics about the reading rate of digital bulletins are listed in Table 5.

The first information security bulletin reached 990 people. Out of 990 people, 500 opened the e-mail they received. The number of personnel who clicked on the link in the e-mail to access and read the newsletter was 27. The rate of those who read the e-mail and access the newsletter was 5.40%.

After the first phishing exercise and awareness training, the reading rate of another bulletin shared with the staff was 8.55%. The reading rates of subsequent bulletins were 8.46%, 18.14%, 8.20%, 11.69%, 11.16% and 7.56%, respectively. After eight information security bulletins were published, the second phishing exercise was performed, and then the second information security awareness training was

Table 5. Statistics about digital bulletins

Bulletin	Number of personnel who were sent e-mail	Number of personnel reading the e-mail	Number of personnel reading the bulletin	Percentage of personnel bulletin readers among reading the e-mail (%)
1	990	500	27	5.40
2	1,015	550	47	8.55
3	1,029	603	51	8.46
4	1,029	667	121	18.14
5	1,023	622	51	8.20
6	1,027	616	72	11.69
7	1,027	636	71	11.16
8	1,023	688	52	7.56
9	1,021	728	246	33.79
10	1,030	704	87	12.36
11	1,033	724	112	15.47

shared online with all personnel. The reading rate of the ninth information security bulletin, which was prepared after these activities, was 33.79%. The reading rates of the other two bulletins that followed were 12.36% and 15.47%.

Comparison of Results

The chronological order and results of the methods applied in accordance with the methodology specified are listed in Table 6.

It can be observed that the rate of reading the bulletins and participation in trainings increases over time. Due to the increase in awareness, a decrease was observed in the rate of personnel who clicked on fake links in the phishing attack in the long term.

On the other hand, the reading rate of the fourth bulletin increased significantly compared to the rate of the previous bulletin. Although no awareness-raising work was carried out, this change has been observed. The reason for this change could not be determined.

CONCLUSION

Survey studies conducted on different groups, such as employees from the private and public sectors, form the basis of the theses in the literature.

Table 6. Comparison of results

Results of Methods	Bulletin Reading Rate (%)	Phishing Drills Rate (%)	Awareness Trainings Rate (%)	Date
Bulletin 1	5.4			January 2021
Phishing Drill 1		28.14		February 2021
Awareness Training 1			57.00	April 2021
Bulletin 2	8.55			June 2021
Bulletin 3	8.46			July 2021
Bulletin 4	18.14			August 2021
Bulletin 5	8.20			September 2021
Bulletin 6	11.69			October 2021
Bulletin 7	11.16			November 2021
Bulletin 8	7.56			December 2021
Phishing Drill 2		28.83		January 2022
Awareness Training 2			70.27	February 2022
Bulletin 9	33.79			January 2022
Bulletin 10	12.36			February 2022
Bulletin 11	15.47			March 2022
Phishing Drill 3		8.61		March 2022

Based on the information received as a result of the surveys, the aim was to determine the level of cyber security awareness in organizations. Suggestions on which methods should be followed to increase awareness and keep awareness at a high level are also included in these studies.

We studied a dataset from a public institution with more than 1,000 personnel. Three methods had been determined to increase the awareness level of the institution's personnel. For phishing exercises, the aim was that users be aware of cyberattacks, while keeping personal and corporate information safe thanks to information security awareness training, and having information about current cyber events with information security bulletins. In addition, as a result of these studies, the change in the level of information security awareness was examined.

As conclusion, we first present our findings followed by discussions that emerged as a result of the study in the items. Then, we include the limitations of our study at the point of revealing the findings. How the limitations can be overcome and research that can be done in future studies are included in the "Future Works" sub-section. Finally, we recall our research question and give an answer in light of these findings.

Findings

Finding One: Organizations Do Not Always Disclose Cybersecurity Breaches to the Public

It is understood that the level of information security awareness of employees not only in Turkey but also abroad is at a low level. However, when we consider public institutions and private sector organizations all over the world, it is observed that the reported information security violations are limited. It is known that many DDoS attacks are made instantly and there are many ransomware attacks. Valuable information on which vulnerabilities have emerged against which attack is not shared by organizations. Moreover, it is not reported whether this vulnerability is caused by the user, and if in fact caused by the user, how to take precautions and train the user.

Finding Two: The Research on the Benefits of Suggestions Presented for Increasing Information Security Awareness Is Limited in the Literature

Studies in the literature generally include questionnaires. Survey questions focused on user habits and the importance of information security. In many studies, survey questions were sent to users by e-mail and their answers were requested via e-mail. According to the results of the survey, the information security awareness levels of users were determined. Moreover, similar to the studies carried out in Turkey, it is observed that phishing exercises and training are organized in order to measure and improve the level of awareness in the studies carried out abroad. The methods we used in our study are global. It is generally suggested that phishing drills and information security awareness training will be beneficial in increasing information security awareness. These methods have been implemented within the scope of information security management system policies. However, a second survey was mostly not applied to confirm these recommendations.

Finding Three: As a Result of Phishing Exercises and Information Security Awareness Training, The Rate of Reading Information Security Bulletins Has Increased in the Long Term

Although written information is more reliable and more permanent, it is understood that the practices regarding the publication of digital bulletins are seen as a supporting element rather than the main activity. On the other hand, it was observed that the digital bulletin first achieved a low average reading rate. In the institution where we conducted the case study, the reading rate of the last bulletin sent to the users before the phishing exercise and awareness training was slightly higher. After the phishing exercise and training activities, it was understood that the rate of reading the bulletin had increased by a factor of 1.5 in the long term. This change is an indication that phishing drills and awareness training increase users' information security awareness and interest in the bulletin.

Finding Four: As a Result of Information Security Awareness Training and Information Security Bulletins, Fewer Employees Believe in Phishing E-Mails

The rate of personnel in the organization who clicked on the malicious link in a fake e-mail in the first phishing exercise was 28%. Furthermore, 50% of the people downloaded the malicious file after being redirected to the fake website. On the other hand, similar results were obtained in the second phishing exercise, which was carried out after awareness training and a few bulletins. After two phishing exercises, two awareness training sessions and the publication of eight different monthly bulletins, a third phishing exercise was held. As a result of the last exercise, the rate of those who clicked on the harmful link decreased to 8%. It is understood that regular activities on information security increase users' awareness of phishing attacks.

Finding Five: As a Result of Phishing Drills and Reading Information Security Bulletins, More Employees Have Attended the Information Security Awareness Training

The participation rate in the first information security awareness training was 57%. Afterward, bulletins were published; these bulletins contained the consequences of current cyber dangers and also news about the importance of information security. In addition, a phishing exercise was performed. After the aforementioned activities, the completion rate of the training, which was open to the online participation of the users, exceeded 70%, indicating an increase in the users' interest and need for training in information security.

Finding Six: Findings Depend on the Literature Review and the Dataset We Had Obtained

In many countries, including Turkey, information security violations in public institutions must be reported to a competent authority. On the other hand, cyberattacks between countries can be tracked by regulatory agencies and service providers over Internet traffic. However, due to institutional sensitivities, this information is not shared in open sources. For this reason, it was observed that the number of studies in this field in the articles and theses in the literature is limited. In addition, no interviews were conducted with the people who performed phishing exercises or the personnel who were exposed to phishing in the institution where our study was made. Nevertheless, partially personal experiences are included.

Limitations

Limitation One: Verification of Users and Individual Assessments

Phishing drills, awareness training content, and digital bulletins within the scope of the study were sent to the e-mail addresses of the users. Users opened their e-mail clients and logged in with a username and password information. Methods such as camera use and static IP address tracking were not followed to verify the personnel. All evaluations were carried out by considering the participants as a homogeneous cluster. This limitation relates to Findings 3, 4, 5 and 6.

Limitation Two: Sociological Effects

While making comparisons between repetitive activities within the scope of the study, sociological changes for users were ignored. There may have been one or more factors in their respective social lives that would have positively or negatively affected the awareness level of users. This limitation relates to Findings 3, 4, 5 and 6.

Future Work

Future Work One: Make User-Specific Evaluations

Within the scope of our study, all employees were handled homogeneously and an overall evaluation was made. The interactions of the personnel regarding phishing, training, and bulletins may be detected following a method involving a user name and IP address. In this way, changes in the level of information security awareness in organizations could be examined individually. Such research may be a solution to Limitation 1.

Future Work Two: Effects of Social Events

While evaluating the results of our study, the experiences of the users in their social lives were not included. There may have been changes in awareness levels due to different social events. In future studies, external factors as well as activities within the organization could be included in the evaluation. In addition, studies on the level of cyber security awareness may be carried out in the field of social sciences. For this reason, research could be conducted in the field of social sciences to compare with the studies that include information security analyses on the social lives of personnel. Such research may address Limitation 2.

Future Work Three: Other Methods of Research

Information security breaches are seldom reported. Face-to-face meetings can be held with officials from many different organizations. The responsibilities of the researcher regarding the use of data shared by companies can be determined. In this way, a number of organizations could also share their data on cyber security incidents. Thus, more data resources pertaining to the awareness of the users can be accessed. In addition, face-to-face interviews may be conducted with the people who organize information security activities, training, and phishing exercises within the organization. Users affected by phishing may also be interviewed. The method of rewarding users with high awareness can also be determined as the fourth method in addition to our study. Such research may be a solution to Limitations 1 and 2.

Answer to the Research Question

The research concludes that it is a useful set of practices that can be applied to increase the cyber security awareness of personnel, providing information about current cyber events by preparing information security bulletins in public and private sector organizations, having users experience real attacks by performing phishing drills, and learning basic information through training.

Returning to the research question:

Can conducting phishing drills, doing awareness training, and preparing information security bulletins increase the cyber security awareness of personnel in the long term?

We can present a sound answer; through the light of the findings and by considering the limitations, organizations can systematically conduct phishing drills, provide awareness training and regularly prepare information security bulletins to increase the cyber security awareness of users.

REFERENCES

Akal, M. (2022). *Information Security Awareness in IT Companies Ufuk University*].

Altiner, I. (2021). *Evaluation of Teachers' Personal Cyber Security Awareness Levels According to Different Variables*. Ankara University.

Arslan, Y. (2021). Phishing Attacks Awareness Exercise Example. *Düzce Üniversitesi Bilim ve Teknoloji Dergisi, 9*(3), 348–358. doi:10.29130/dubited.832862

Aslay, F. (2017). Cyber Attack Methods and Current Situation Analysis of Turkey's Cyber Safety. *International Journal of Multidisciplinary Studies and Innovative Technologies, 1*(1), 24–28.

Chatchalermpun, S., Wuttidittachotti, P., & Daengsi, T. (2020). *Cybersecurity Drill Test Using Phishing Attack: A Pilot Study of a Large Financial Services Firm in Thailand.* 10th Symposium on Computer Applications & Industrial Electronics (ISCAIE), Malaysia.

Erdogan, S. E. (2020). *Building an Information Security Management System, Implementation of IEC / ISO 27001 Standard in A Civil Aviation Organization.* Istanbul Kültür University.

Eroglu, C. (2023). The Size of Cyber Crimes that Businesses Are Exposed. *Journal of Security Sciences, 12*(1), 69–96. doi:10.28956/gbd.1264593

Gulhan, B. (2021). *Awareness of Information Security in Higher Education Institutions: The Case of Bahçeşehir.* University Bahçeşehir University.

Gunduzalp, C. (2021). University Employees' Awareness of Digital Data and Personal Cyber Security (A Case Study of IT Departments). *Journal of Computer and Education Research, 9*(18), 598–625. doi:10.18009/jcer.907022

Hwang, I., Wakefield, R., Kim, S., & Kim, T. (2021). Security Awareness: The First Step in Information Security Compliance Behavior. *Journal of Computer Information Systems, 61*(4), 345–356. doi:10.1080/08874417.2019.1650676

Ileri, Y. Y. (2017). Information Security Management in Organizations, Enterprise Integration Process and A Case Study. *Anadolu University Journal of Social Sciences, 17*(4), 55–72. doi:10.18037/ausbd.417372

Ileri, Y. Y. (2018). Security in Accessing Enterprise Information Resources: A Research on Password Management of Physicians. *International Journal of Health Management and Strategies Research, 4*(1), 15–25.

Mursul, D., & Kaya, A. (2019). A Review of Public Institutions in Terms of National Information Security Policies: Sample of Kayseri Bar Association. *ASSAM International Refereed Journal*(Special Issue), 331-343.

Naci Unal, A., & Ergen, A. (2018). Cyber Security Behaviour: A Research Conducted in Istanbul. *Manisa Celal Bayar University Journal of Social Sciences, 16*(2), 191–216. doi:10.18026/cbayarsos.439489

Nezgitli, S., & Gokcearslan, S. (2022). Review on Information Security Awareness for Public Institutions and Private Sectors. *Instructional Technology and Lifelong Learning, 3*(1), 19–44. doi:10.52911/itall.1115701

Ozdemir, A. & Uluyol, C. (2021). Information Security Awareness in Public Organizations. *The Journal of Turkish Social Research, 25*(3), 649–666. doi:10.20296/tsadergisi.815635

Ozdemir, D., & Aslay, F. (2016). *Examination of The Effects of Information Security Awareness Training on Staff of Erzincan Public Health Administration.* International Erzincan Symposium, Erzincan, Turkey.

Taner, E., & Kilic, I. (2019). A Study on Determining Information Security Awareness of Security Forces. *The Journal of Security Sciences, 8*(2), 253–269. doi:10.28956/gbd.646321

Telecommunication Standardization Sector of ITU. (2009). Series X: data networks, open system communications and security. In *Overview of cybersecurity*. ITU.

Tugal, I., Almaz, C., & Sevi, M. (2021). Cyber Security Issues and Awareness Training at Universities. *Journal of Information Technology, 14*(3), 229–238. doi:10.17671/gazibtd.754458

Yildirim, E. Y. (2018). *Cyber Attacks Directed Information Systems (IS) and Maintenance of Cyber Security.* 2nd International Vocational Science Symposium (IVSS), Antalya, Turkey.

KEY TERMS AND DEFINITIONS

Digital Cyber Security Bulletin/Newsletter: Content that is shared online through web pages, in which recent developments in the field of cyber security are compiled. In general, they are regularly published

Information as Asset: The sum of the information infrastructure, applications, services, and information transmitted and/or stored in the cyber environment.

Information Security Awareness Training: Training courses organized by the workplaces; such courses aim to increase awareness of the end-users by teaching them type of attacks and how to dodge them.

ISMS: Information Security Management System, a systematic series of policies that enforce/guide the workplace in operating corporate information security processes.

Phishing Drill: Attacks that are permitted by the workplace; they are similar to phishing attack scenarios and are performed to determine the level of readiness of human resources.

Phishing E-Mail: A specially crafted e-mail by an attacker; it may appear as a genuine e-mail. With the help of the prepared e-mail, computer users are directed to fake screens and they are requested to log in with a password, usually.

Chapter 10
Network Security Breaches:
Comprehension and Its Implications

Yash Bansal
The NorthCap University, India

Shilpa Mahajan
(iD) https://orcid.org/0000-0001-8975-8114
The NorthCap University, India

ABSTRACT

Threats on communication networks are proliferating. It is surprising that more than half of Indian organizations have experienced a breach in 2022. These breaches can occur due to various reasons, such as human error, software vulnerabilities, or malicious attacks. Tech executives frequently commit the sin of being unready for network data breaches despite the early indications. The consequences of a network security breach can be severe, including financial losses, damage to reputation, and loss of sensitive data. The punitive damages are severe, and it is easier to lose and much harder to restore the faith of customers. This chapter gives an overview on the evolution of network architecture, the grounds behind data leaks and violations. It also discusses the best approaches to counter the threats, governance structure and emerging developments in network security. This chapter also gives you the analysis on the methodologies adopted for network forensics and explores the viewpoint on where network forensics could be applied.

DOI: 10.4018/978-1-6684-8218-6.ch010

INTRODUCTION

For many years, firewalls, VPNs, and antivirus software served as the foundation for network security. Nevertheless, this defence trinity has reached its breaking point. Why? Firstly, Online services are now frequently employed by companies to drive sales, enhance interactions, and streamline workflows, but because network layer standards are protected by today's security architecture, Web platforms are fundamentally unprotected.

Evolution of Network Architecture

- It is an enormous task for the data security operations to stay up with pace—and even be forward so as never to behave hastily—as new ways to breach technology infrastructure are developed daily. Due to its ability to prevent impulsive attempts on the systems from both the standpoint of the network and by infecting a machine directly, the installation of an effective IDS strategy was essential in a network architecture but with passing time security managers stress that they are perpetually playing tag when it comes to dealing with security governance, market needs, and evolving threats. Although the IDS system are effective as any, predictive analytics and analytics are still laborious manual processes even after regular adjustment (Kozlosky, 2021). Experts and clients claim that signature-matching IDS is gradually losing its relevance in the face of a constantly reducing window of opportunity betwixt weaknesses and exploitation. They gripe about how IDS behaviour is greatly increased by automated attacks and suggests how acquiring all the information and screening the event logs automatically will help us identify the pertinent facts.
- In a blockchain system, each transaction is saved in a block and added to the chain of last blocks, creating a permanent and tamper-proof record of all transactions. Around 2017 several have seen the most recent stateful inspection firewalls deployed as a new barrier to defend against TCP assaults. This method offered layer 4 protection, but since many modern attacks—like buffer overflows, SQL injections, and cross-site scripting—occurred at the application layer, it was no longer effective to follow those ideals. Just as a workaround for the holes was being developed, it was easy to overlook about the stress it would bring. The inclusion of additional defence from various vendors, as well as application firewall and malware gateways to the perimeter, overburdened the employees in charge of administering it (Qureshi et al., 2021). As if this weren't enough, many businesses believed they were safe from loads of rising attacks. But those assaults may end up being the

favoured method used by today organizations or felonious syndicates to impede commerce or drag down an enterprise.

Grounds Behind Violation

Business requirements necessitate the creation of fresh servers, applications, and channel layout that require protection, and advancements like Wireless LAN, instant communications, and IP telephony introduce insecure network protocols, nullifying the efficacy of current protection measures. We have met the enemy and it is us, rather than encouraging teamwork and collaboration network groups often concentrate on uptime and time-between-failure. This frequently leads to increased conflict and blame-laying. Without the proper restrictions, it's possible that both groups will end up working against one another. Moreover, handoffs and acquiring and integrating the product could be difficult and ineffective because security and network operators frequently use manual approaches. The fact that IT and protection groups typically use distinct technologies and depend on various data collections as their sources of truth is another factor. Therefore, the IT and defence teams have different perspectives on the network's reality, which prevents them from sharing crucial information necessary to maintain the network's security and performance of products. Firms must resolve the conflict between the two groups if they want to guarantee the effectiveness of activities and protection. This entails making sure they share similar technologies, uniform processes, and clear goals for them to collaborate peacefully (Oltsik, 2021). There also exists the dilemma for small organizations or new foundations. It's crucial to have a team of computer specialist on standby to assist when things go wrong since network problems are unavoidable. But for many organizations, particularly those that are smaller, having an on-site crew is simply not practical. They must be informed that virtually all business types can benefit from mobile IT support. They offer remote management and maintenance for virtual computer networks in addition to assistance. Network security is the main area of focus for remote IT businesses, and they will notify you of any weaknesses (Qureshi et al., 2021).

Popular Breach Vectors

1)   1 in 131 emails contain malwares.
2)   45% of companies are not using a dedicated patch management solutions to distribute updates.
3)   81% of data breach victims do not have a system in place to self-detect data breaches.
4)   30% of companies are not dedicated to resolve security misconfigurations.

5) Around quarter of the companies have poor encryption.
6) 50% of resources are sensitive and exposed.

LITERATURE REVIEW

Review

Experts expect various trends that will influence the future of network security, which is an area that is always developing. From zero security in which network users and devices are presumed to be unreliable until they can demonstrate their identity and access rights to new encryption methods that are based on quantum physics which can overcome the threats of quantum computing that is easily able to defeat conventional cryptographic schemes, there are a lot of new possibilities. But, despite the seemingly unlimited possibilities, we must constantly be on the lookout for dangers that are always one step ahead. There are many studies that have been conducted about network security, according to the Insights by ESG group one of the more recent strategies to combat network vulnerabilities is the switch from a private hub to a mesh design employing an MPLS-based VPN. With this change, communication between stores is now possible, inventory can now be managed geographically, and service quality has greatly improved. While it will expand network security duty from the main office to all retail outlets, it is still regarded as a viable option in contrast. Another study by Cisco focuses on process control and describes how everything starts with determining the network's risk and assembling a team to handle it. To maintain the program, a security shift administration procedure must be put in place, and the network must be watched for security flaws (Cisco, 2021). There are bound to be employees who contact a competitor to sell the intellectual property, doing behaviour surveillance will assist in detecting unexpected and perhaps harmful conduct. It has been discussed before, and warning signs have frequently been cited. The insights provided by Constella in the article titled "The Data Breach Era" is one ideal illustration of it. The study examines an alternative strategy and demonstrates why, despite what many people may believe, constructing a wall around a company's network to keep thieves out is insufficient. Many security concerns are caused by ineffective processes or configuration flaws (Constella, 2021). A sys admin installs the incorrect patch on a database server, a security manager frequently makes undocumented changes to the firewall, and new parts is added to the network without notification to the security team, or upgrading configuration management documents, to name a few examples. These seemingly simple blunders can lead to security flaws that are difficult to detect and expose businesses to hackers or automated attacks. The document by LeadingAge, which assists geriatric companies in understanding vulnerabilities, how to reduce them, and how to react if attacked, provides another excellent example. The study also focuses extensively on policies and management strategies, which is why it serves as a useful example. It is interesting how businesses often base their security operations on IT governance standards like the Management Standards for Information and Associated Technologies, IT Service Operations, or the IT Infrastructure Library to address the flaws. These Standard policies, practices, and documentation works well for crucial risk activities like modification and setup management by these tried-and-true governance models.

More About IT Governance and Management Standards

IT Governance and management standards can be defined as organization or personnel responsible for controlling and directing both the information security and its system and responsible for suggesting business measures that align with the security goals. In within its parameter, we can observe how it focuses on four vital components and how it works its way to integrate them with each other as shown in table 1.

Table 1. Four components of management standards

Security Regulations and practices
Specific Business requirements
Organizational Requirement
Security Approach

1) Security Regulations and practices- Application Filtering, Constant Scanning, Patches and Protocols.
2) Specific Business requirements- Security Documents, Cobit, Host Protection, Behaviour modelling, Business modelling.
3) Organizational Requirement- Meetings, Management, Maintenance, Monitoring, Audits, User Training.
4) Security Approach- Immediate actions based upon the nature of vulnerabilities, Recommended Controls.

SECURITY METHODS

End-to-End Security

When talking about network security, end-to-end security is a key component that shouldn't be disregarded. It is a crucial issue that every business must prioritize and deal with. End-to-end security guarantees that confidential data is safeguarded during its entire route, from source to use. A security compromise may significantly slow down daily operations. End-to-end security guarantees that crucial programs and networks continue to function even in the case of a security incident. Its implementation can reduce the cost of data breaches, legal problems, and other stability mishaps (Grigas, 2023). Also, it saves firms the expense of having to restore broken systems and recover lost data. But how does everything operate and how do the parts fit together to form the puzzle we are trying to solve? Basically, the communication is encrypted by the originator using a public key, which only the intended receiver can decipher. The recipient decrypts the message using their private key, which is accessible only to them, to prevent listening, manipulation, or interception of the data is further then delivered across an encrypted connection, like HTTPS or SSL (Laliberte, 2023). From here on out, access controls measures are adopted allowing who can send what and how as shown in Figure 1.

Figure 1. End-to-end security

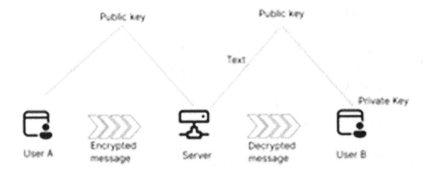

Table 2. DLP stages

Stage 1	Safety and Security Risk Management
Stage 2	Audit and Survey
Stage 3	Master Planning
Stage 4	System Design
Stage 5	Policies and Procedures
Stage 6	Training and Awareness

Data Loss Prevention

The data loss prevention system works just like end-to-end security, the difference lies in the extend it takes and incorporates those measures for defence against network attacks. To discover critical data and implement security standards, DLP systems often include techniques like content examination, contextual analysis, data classification, encryption, and access controls. They might also have features like user activity tracking, data masking, audit trails, and backup and recovery of data. The possibility of data breaches, compliance violations, and reputational harm can be minimized using a DLP plan and solutions as shown in Table 2.

Security Prioritisation

In order to efficiently handle and dedicate assets to minimize risks and threats according to their degree of severity, it is crucial to develop a priority order in security. Prioritization enables a better organized security strategy, making it simpler to identify which dangers are most important and call for quick intervention. Organizations can use their limited protection resources more wisely and handle the most serious risks while limiting the effects of less serious ones by prioritizing security threats in order of importance (MITRE, 2019). To ensure that an organization's security precautions are strong and effective throughout time, a security priority order is crucial. In general terms, these tactics should start with a merit based hierarchical evaluation of the network's resources, from user workstations down to task critical utility hosts as shown in Figure 2.

1) Series 1- Refers to the top levels, which attackers attempt to strike initially in order to infiltrate the organisation and compromise it
2) Series 2- Refers to the procedure that aids in safeguarding the organization's top layers and other assets

Figure 2. Observed priority hierarchy in industries

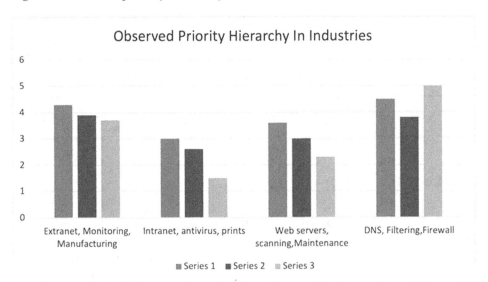

3) Series 3- Refers to the documents and protocols that are required for the procedures to function.

NETWORK FORENSICS

In addition to helping to detect and capture subversives and intruders, network forensics has an essential purpose in expanding the safety paradigm within a network. With the help of network forensics, a felony can be solved. Network forensics is a systematic method for finding and retrieving material having probative value. Hard drives, modems, ports, wireless connections, and digital hubs are just a few examples of the network and computer clusters from which the evidence is obtained. Network forensics is different from intrusion detection in the aspect that obtained data must meet both statutory and technological standards in order to be acceptable in court (Rode et al., 2022). While intrusion monitoring aids in bolstering and enhancing a computer channel's safety, network forensics is more commonly related to the requirement to locate the proof of a breach of security. It is also crucial to remember that companies might utilize network forensics to guarantee the uptime and continuation of their main services. In this circumstance, network forensics aids in locating corporate network flaws that make it simple to deploy the required security upgrades. There are numerous methodologies used and observed in industry to ensure that forensic information is precise and reliable.

1) Strategize and collect: This methodology calls for the creation of a thorough plan for conducting the investigation. This is crucial because it enables the forensic to determine the urgency of allocating the necessary staff and assets, which is essential since varying pieces of evidence from diverse sources have differing degrees of uncertainty. The next step is to gather data using the defined urgency from the sources that have been selected. Thus, there are vital components that are considered afterwards which are documentations and transit. In essence, it entails that every action, particularly a list of every resource, framework, and data, should be meticulously recorded. In order to uphold the chain of custody, it also indicates that documentation should be kept in a secure location. By using this methodology, a company can assess the worth and expense of getting evidence, calculate the purpose of the research and the time period, create reliable duplicates of the evidence, and acquire evidence as quickly as practicable.

2) Extended model: This methodology outlines a comprehensive framework with distinct phases for the study of network forensics. It contains numerous moves that must be completed through most of network forensic investigations, including recognition, permission, preparation, alerts, searching for and identifying substantiation. Although the extended model adheres to the entire set of acts agreed upon by researchers, it contains parts that are unrelated to digital forensics, which implies that many parts of the model can be done by non-digital forensic practitioners. For example, in a crime prosecution, police officers can complete the authorization process to provide forensic investigators legitimate evidence to work with. So, the extended model is a good model, but it is unknown how closely it might be tied to forensics and its process.

3) Honeypots for attack patterns: This methodology calls for the creation of honeypots to aid in network forensics and to keep an organization one door away from danger. A unit that is broadcasted on a network with the intention of luring unsanctioned activity is known as a honeypot. A circuit of these pots with a gateway attached makes it complete. After an intrusion has infiltrated the network, it gathers information based on the illegal access which allows for its investigation, which can reveal the newest techniques and malwares used by an attacker and aid in reconstructing the flow to the attacking machine. It's crucial to keep in mind that the sort of honeypot used depends greatly on the goals of the organization. In the interest of recovering the attacker's objectives and identities, investigators need a lot of data; hence, an elevated interaction honeypot may be used. Also, the investigators should possess the tools required to fund and support this system. Companies might not require as much information gathering, thus a limited honeypot that is simple to put up might be recommended. Overall, honeypots offer a wide range of advantages

Figure 3. Honeypot and its placement

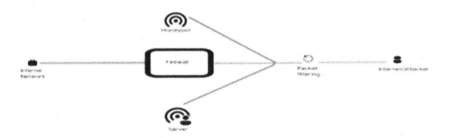

to the company that uses them. With the aid of honeypots, a company can investigate potential hazards. As a result, inquiries into the identity of the perpetrator and the instruments they employ in their attempts can be resolved (Rode et al., 2022). This will help the company's IT security department better assess its probable risks, hence enhancing its readiness and defense systems. It serves as a simple target, deterring the intrusive party from targeting the actual organization's system. As long as a possible intruder just targets the more alluring honeypot and leaves the organization's system unharmed, this provides a certain amount of safety as shown in Figure 3.

4) OSCAR: This approach calls into consideration the investigator's access to the internal assets of the impacted entity, the objectives and timescale put forth by the company for a probe, as well as other elements that are highly reliant on the companies engaged. Finding and examining information from network sources must be done in stages, just like other diagnostic tasks, for the outcomes to be credible. The operations of forensic analysts should be carried out inside an orderly structure. According to Oscar methodology the recommended way to get digital information for forensic purposes is to obtain information, strategize, collect evidence, analyse, and report as shown in Figure 4. At the very first step of this methodology analysts must gather specifics regarding the occurrence and the network architecture in which it transpired, how it took place, why it happened, who was responsible, which databases and systems were impacted, and what steps were previously done in response to the incident. After the first step is done the strategize and collect methodology

Figure 4. OSCAR methodology

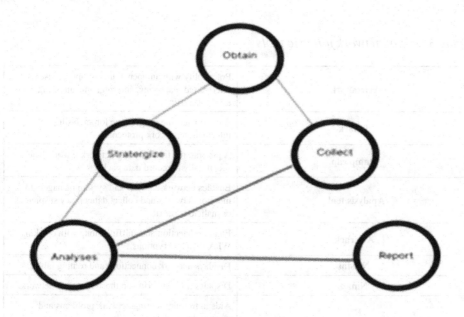

mentioned earlier in pointer one of network forensics is adapted(Lee & Liles,). This methodology's final stages entail connecting information like date and time stamps and Internet Protocol (IP) addresses. Following the correlation of the proof, the investigator can create an outline of what happened, detailing who did what how and on which machines. Initial discoveries necessitate the gathering of further pieces of data, which in turn spurs additional analysis. But indeed, all of the aforementioned efforts will be useless if you are unable to share the outcomes with people. Presentations need to be adapted for those watching, which could be less technically minded. It's crucial to be ready to effectively communicate to an extensive group about what transpired amid an incident involving security and how it can be proved that the events occurred as recounted.

NETWORK FORENSICS TOOL

Table 3. List of network forensic tools

Wireshark	Practicality with an open-source graphical user interface for capturing, filtering, and analysing network traffic.
Tshark	A command-line programme for analysing information network protocols.
Dump cap	A programme for analysing network traffic made specifically to record data packets.
Analysis tools	Enables network analysts and system managers to monitor networks and collect data on any suspicious or malicious traffic.
Aircrack	Enables detection and sniffing along with cracking WPA/WEP encryptions.
Netstat	Displays network connections and routing tables.
Nmap	Displays host and ports on those host on a network.
Dig	Aids in troubleshooting network problems and queries dns.
Nessus	Scans for vulnerabilities.
Burp Suite	A vapt java-based tool.
Angry Ip Scanner	Scans for ports and Ip address.
Ettercap	Enables sniffing connections and filtering.
Network Stumbler	Detects wireless lans.

DEVELOPMENTS IN NETWORK SECURITY

The network security league has dramatically evolved during the last few years. Corporate firms require a thorough defence and security infrastructure to secure rising network-based assets instead of relying solely on firewalls and other border defences. Network security now combines technology with the appropriate rules, processes, and objectives because technology by itself is inadequate. Yet without a doubt, having stated that, we simply cannot overlook the new sectors that are becoming more important as time goes on. As more businesses shift their data and apps to the cloud as shown in Figure 5, data and application security is becoming a critical responsibility. Cloud security mechanisms should continue to evolve to protect against cloud-specific assaults such data theft, malicious insiders, and configuration issues. Not only that, IoT devices that were already susceptible to cyberattacks over

Figure 5. Implementation of cloud in organizations

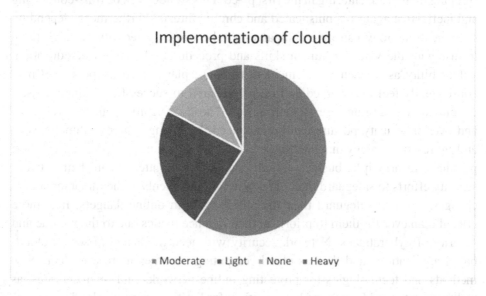

past few years, their number is anticipated to rise sharply (Bienkowski, 2022). As a result, protecting them is turning into a crucial problem that will call for additional study and the creation of security protocols. To figure out and break this pattern, there is a clear beginning point; it is critical to teach workers about cybersecurity best practices to help mitigate the risk of security breaches and that is happening. But what's more? Companies are steadily coming to understand how crucial it is to follow data protection laws to preserve confidential data and uphold user confidence. They are aware that breaking these rules could lead to severe monetary penalties, a loss of trust, and harm to the image of the company. This awareness is a good sign for the future. Many may argue that this knowledge provides little reward because dealing with data breaches and cyber-attacks manually is problematic. This is where AI and ML can assist in identifying and preventing security threats by continually acquiring information from previous incidents. One well-known tool that has gained worldwide attention, chatGPT, can help explain its potency. Whether it's a general code or steps for developing a website, experience the power of ai and ml.

SUMMARY

A proactive, technology-focused network security policy might stress and frustrate security personnel. This is due to the fact that such tactics rely mainly on recognizing and reacting to threats that have already infiltrated the system, rather than actively stopping them from entering in the first place. This strategy can be time-consuming and ineffective against sophisticated and chronic threats. Furthermore, depending only on technology can give companies a false sense of security, causing them to disregard the value of human skills and procedures like employee education, vulnerability assessments, and incident response plans. Security personnel may consequently feel overworked and under-prepared to successfully fight against a continuously changing threat environment. These difficulties can be overcome, and overall security posture can be enhanced, by taking a more comprehensive and aggressive strategy that integrates technology with human experience and best practices. In principle, businesses will continue to be attentive in their network security efforts to safeguard their data, systems, and clients as the threat landscape changes. To help safeguard their digital assets from online dangers. It is more crucial than ever for them to prioritize their defence tactics due to the volume and severity of cyberattacks. Network security will need to change to secure cloud-based applications and data as cloud computing becomes more prevalent. New methods and technologies for protecting online networks and microservices, as well as updated guidelines and best practices for limiting access to cloud resources, will probably be required. Corporations in specialized sectors, such as healthcare and finance, will be under greater strain to adhere to data protection and security rules. Network security experts will need to stay up to date on the newest regulatory requirements and industry standards as these policies evolve. The usage of ai and ml, which integrate with security technologies like gateways, intrusion detection systems, and antivirus software, streamlines the risk identification response, and is one of the ongoing trends in network security. They also give security personnel real-time insights about potential security concerns, which helps them make quicker and more accurate judgements.

FUTURE WORK

Threats from harmful schemes as well as other sources are getting stronger and doing more harm. Companies must adopt a more aggressive security standpoint since they have limited opportunity for reaction. Responsive security won't be effective anymore. Hence, in order to be more capable of keeping their enterprises as safe as feasible, companies need to fully grasp the patterns, dangers, and challenges of the coming. The future work involves developing a toolkit that simplifies and automates the process of implementing the most known encryption methods and monitors the network traffic to generate reports at regular intervals. This future work will help in eliminating any sensitive data that is exposed globally and will motivate the culture of security intensively.

CONCLUSION

As the use of the internet expands, safeguarding networks is a crucial idea that is receiving greater emphasis. To preserve network security, several protocols and dangers to security should be examined. The sophistication of an attack mirrors the sophistication of the means. The most significant obstacle currently is the implementation of measures to counter these risks. In the end, it enables the capture of a criminal on the internet or any individual accused of perpetrating a digital crime. Thus, it is crucial for a network forensic analyst to think about implementing and employing a forensic network study methodology that is effective and resilient and eventually serves streamline the inquiry workflow.

REFERENCES

Bienkowski, T. (2022). 4 trends in network security to keep an eye on. *CSO INDIA*. https://www.csoonline.com/article/3649754/4-trends-in-network-security-to-keep-an-eye-on.html

Cisco. (2021). *Encrypted Traffic Analytics* (White Paper). Cisco Enterprise Security. https://www.cisco.com/c/en/us/solutions/collateral/enterprise-networks/enterprise-network-security/nb-09-encrytd-traf-anlytcs-wp-cte-en.html.

Constella. (2021). The Data Breach Era. Constella. https://info.constellaintelligence.com/white-paper-the-data-breach-era.

Grigas, L. (2023). *What is end-to-end encryption and how does it work?* Nord Pass. https://nordpass.com/blog/what-is-end-to-end-encryption/

Kozlosky, B. (2021). *5 Challenging Network Security Problems & the Fixes*. TechSling Weblog. https://www.techsling.com/5-challenging-network-security-problems-the-fixes/.

Laliberte, B. (2023). *End-to-End network visibility and management?* Enterprise Strategy Group. https://www.esg-global.com/research/topic/networking

Lee, B., & Liles, S. (2013). *Applying the OSCAR Forensic Frameworkto Investigations of Cloud Processing*. CERIAS. https://www.cerias.purdue.edu/assets/symposium/2013-posters/23E-5F7.pdf,2013

MITRE. (2019). *Mitre att & ck*. MITRE. https://attack.mitre.org/

Oltsik, J. (2021). *Network Security Without Borders*. NetScot. https://www.netscout.com/ whitepaper/esg-wp-network-security-without-borders, August .

Qureshi, S., Tunio, S., Akhtar, F., Wajahat, A., Nazir, A., & Ullah, F. (2021). NetworkForensics: A Comprehensive Review of Tools and Techniques. *International Journal of Advanced Computer Science and Applications*, *12*(5), 879–887. doi:10.14569/IJACSA.2021.01205103

Rode, K., Patil, S., Patil, A., & Dahotre, R. (2022). Network Security and Cyber Security. *International Journal of Research in Applied Science & Engineering Technology*. https://www.ijraset.com/research-paper/network-security-and-cyber-security, June

Chapter 11
Penetration Testing Building Blocks

Abhijeet Kumar
Independent Researcher, India

ABSTRACT

Penetration testing is an art, and the path of its mastery starts from having a good grasp on fundamental knowledge. Throughout this chapter, readers will be made familiar with the building blocks in an easy-to-interpret manner. During the course of this chapter, readers will learn about topics such as penetration testing types, phases, dos & don'ts, and types of OSINT (open source intelligence) methods including image OSINT, email OSINT, Google dorking and social media intelligence (SOCMINT). Please note that this chapter will not be very comprehensive. Readers are recommended to do their own research on such topics in order to gain more in-depth knowledge. Links to additional reading materials will be added in reference section at the end of the chapter. Readers are advised to go through them at least once to learn more comprehensive information about some chapter topics.

1. INTRODUCTION

Information Security a.k.a. InfoSec is the practice of protecting sensitive information from threat actors. Some things included in this practice are tools, techniques, standards, regulations, and policies. Please note that InfoSec is a very broad domain which spans outside the Internet realm and covers physical security, backups, people, processes, and technology among others.

Cybersecurity is a subset of Information Security which deals with information in Cyberspace. It encompasses Data, Information, Networks, Devices, Channels,

DOI: 10.4018/978-1-6684-8218-6.ch011

Personnel and others. CIA (Confidentiality, Integrity, Availability) triad is widely used as a base when implementing security within an organization.

Here are some terms which every Penetration tester should be familiar with:

- **Asset:** Asset is a something that provides value and should be kept safe.
- **Threat:** Threat is the danger to an asset.
- **Vulnerability:** Vulnerability is a weakness that could be leveraged to make that threat successful.
- **Risk:** Risk is the potential chance of damage to an asset and asset owners/ users.
- **Exploit:** Exploit is the method to leverage a vulnerability. For e.g: an exploit could be considered as a syringe.
- **Payload:** A payload is a code which could runs after exploitation and performs the actual intention behind exploit. For e.g: A payload could be considered as medicine inside the syringe.
- **Authentication:** To verify whether someone is who they claim to be.
- **Authorization:** To verify whether someone has access to perform certain action.

1.1 CIA vs. DAD

In Information Security domain, more specifically in the realm of Cybersecurity there are two sides i.e, Blue Team and Red Team. These two sides are two sides of the same coin and together the help strengthen the security of an organization.

Blue Team act as defenders with following main objectives:

- **Confidentiality:** To ensure that a communication is happening only between authorized persons/devices.
- **Integrity:** To ensure that a data is intact and hasn't been modified by any parties.
- **Availability:** To ensure that data/resources can be available to relevant parties.

Whereas, **Red Team** act as attackers with following main objectives:

- **Disclosure:** To ensure that confidential communication can be broken.
- **Alter:** To ensure integrity of data has been damaged.
- **Denial:** To ensure data/resource cannot be available to relevant parties.

1.2 Operational Security

As a penetration tester handles lots of confidential information during assessments, it is of utmost importance to keep it safe & secure. In recent years it has been observed that threats actors target pentesters to gain access to those secrets. Some of the techniques to prevent this are as follows:

- Use Firewall with strict rules. Deny all incoming connections and allow only necessary services. Additionally whitelisting & blacklisting along with other methodologies should be considered as per use case of a penetration tester.
- Harden the machines & uninstall/remove/block all unnecessary services. Keep system(s) updated and isolated from open networks.
- Setup logging & monitoring on pentest machines.
- Encrypt hard drives & all removable media.
- Since online services could be attacked, storing confidential notes and files should be stored locally. In case data needs to be kept online, strict protection measures like MFA, Allowed IP, etc should be implemented. Additionally that data should be encrypted and key should be stored safely.
- Pentester should be carefully implementing the usage of IP scrambler services in order to confuse any adversary. Please note that a skilled adversary can still bypass this however defense-in-depth will slow them down if not stop.
- VPN, Proxies & TOR are some of most widely used IP scramblers (Kumar, 2022a).

2. PENETRATION TESTING 101

Penetration testing a.k.a. pentesting is the practice of cyber attack simulation on a web application, computer system, network, in order to identify vulnerabilities that could be exploited by a threat actor. The goal of penetration testing is to identify and assess the security of a system or network by attempting to exploit vulnerabilities, and to provide a detailed report of the vulnerabilities found, along with recommendations for remediation. Please note that while a penetration tester should not be in the one to implement these remediation. It should be done by either client's IT/Security teams or a separate entity.

2.1 Types of Penetration Testing

There are three main types of penetration testing as follows:

- **Black box testing:** In this type of testing, penetration tester is unaware of the architecture of the infrastructure to be tested. This is to simulate the test from the perspective of an external threat actor residing outside the client infrastructure. While this type of testing has its perks it has some downsides which should be taken in account while performing black box testing. One downside is unless the tester penetrates outside perimeter, they can't test out internal systems. Another downside is since a penetration test has time constraints, a tester could miss some vulnerabilities that could be leveraged by a threat actor who spends a great deal of time to recon the target.

- **White box testing:** In this type of testing, penetration tester is made aware of the client infrastructure to a great deal. They are provided with full access to network diagrams, source code, access credentials, internal documents, and infrastructure. This type of testing often yields the most results and can led to discovery of most vulnerabilities. However since this type of testing is very comprehensive (covering both external and internal infrastructure) it tends to be expensive as well as resource intensive. It also depends upon the skill and experience of a tester to notice some hidden vulnerability which was not discovered earlier, such a vulnerability is known as Zero day. A Zero-day is the vulnerability which is previously unknown to the vendor.

- **Gray box testing:** It is the combination of black box and white box testing and penetration tester is given access to a limited number of resources. This type of testing is often more efficient than black box due to internal information. This leads to less time spent of recon phase and more of testing phases.

2.2 Penetration Testing Stages

1. **Pre-engagement:** This stage involves activities such as interaction with client, discussion on rules of engagement, establishment of scope, allowed tools & techniques, engagement period, point of contact, and legal documentation. It is very important to have a legal advisor go through all documents for the safety of all parties involved. Furthermore all required approvals should be obtained at this stage including from client decision makers and law enforcement authorities.

2. **Information Gathering & Recon:** Information gathering is the process of gathering information about the target from a multitude of sources. While this process could be time consuming, it is always worth it. This could be performed by both manually and automatic tools. Gather as much data about the target as possible in this stage and use it in later stages.

Information gathering can be divided into two main categories as follows:

- Active Recon: It involves interacting with target(s) directly.
- Passive Recon: It involves data collection via passive means such as OSINT (Open Source Intelligence), HUMINT (Human Intelligence) or covert channels.

3. **Enumeration:** This stage involves direct interaction (Active Recon) with target(s) to perform activities like port scan, service enumeration, accounts enumeration, vulnerability assessment and eventually leads to planning attack paths. Targets could range from an internet facing web application to complex systems and everything in between. The findings from this stage should be compared with previous ones to determine vulnerabilities. These vulnerabilities should then be studied in order to plan an efficient attack.

4. **Exploitation:** Exploitation is the process of leveraging identified vulnerabilities to gain access to target infrastructure (systems, networks, or applications). Some methods used in this stage include exploitation of publicly disclosed vulnerabilities, use of custom exploits, social engineering, or physical access to system.

5. **Persistence & Pivoting:** After successful exploitation, pentesters need to further perform different activities such as elevating privileges, maintain access, pivot to different systems or access critical information. One portion of these activities are termed as persistence which refers to techniques for maintaining access to exploited targets. Another portion of these activities is known as pivoting which deals with techniques for moving within the target infrastructure. Another portion of these activities is classified as exfiltration which refers to extracting sensitive data from target infrastructure to penetration tester controlled systems.

6. **Cleanup & Reporting:** At the last stage, a penetration tester need to perform two final tasks i.e, cleanup (removal of all implants/tools/scripts and reverting all systems to original state), and reporting (providing a comprehensive report to the client). These two phases should be done very carefully and client should be properly briefed after report submissions. Readers please note that the goal of a penetration test is to provide client organization with a thorough report and not just owning the machines. The mindset of a penetration tester should be to help secure the organization not just hack n tell. Special care should be taken while preparing this report and every vulnerability should be mapped with Common Vulnerability Scoring System (CVSS) to find severity score. Additionally not every vulnerability is exploitable and while it could do damage in one environment it could be benign in another. These points should be mentioned both in the report and during the briefing.

For a more comprehensive information on these phases please refer to Penetration Testing Execution Standard (Nickerson et al., 2014) and NIST SP 800-115 (Scarfone et al., 2008).

2.3 Categories of Penetration Testing

Penetration testing can be divided into multiple categories based on the target. Some of most common types are as follows:

- External network penetration testing: This type of assessment deals with testing internet facing perimeter of an organization. This assessment simulates the activities of an external attacker and mainly involves testing network infrastructure like routers, firewalls, servers.
- Internal network penetration testing: This type of assessment takes place within the internal network of target organization. This scenario assumes that an attacker has successfully breached the outside perimeter and infiltrated the network. This assessment involves the testing of Active Directory (AD), routers, firewalls, servers, Virtual LANs(VLANs). Subnets, workstations, web applications, and other internal devices.
- Web application penetration testing: This type of assessment involves the source code analysis, logic flow checking, and identifying vulnerabilities within various components of a web application.
- IOT (Internet Of Things) penetration testing: This type of testing deals with finding vulnerabilities in internet connected devices or networks like smart watches, smart homes, security cameras, or baby monitors.
- ICS (Industrial Control System) penetration testing: This type of assessment takes place within an industrial organization/network and deals with Operational Technology (OT) and Critical Infrastructure (water, electricity, oil, etc.).
- Mobile application penetration testing: This type of assessment deals with finding vulnerabilities in the mobile devices like Android phones, IOS devices, etc. This type of assessment focuses significantly focuses on source code analysis and logic flows.
- Application Programming Interface (API) penetration testing: This type of assessment deals with finding vulnerabilities in Application Programming Interface used to connect different pieces of software.
- Automobile penetration testing: This type of assessment focuses on finding vulnerabilities in connected cars, or AI cars. With the evolution of Cloud Computing and Artificial Intelligence, car manufacturers have been

implementing computing devices in cars to provide the owners with easy of access.

In next chapter, readers will be introduced to Network penetration testing and Web application penetration testing. Other penetration testing types mentioned above are geared towards advanced skillset and thus are outside the scope of these two chapters.

2.4 Penetration Testing Do's and Don'ts

While performing any activity it's very important to know what is allowed and what is restricted. Same principle applies to penetration testing as well. Below is collection of some of the most crucial guidelines that will ensure the integrity and security of all parties involved in a penetration test.

Some important do's are as follows:

- Make sure you are legally covered. In case you are doing physical penetration test make sure to notify local authorities including Law Enforcement in advance.
- Do ask about their previous penetration test and identify critical systems (systems that are crucial for operations, systems that crashes often, old systems, etc).
- Make sure you have proper authorization to test a particular IP/Machine/ Domain.
- Do check that the target list you got is correct. There has been instances where a penetration tester hacked the wrong organization due to incorrect target list.
- Define the scope clearly in ROE (Rules of Engagement) document.
- Make sure to have records with screenshots of attacks/vulnerabilities/exploits with proper logs.
- Share the final report only with the authorized personnel and redact any sensitive information (passwords, secret keys, API keys, etc.).

Some important don'ts are as follows:

- Don't run untested exploits against system. Before running any critical attacks like Kernel attacks/DoS/DDoS/ make sure these attacks are important for penetration test and you know the damage they could cause. Furthermore notify your point of contact before running such attacks. As an additional prevention measure run these exploits on a simulated lab environment mimicking that of target system.

- Don't test any systems/networks/IP/attacks not defined in ROE.
- Don't share confidential information with unauthorized personnel.

3. OSINT (OPEN SOURCE INTELLIGENCE)

OSINT is the process of obtaining data available in public domain and analyzing it to extract useful intel. This information can be from sources such as websites, newspapers, articles, or social media platforms.

Please note that the operating word in OSINT is intelligence. This is due to the fact that data can be collected however without applying intelligence that data would not yield any meaningful intel. An investigator should be very careful while performing OSINT and should use multiple sources to verify the extracted intel.

In this chapter we will focus on some of the most common types and learn about how to gather information in that particular segment. A good OSINT mindmap can be found at YOGA (Your OSINT Graphical Analyzer) (Hoffman, 2018). Additionally there are good web pages like OSINT Framework (Nordine, 2022) or The Ultimate OSINT Collection (Glynn, 2023) containing links to hundreds of resources.

3.1 Image OSINT

Intro: This field deals with collecting, analyzing, and disseminating information from image sources that are publicly available and legally obtainable. Some sources are social media images, the internet, public archives, and other sources.

Some things that could be determined by image OSINT are as follows:

- Geo-location
- Image authenticity
- Identification of the photographer or the image source
- Date and time of image capture
- Identification of equipment
- Identification of objects, people or places visible in an image

Two main methodologies to obtain information from image are as follows:

3.1.1 Metadata Analysis

Intro: Metadata refers to data about data. In our case, every image contains some data about its origin, characteristics, and other details. These details when properly analyzed could provide useful insights in an investigation.

One of the widely recognized standard is EXIF (Exchangeable Image File Format) which deals with formats of image, audio and other media files.

This is worth pointing out that some of the important EXIF fields in an image are Device information fields and geo-location fields (Latitude, Longitude, Altitude). However most social media websites strip EXIF data before uploading, its always worth a shot. Additionally EXIF data can also be manipulated or removed with ease. Make sure to co-relate obtained EXIF information with other sources to ensure its accuracy.

Steps:

I.   **Obtain the image:** The first step is to obtain the image that is to be analyzed. This can be done by downloading the image from the internet, or by taking a screenshot of the image. Other methods could be to obtain it from a co-worker, public archives, a gallery among others.

II.   **Extract the metadata:** After acquiring the image, we need to extract the metadata. This can be done using This could be done locally via tools like exiftool or with websites like onlineexifviewer or exifdata.

III. **Analyze the metadata:** After extracting the metadata, we can analyze it to gain insights into the image. This can include information such as the date and time the image was taken, the camera make and model, the lens used, the exposure settings, and GPS coordinates if the device had GPS enabled.

IV.   **Interpret the results:** At last, we can interpret the results of the metadata analysis to gain insights into the image. This result could be co-related with other sources, data could yield some meaningful connections. Additionally it could also point out in direction of next phase of investigation.

Some use Cases:

- Metadata can contains information such as File name, File size, Access/modification timings, GPS coordinates among other things. Combined with other techniques this could be leveraged to track the movements of people/groups by analyzing multiple images and establishing links among them.
- It could establish the authenticity of a given image. For e.g, nowadays AI based systems are used to generate deep-fakes which could then be used in a multitude of ways. By analyzing the image we could establish whether its real or fake.
- Now let's imagine a scenario where we are penetration testing XYZ Company and manage to acquire an image clicked by one of their employee. After analysis we discovered some juicy information, it could then be used to further strengthen Social Engineering campaign (if in scope).

3.1.2 Reverse Image Search

Intro: Reverse image search is an OSINT technique which refers to collection of activities performed on an image such as finding the source, tracking its spread across the internet, altered versions or plagiarism detection.

In reverse image search image first go through feature extraction where its compressed and then converted into a unique hash. These extracted features are then used to compare the image across multiple databases. Once a match is found the result is returned to the user.

Some techniques involved in feature extraction are as follows:

- CNN (Convolutional Neural Networks)
- ORB (Oriented FAST and Rotated BRIEF)
- SIFT (Scale-Invariant Feature Transform)
- SURF (Speeded Up Robust Feature)

After the feature is extracted then some other techniques could be performed:

- Word and watermark detection: If there is a word or watermark in given image, it is then extracted and matched with other images in database. It helps in reducing the search time and could provide close if not accurate results.
- Indexing: Algorithms such as k-d tree, locality-sensitive hashing (LSH), product quantization (PQ) are used to index images in a database so that they can be searched efficiently.

Here's a reverse image search result for a famous OSINT challenge named Find Satoshi. This search returned six webpages where this image is hosted:
Steps:

I. **Obtain the image:** The first step is to obtain the image that is to be analyzed. This can be done by downloading the image from the internet, or by taking a screenshot of the image. Other methods could be to obtain it from a co-worker, public archives, a gallery among others.

II. **Submit the image:** After acquiring the image, we need submit it to an image search engine in order to find matches. This can be done by either uploading the image or entering the URL of that image. The search engine then applies a series of techniques on that image and returns the matches. Some of the most popular reverse image search engines are Google Image Search, Yandex, Bing Image Search, TinEye, and PimEyes.

III. **Analyze the results:** Once search engine reverts with the result, we need to comb through to find ones that we are after. Please note that these results will not be accurate all times and could lead to false positives. In some cases, the returned images may not be the ones we want. This step should be performed very carefully to get the most out of results. Additionally the use of multiple search engines is recommended as one could yield incomplete results. Once the desired results are found, we can then focus on extracting meaningful information.

IV. **Interpret the results:** After finding the desired images we will perform activities to find more about them. In this stage we will co-relate multiple sources, use different tools and techniques, check the spread, or find out about the sites where these images are hosted.

Some use cases:

• Reverse image search could be used to detect image plagiarism. This one is particularly beneficial to content creators and photographers.
• It can find people, landmarks, or other objects to a great extent. This is particularly useful for Law Enforcement Agencies for tracking a suspect.
• In other instances, it could lead to websites hosting the concerned image and thus be used to track the spread.

3.2. Email OSINT

Email OSINT is the process of email analysis, origin tracing and attribution from public sources. A successful email investigation depends on sever factors including age of email address, spread across internet and social media sites and the provider.
    One methodology to performing email OSINT is as follows:

3.2.1 Email Header Analysis

Intro: Every email sent contains some information in the header to aid email servers in routing and email delivery. This information format is called Internet Message Format (IMF) which was introduced in RFC 5322 (Resnick, 2008). The IMF format includes a set of standard fields, such as the "From", "To", "Subject", "Date", and "Message-ID" fields, which are used to describe the email message. Additionally, it includes some fields to convey additional information, such as the "CC" (carbon copy) and "BCC" (blind carbon copy) fields.
    Here's the example of an email header:
    Steps:

Figure 1. Example of a simple email header

Delivered-To: recipient@example.com
Received: from mail.example.com (mail.example.com [192.0.2.1]) by example.com (8.14.5/8.14.5) with ESMTP id x0EKb1I3007835 for <recipient@example.com>; Sat, 31 Dec 2022 12:34:56 -0530
Received: from user by mail.example.com with local (Exim 4.94) (envelope-from <sender@example.com>) id 1l2Kb1-0007HI-9x; Sat, 31 Dec 2022 12:34:56 -0530
To: recipient@example.com
From: sender@example.com
Subject: Test Email
Date: Sat, 31 Dec 2022 12:34:56 -0530
Message-ID: <abcdefghijklmnopqrstuvwxyz@mail.example.com>
Content-Type: text/plain; charset=us-ascii
Content-Transfer-Encoding: 7bit

I. **Obtain the header:** As the first step goal is to acquire the email header from the message. While every provider has different options to view this header, below are steps for some widely used providers.
   ◦ **Gmail:** Open the email, Click "More", select "Show original" and click "Download original"
   ◦ **Yahoo! Mail:** Select the email, click "More" and then "View Message Source"
   ◦ **Hotmail:** Right click on the email, click "View Message Source"
   ◦ **Outlook:** Open the email, Click "File" and then "Properties"
   ◦ **Apple Mail:** Open the email, Click "View" then "Message" and finally "All Headers"
   ◦ **AOL:** Open the email, in "Action Menu" select "View Message Source"
II. **Identify the key fields:** After obtaining the email header, the next step is to find out which fields contains information related to our investigation. This is done because a typical email header contains about 10-12 fields and some of them might not be relevant to the nature if investigation. Some of the fields worth looking are as follows:
   ◦ **From:** It contains sender's email address.
   ◦ **To:** It contains recipient's email address.
   ◦ **Received:** It contains routing information, details of intermediate mail servers and the timestamps of when the mail reached these servers.
   ◦ **Reply-To:** This field contains the email address to which replies to the email should be sent
   ◦ **Subject:** It contains the subject of the email.
   ◦ **Date:** It contains the timestamp of when the message was sent.
   ◦ **Message-ID:** It contains a unique identifier of that email.

- ◦ **Return-Path:** It contains the email address to which delivery status notifications will be sent.
- ◦ **Received-SPF:** It contains the result of Sender Policy Framework (SPF) authentication.
- ◦ **X-Originating-IP:** It contains the IP address of originating computer.
- ◦ **X-Mailer:** It contains the name and version of the email client.
- ◦ **X-Forwarded-For:** It contains the IP addresses of the intermediate email servers.

III. **Analyze the results:** After identifying the important fields, the next step is to perform some analysis and obtain useful information. Some methods to obtain this information includes

- ◦ **IP Geolocation:** IP addresses are primarily assigned by the Internet Assigned Numbers Authority (IANA). IANA assigns these address to one of five regional authorities Asia-Pacific Network Information Centre (**APNIC**), American Registery for Internet Numbers (**ARIN**), Réseaux IP Européens Network Coordination Centre (**RIPE NCC**), Latin America and Caribbean Network Information Centre (**LACNIC**), and African Network Information Center (**AFRINIC**). These regional authorities then assigns these address to ISPs and organizations which in tuen made them available to the end users. A detailed record of every IP is kept which can be queried from databases like IPinfo or IP2Location. Every IP address is tied to a particular location finding it could give insights on the user and email servers.

Here's an example of an IP lookup of Google's DNS server from IPinfo website:

Figure 2. IP Lookup example

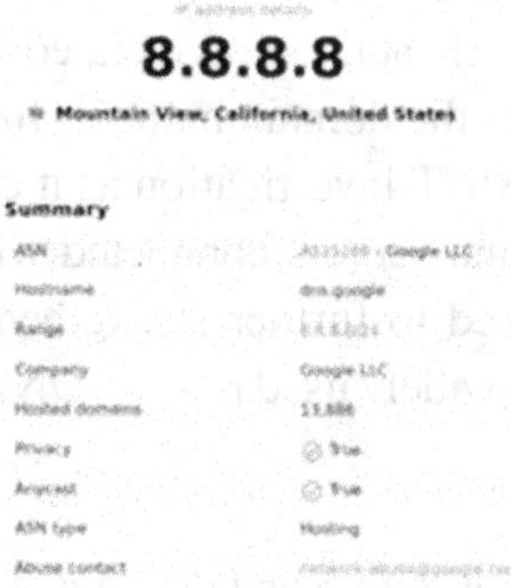

IP address details

8.8.8.8

📍 Mountain View, California, United States

Summary

ASN	AS15169 - Google LLC
Hostname	dns.google
Range	8.8.8.0/24
Company	Google LLC
Hosted domains	13,886
Privacy	✓ True
Anycast	✓ True
ASN type	Hosting
Abuse contact	network-abuse@google.com

Figure 3. Geo-location part of IP lookup

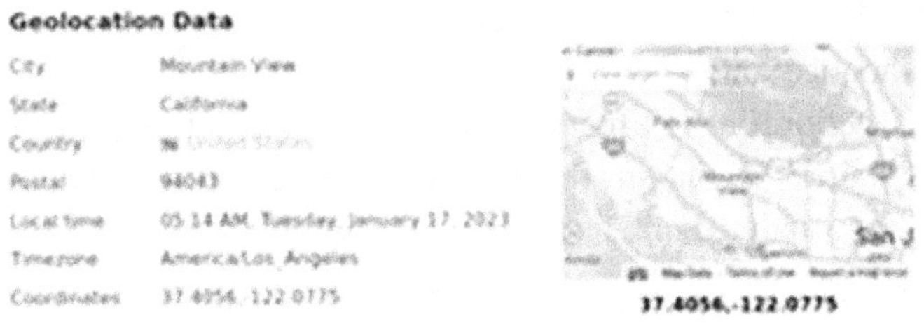

WHOIS Lookup: WHOIS is one of the largest public databases which contains records of every IP address, domain names and DNS records. A WHOIS search of an IP returns a lot of details about that address and the organization which controls it. There are tools like whois (linux) and nslookup (Windows) which can perform this lookup. Additionally there are many websites like Whois.com, Who.is, or Whois.net which can be queried via web browser. However please note than with exception to few specific Top Level Domains (TLDs) almost all offer Whois Privacy. When enabled it replaces the Registrant records with proxy records of domain registrar. In this case one should search for historical records as they can be unproxied and thus reveal what is hidden.

- **Reverse DNS Search:** Domain Name System (DNS) is considered as the Internet phonebook. Just like the phone numbers, every domain name connected to Internet is given an IP address so that other devices can communicate with it. This is due to the fact that web browsers interact via IP addresses which is very difficult to memorize for a significant number of sites. DNS aids in this scenario by providing an effective way to translate these domain names to IP address and vice verse. While a simple DNS lookup results translates the IP address associated with a domain name, reverse DNS lookup translates the domain name(s) from an IP address. This is particularly useful in an OSINT investigation as it could yield information about the different websites and servers connected with an IP address. This information could then be used to further strengthen the investigation and assist in next phases. Some of widely used reverse DNS sites are DNSchecker and MXToolbox.

Readers looking for technical information on DNS may visit this blog by Internet Engineering Task Force (IETF) (Internet Engineering Task Force, 2019).

- **Email Reputation analysis:** Email reputation analysis focuses on finding whether the email has arrived from a legitimate source or not. Additionally it also checks for whether the domain and IP connected have been used in any malicious activities or not. Please note than most IP address are assigned dynamically and thus they change very often. This could lead to false positives (instances where a previously flagged IP is assigned to someone clean) and should be checked carefully. In addition to this the email content should be carefully analysis along with the SPF/DKIM/Anti-Spam header fields. Additionally websites like Emailrep.io, or Trumail.io are a good starting point to check the reputation of a email. Please note that these site could provide false positives so use of multiple sources is recommended.

- **Co-relation:** This one is performed mainly for attribution. Email address and other extracted data (IP address, mail servers, etc) should be searched on various sources like social media platforms, forums, websites, logs and threat intelligence databases such as AlienVault OTX, Pulsedive, CISCO Talos, or AbuseIPDB to see whether are associated to any user or group. Additionally there are databases such as Hunter.io, Haveibeenpwned.com, or dehashed.com can provide details such as if email address was found in any data breach, leaked hashes, common address format among other things. If found, those details should be added in the notes and investigated further.

Some use cases:

- Email header analysis is very useful in spam and phishing investigations. Since it can attribute the person behind an email in most cases, it could be used to track down the threat actor behind a malicious campaign.

- It is very helpful while analyzing potential security risks. This is due to the fact that after analyzing an email, certain patterns could be extracted which co-related with other sources can be converted into some rules. These rules can then be applied to further identify such emails and reject them. This could be done with the help of Artificial Intelligence (AI) like Natural Language Processing (NLP) systems which can identify the content pattern among other things.

3.3. Website OSINT

Intro: Website OSINT focuses on passively collecting information about a website and connected assets such as Domain names, IP addresses, email addresses, or software. Some important information that can be found are DNS records, technology stack, infrastructure, misconfigurations, or potential vulnerabilities. Additionally details of organization or individuals behind the website can be found out.

Here's a sample domain name: www.example.com

Where,

"www" is the subdomain

"example" is the Second Level Domain (SLD)

"com" is Top Level Domain (TLD)

" . " is the separater

Steps:

I.   **WHOIS Lookup:** As mentioned in the Email Header Analysis section, Whois records contains ownership records of a domain and are publicly available. The techniques here will be similar to those in previous section.

II. **IP Lookup and asset inventory:** Every website resolves to at least 1 IP (multiple in case of large websites). Since we are focusing on avoiding direct interaction with the website, Who.is provides DNS information along with WHOIS records. This is a great website to retrieve IP addresses associated with a website.

Now there are two types of IP addresses i.e, IPv4 and IPv6. IPv4 is a 32bit address spread over 4-blocks of 8bits separated by dot (.). IPv6 is a 128bit address spread over 8-blocks of 4 hexadecimal digits separated by colons (:). In DNS records, IPv4 is present in "A" field while IPv6 is present in "AAAA" field.

After extracting IP address of website, at first it should be search on IP services mentioned in Email Header Analysis section. Afterwards there are some platforms like Shodan.io, Zoomeye.org, Netlas.io, or Censys.io which provide information about services running on these IP addresses. This information includes open ports, web server information, software versions, SSL/TLS certificates, or IP geolocation.

Here's a sample result from Who.is showing DNS records:

III. **Subdomain Enumeration:** Subdomain is a prefix which is often use to create multiple sections of a website. Each subdomain is unique to that website and points to a webpage. A website can have as many subdomains as needed and they often contain information useful for a penetration test assessment. However since this section is focused on OSINT, actively browsing these subdomain is out of scope. These subdomains can be discovered by many methods one of which is Crt.sh which contains the SSL/TLS. SSL stands for Secure Socket Layer and it was the predecessor to TLS which stands for Transport Layer Security. SSL is now depreciated and its last version v3 was released in 2015. TLS is the modern standard is currently at v1.3 and uses Public Key Infrastructure

Figure 4. Example of DNS records

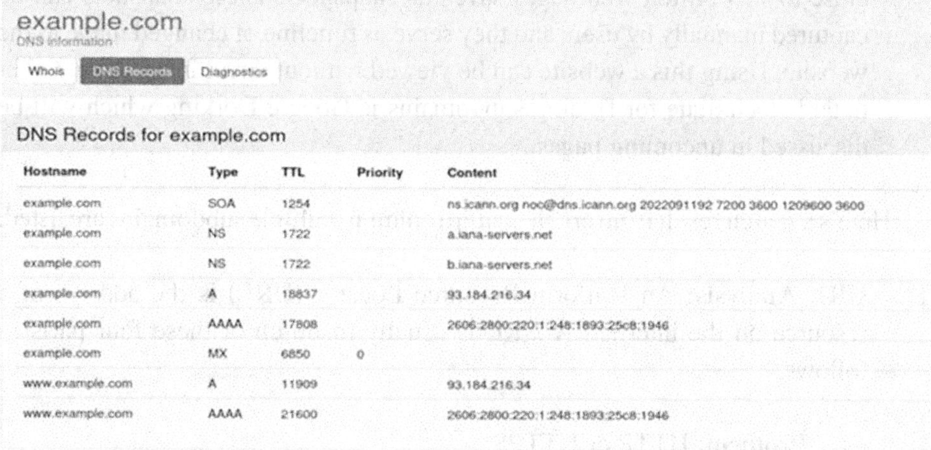

example.com
DNS Information

| Whois | DNS Records | Diagnostics |

DNS Records for example.com

Hostname	Type	TTL	Priority	Content
example.com	SOA	1254		ns.icann.org noc@dns.icann.org 2022091192 7200 3600 1209600 3600
example.com	NS	1722		a.iana-servers.net
example.com	NS	1722		b.iana-servers.net
example.com	A	18837		93.184.216.34
example.com	AAAA	17808		2606:2800:220:1:248:1893:25c8:1946
example.com	MX	6850	0	
www.example.com	A	11909		93.184.216.34
www.example.com	AAAA	21600		2606:2800:220:1:248:1893:25c8:1946

(PKI) to provide encryption between browser client and web server. Now all granted and revoked certificates by a Certificate Authority (CA) to a website can be found on Crt.sh. These records often contains different subdomains these certificates were granted for (with the exception for wildcard certificates). These records are a great place to find existing subdomains discretely. Subdomain enumeration can also be performed by tools such as Sublist3r, Dnsrecon, or Dnsdumpster.com. Additionally there is one website in particular i.e, Wayback

Figure 5. Subdomain enumeration using Crt.sh

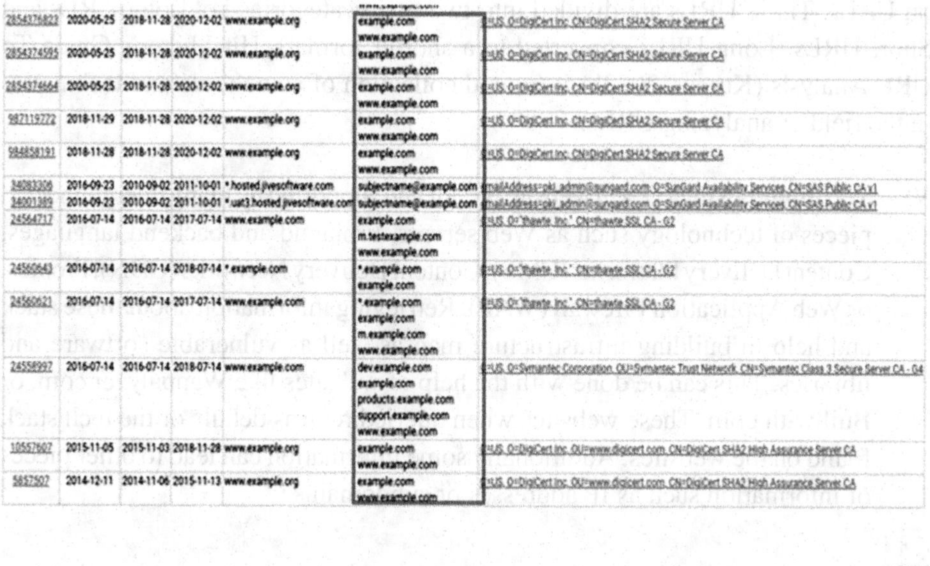

Machine which crawls and stores the snapshots of a website. Currently it has close to 778 billion web pages saved as snapshots. These snapshots can be captured manually by users and they serve as timeline of changed made to the website. Using this a website can be viewed without direct interaction. Other notable technique for finding subdomains is Google Dorking which will be discussed in upcoming pages.

Here's a search result from crt.sh, in third column multiple subdomains are listed.

IV. **URL Analysis:** An Uniform Resource Locator (URL) is the address of a resource on the internet. A URL is usually made up of these four parts as follows:

  ○ **Protocol:** HTTP or HTTPS
  ○ **Domain name:** Website's domain name
  ○ **Path:** Resource's location on web server
  ○ **Query string:** Parameters associated with resource

Here's an example of URL: https://www.example.com/abc.html?q=nil
Where,
Protocol is *"HTTPS"*
Domain name is *"example.com"*
Path is *"abc.html"*
Query string is *"?q=nil"*
From this example, it can be determined that much can be glean from a single URL. Please note that most of the previously discussed techniques can be performed on URLs. These URLs are divided into two main categories i.e, Long URLs and Short URLs (Long URLs converted to a shorter format). Hitchhiker's Guide To URL Analysis (Kumar, 2022b) is a good collection of some tools and techniques in the field of analyzing URLs.

V. **Technology Stack:** Each website is built and supported by many interconnected pieces of technology such as Web servers, Frontend and backend languages, Content Delivery Platforms (CMS), Content Delivery Networks (CDN), Proxies, or Web Application Firewall (WAF). Retrieving information about these stack and help in building infrastructure map as well as vulnerable software and libraries. This can be done with the help of websites like Wappalyzer.com, or Builtwith.com. These websites when queried returns details of the tech stack found on the websites. Additionally some information can lead to other pieces of information such as IP addresses or subdomains.

Some use cases:

- Website OSINT can help find vulnerabilities and attack paths in early stages of penetration testing assessment. This can save time as well as effort required. However please note that the all findings might not be accurate and should be verified in enumeration phase.
- It is widely used in the field of Business Intelligence where information about a company can be easily found out.

3.4. Google Dorking

Intro: Google is a name that almost anyone owning a smartphone/computer knows of. Its Search Engine is used by millions of people on daily basis making it the most used search platform in the world. This is due to the fact that Google has one of the best web crawlers and an enormous collection of crawled web pages among other things. Since Google has so much data, finding relevant information could be an overwhelming task. For this reason there exists a special branch of OSINT knows as Google Dorking which uses the search operators to find information which could be difficult to obtain otherwise.

Below are some of the most common search operators:

- **Exact match operator:** Anything quoted between double quotes " " will be searched exactly as it is. This one is particularly useful while searching some unique strings.

    e.g: *"wubba lubba dub dub"* will search for this exact phase which is a catchphrase of a popular TV Series Rick and Morty.

- **AND Operator:** It has at least two keywords and finds instances where both of them appear in same manner.

    e.g: *quantum AND superposition*

- **OR Operator:** It returns results if one or both keywords are found.

    e.g: *quantum OR superposition*

- **Pipe (|)Operator:** It works similar to the OR Operator.

    e.g: *quantum | superposition*

- **Negation (-) Operator:** It excludes any keyword on the right side of itself.

  e.g: *quantum - superposition* will search for only quantum while excluding superposition from results.

- **Range (..) Operator:** It search for numbers in a particular range.

  e.g: *1900..2022*

- **Wildcard (*) Operator:** It will match any keyword and acts as a placeholder.

  e.g *quantum * entanglement*

- **Parenthesis () Operator:** It isolates the keyword contained within and often used with combination of other operators.

  e.g: *(quantum superposition) entanglement*

- **in Operator:** It search for one keyword in another.

  e.g: **entanglement in quantum**

- **Site Operator:** It search for a keyword in given site.

  e.g: *site:example.com quantum entanglement* will search for keyword quantum entanglement in only example.com

- **Define Operator:** It searches for keyword in Google's dictionary.

  e.g: *define:quantum*

- **filetype Operator:** It returns results to specified filetypes only.

  e.g: **filetype:pdf**

- **cache Operator:** It returns the cached version of requested webpage.

  e.g: **cache:example.com**

- **intitle Operator:** It returns the results where given keyword is found in title.

e.g: **intitle:quantum**

- **allintitle Operator:** It returns the results where all keywords are found in title.

e.g: **allintitle:quantum entanglement**

- **intext Operator:** It returns the results where given keyword is present in webpage.

e.g: **intext:quantum**

- **allintext Operator:** It returns the results where all given keywords are present in webpage.

e.g: **allintext:quantum superposition**

- **inurl Operator:** It returns the results when given keyword is present in URL.

e.g: **inurl:superposotion**

- **allinurl Operator:** It returns the results when all given keywords are present in URL.

e.g: **allinurl:quantum superposition**

- **insubject Opertor:** It returns of keyword present in Google groups.

e.g: **insubject:quantum**

Now that readers are familiar with some common search operators, let's focus on building queries. Queries are the especially crafted collection of keywords combination of one or more search operators to further fine tune the results. One alternative to these queries is Google's Advanced Search feature which offers graphical interface. While that interface is easy to use, it is recommended to learn how to build queries manually.

Here are example of some queries

- "Quantum superposition theorem"
- site:example.com inurl:quantum
- intitle:"Index of" inurl:superposition

- site:example.com filetype:pdf inurl:entanglement
- cache:example.com filetype:log intitle:quantum

Now let's see how we can build these queries for a penetration testing engagement.

- **Queries with results related to FTP servers:**

Figure 6. intitle: "index of" inurl:ftp

Figure 7. intitle: "index of" "/ftp.txt"*

- **Queries with results related to web servers:**

Figure 8. intitle::WAMPSERVER Homepage

Figure 9. Re: intext: "index of /" "server at"

A much comprehensive collection of these queries can be found at Google Hacking Database (Offensive Security, n.d.).

3.5. Social Media OSINT (SOCMINT)

Intro: Social Media has been rapidly embedding itself in human life for past decade. Very few are immune to its charms and most users share stuff without considering the consequences. This had made social media platform one of the biggest sources for gathering personal information. This section will briefly touch upon some methods to extract this information. Readers are recommended to research more tools & techniques in order to stay updated.

Some Social Media platforms are as follows:

I.   Facebook
II.   Twitter
III. Instagram
IV.   Snapchat
V.   Tiktok
VI.   LinkedIn
VII. Quora
VIII. Reddit
IX.   Pinterest
X.   Youtube
XI.   VK

Now let's see some generic steps of gathering this data:

Step 1: These platforms often have inbuilt advanced search features which could act as a great starting point while investigating. Some platforms include UserID (unique identifier) in the source code of profile page which can be leveraged to find accounts and related content easily. Some of the information that can be extracted is friends, address, likes/dislikes, travel details, or devices own.

Step 2: Once these details have been extracted tools like Maltego, websites like Intelx.io, or Inteltechniques.com can further scrape more useful information. Additionally Google dorking can also be utilized to filter the results from a social media platform.

One such example of a query is:

site:facebook.com "quantum entanglement" 2022..2023

Step 3: After obtaining information, it should be properly analyzed. However before doing just that, please note that these methods are not foolproof and many people may share same details (name, city, company, etc). So its the task of the investigator to rule out false positives. After doing so this information should be compared with other sources and utilized as per needed.

4. CONCLUSION

This chapter is designed to give readers an overview of how to make preparations before actively engaging a target. Readers are expected to make themselves familiar with different methods mentioned during the course of this chapter. Precautions were taken to make sure the content provided is up-to-date and accurate. However if readers find any issues with it feel free to contact the publisher. The next chapter will focus usage of various tools & how-to techniques.

REFERENCES

Glynn, G. (2023). *The Ultimate OSINT Collection*. Startme page. https://start.me/p/DPYPMz/the-ultimate-osint-collection

Hoffman, M. (2018). *Your OSINT Graphical Analyzer*. Myosint. https://yoga.myosint.training

Internet Engineering Task Force. (2019). *Learning About the Domain Name System (DNS) from its Terminology*. Internet Engineering Task Force (IETF). https://www.ietf.org/blog/dns-terminology/

Kumar, A. (2022a, September 16). Deep dive into VPN & Proxies: How to stay safe online. *Medium*. https://medium.com/@abhijeet-secops/deep-dive-into-vpn-proxies-how-to-stay-secure-online-79731b9b654

Kumar, A. (2022b). *Hitchhiker's Guide To URL Analysis*. Github. https://github.com/wand3rlust/Hitchhikers-Guide-To-URL-Analysis

Nickerson, C., Kennedy, D., Riley, C. H., Smith, E., Amit, I. I., Rabie, A., Friedli, S., Searle, J., & Knight, B. (2014). *Home*. Penetration Testing Execution Standard. http://www.pentest-standard.org/index.php/Main_Page

Nordine, J. (2022). *OSINT Framework*. OISNT. https://osintframework.com

Offensive Security. (n.d.). *Google Hacking Database*. Exploit DB. https://www.exploit-db.com/google-hacking-database

Resnick, P. (2008). *RFC 5322*. RFC. https://www.rfc-editor.org/rfc/rfc5322

Scarfone, K., Souppaya, M., Cody, A., & Orebaugh, A. (2008). *SP 800-115 Technical Guide to Information Security Testing and Assessment*. National Institute of Standards and Technology (NIST). https://csrc.nist.gov/publications/detail/sp/800-115/final

Chapter 12
Penetration Testing Tools and Techniques

Abhijeet Kumar
Independent Researcher, India

ABSTRACT

Penetration testing is an ever-growing field that deals with a lot of products and services. This chapter will begin introduction of networking, Linux, and Bash. Furthermore, the common tools of the trade (Nmap, Nessus, Metasploit, Burp suite, etc.) for specific type of penetration testing assessments. Readers are recommended to do their own research on such topics in order to gain more in-depth knowledge. Links to additional reading materials will be added in the reference section at the end of the chapter. Readers are advised to go through them at least once to learn more comprehensive information about some chapter topics.

1. NETWORKING 101

Networking is one of the most important subject for anyone interested in mastering the art of Penetration Testing. Every penetration tester should at the very least be familiar with the working of different protocols and how they interact with different software/tools/hardware.

1.1 Open System Interconnect (OSI) Model: A Reference Framework Made up of Seven Layers

Each layer performs a specific function and supports the layer on top of it.
OSI layers with characteristics are as follows:-

DOI: 10.4018/978-1-6684-8218-6.ch012

- **Layer 1 (Physical Layer):**
  § It transmits bits from one communication channel to another
  § No Protocol works in this layer
  § Protocol Data Unit (PDU) is bit
  § Attack method: Sniffing
- **Layer 2 (Data Link Layer):**
  § It implements error checking methids like Cyclic Redundancy Check (CRC),
    and Frame Check Sequence (FCS)
  § Protocols are ARP, PPP, etc
  § Protocol Data Unit (PDU) is Frame
  § Attack method: Spoofing
- **Layer 3 (Network Layer):**
  § It transfers data packets from one network to another
  § Protocols are IP, ICMP, etc
  § Protocol Data Unit (PDU) is Packet
  § Attack method: Men in The Middle (MiTM)
- **Layer 4 (Transport Layer):**
  § It handles end to end communication between devices
  § Protocols are TCP & UDP
  § Protocol Data Unit (PDU) is Segment/Datagram
  § Attack method: Reconnaissance
- **Layer 5 (Session Layer):**
  § It handles sessions (creation, maintenance, termination) between different
    services
  § Protocols are NetBIOS, RPC, etc
  § Protocol Data Unit (PDU) is session
  § Attack method: Hijacking
- **Layer 6 (Presentation Layer):**
  § It handles data operations like compression, encryption, or decryption
  § Protocols are NCP, LPP, etc
  § Protocol Data Unit (PDU) is data unit
  § Attack method: Phishing
- **Layer 7 (Application Layer):**
  § It provides interface between network and the user
  § Protocols are HTTP, FTP, etc
  § Protocol Data Unit (PDU) is message
  § Attack method: Exploit

1.2 Transmission Control Protocol/ Internet Protocol (TCP/IP Model)

It is the networking model used by the systems to communicate with each other. It is made up of 4 main layers (5 if Physical and Data Link are counted separately) and support the current Internet infrastructure.

TCP/IP 5-Layer model is as follows:-

- **Layer 1:** Physical Layer
- **Layer 2:** Data Link Layer
- **Layer 3:** Network Layer
- **Layer 4:** Transport Layer
- **Layer 5:** Application Layer

1.3 Internet Protocol (IP) Address

IP address is an unique identifier of a device connected to any network.

- **IPv4:** IPv4 is a 32bit address spread over 4-blocks of 8bits separated by dot (.).

  Example: *192.168.1.1*

- **IPv6:** IPv6 is a 128bit address spread over 8-blocks of 4 hexadecimal digits separated by colons (:).

  Example: *2001:0db8:85a3:0000:0000:8a2e:0370:7334* or *2001:0db8:85a3::8a2e:0370:7334*

1.4 TCP and UDP

- **Transmission Control Protocol (TCP):** TCP establishes connection between client and server and ensures that data packets are delivered reliably. A 3-way handshake is performed while establishing and terminating connections.
- **User Datagram Protocol (UDP):** UDP is an unreliable connection-less protocol and it sends data packets without checking for errors.

1.5 Ports

Ports are the logical opening in a system through which a connection is established. The main function of a port is to decide where a running service can send/receive data.

Some commonly used default ports are listed below. Please note these ports are TCP unless specified otherwise.

- *21 ® File Transfer Protocol*
- *22 ® Secure Shell*
- *25 ® Simple Mail Transport Protocol*
- *53 (UDP) ® Domain Name System*
- *80 ® Hyper Text Transfer Protocol*
- *123 (UDP) ® Network Time Protocol*
- *139 ® NetBIOS*
- *443 ® Hyper Text Transfer Protocol Secured*
- *445 ® NetBIOS*
- *623 (UDP) ® IPMI*
- *1098 ® Java RMI*
- *3389 ® Remote Desktop Protocol*
- *4786 ® CiscoSmartInstall*

1.6 Common Networking Devices

Some network devices are as follows:-

- **Router:** It routes traffic between different networks.
- **Switch:** It transfers packets within internal network.
- **Firewall:** It allow/deny the data transmission based on configured rulesets.
- **Next-Gen Firewall (NGFW):** It is an upgraded firewall with advanced capabilities like pattern recognition, application awareness, advanced filtering features, etc.
- **IDS:** It monitors network traffic and creates an alert whenever suspicious activity is detected.
- **IPS:** It monitors the network traffic and block suspicious activity once detected.

2. LINUX 101

Linux a.k.a GNU/Linux is an open source highly configurable Operating System (OS) derived from UNIX. To be precise, Linux is a kernel and different OS (called distros) are built on top of it. Readers are advised to get themselves familiar with at least one Linux OS and understand how it works. While some introductory information will be provided whenever necessary, readers are expected to perform additional research to learn more about those topics.

Some of the most commonly used distros are are follows:-

- Debian
- Ubuntu
- Fedora
- Arch
- Red Hat
- SUSE
- Kali Linux
- Parrot Security OS
- Blackarch

The most powerful thing about Linux is the customization if offers to the users. Almost every component of a Linux system can be changed if only the user knows how. From a penetration tester's perspective this means that any Linux OS can be used in a penetration test assessment. While most penetration testers prefer specialized OS like Kali Linux, Parrot OS, etc which come preinstalled with security tools, please note that these tools can be installed in any Linux system and most of them are also compatible with Windows and Mac OS.

It is worth mentioning that while Linux is the OS a penetration tester ends up working most of time, the targets are more often than not Windows/Mac systems. Deep understanding of these OS is a necessity and should be done sooner than later.

Now let's see some of the inbuilt Linux tools that often come handy during an assessment.

- **awk:** Its a pattern scanning and processing tool
- **cat:** It concatenate and display the content of files
- **chmod:** Changes file permissions
- **cp:** Copies files and directories
- **curl:** Transfers data to and from a server
- **find:** search for files in a directory hierarchy
- **grep:** It filters given keyword(s) within a given file/filesystem

- **ip:** show / manipulate routing, devices, policy routing and tunnels
- **jobs:** It displays the status of jobs in the current session
- **kill:** send a signal to a process
- **ls:** It lists directory contents
- **man:** It displays the documentation of command
- **netstat:** It displays network connections and statistics
- **ps:** It displays details of current processes
- **route:** It deals with routing table and its alteration
- **scp:** It transfers files securely

2.1 BASH

Bash or Bourne Again Shell is the command line environment present by default in a vast majority of Linux systems. Bash is very helpful for a penetration tester due to the scripting feature it offers. Bash is one of the easiest language to learn and implement. However it is highly dependent on the system so while it can help in automation and some quick scripting, doing data/compute intensive tasks with it is not recommended. For those use cases consider other languages like Python, Perl, etc.

3. PENETRATION TESTING TOOLS

In this section, readers will be introduced to some of the widely used tools for Penetration Testing.

3.1 NMAP

Nmap a.k.a Network Mapper is one of the most coveted tools of a penetration tester. Nmap is also known as swiss army knife due to the vast range of features it offers. Some of its features are port scanning, service detection, vulnerability detection among others. A GUI version of Nmap is available as Zenmap which helps see reports in an organized dashboard.

One of the most powerful feature of Nmap is the Nmap Scripting Engine (NSE) which uses scripts written in Lua language to perform a multitude of tasks including enumeration, vulnerability analysis, or exploitation. Use cases of NSE include discovering hosts, identifying services, and exploiting vulnerabilities.

Some key features of NSE include:

- **Large library of scripts:** NSE comes with a large library of pre-written scripts that can be used out of the box or modified to meet specific requirements.

Figure 1. Simple NMAP scan

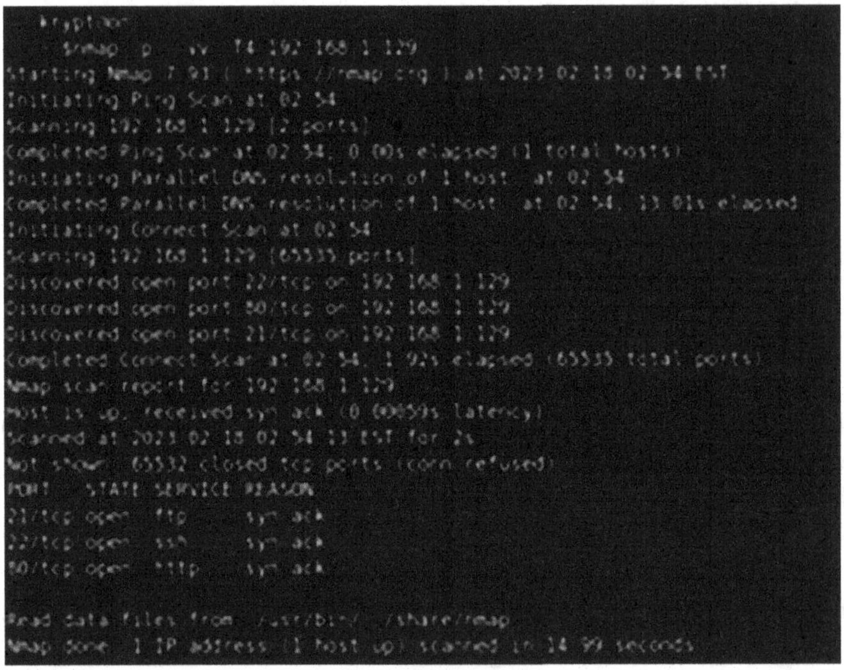

Figure 2. NMAP scan with service versioning, OS version and default scripts flags

- **Customizable:** NSE allows users to write their own scripts in Lua, a lightweight scripting language, making it easy to create customized scripts to meet specific needs.
- **Speed:** NSE is designed to be fast and efficient, with the ability to execute multiple scripts in parallel.
- **Flexibility:** NSE can be used with Nmap for a variety of purposes, including host and port scanning, service detection, and vulnerability detection.
- **Integration:** NSE can be integrated with other tools and platforms, providing a powerful and flexible solution for network security and management.

Readers should take note that any good security solution such as NGFW/IDS/IPS/XDR has the signatures of Nmap scans and thus can detect it easily. For this reason special care should be taken while scanning to avoid triggering these security solutions.

Here are some flags which can help evade this solutions. The effectiveness of these flags however might differ from case to case basis.

Flag 1: **-T<num>** (It controls the speed by which packets are sent)
Flag 2: **-f** (It fragments the packets)
Flag 3: **--data-length** (It appends random data to each packet)
Flag 4: **-S<IP address>** (It spoofs the IP address of Nmap system)
Flag 5: **--randomize-hosts** (It randomizes the target IP addresses)

Now here are some useful tips:

- Nmap doesn't have any fixed order of switches so *nmap <ip> -p- -A* is same as *nmap -A -p- <ip.*
- Only top 1000 ports are scanned in default scan.
- Scanning gateway should be avoided whenever possible.
- Metasploit Framework has an option to directly integrate Nmap scan results in XML format.
- -sS (Stealth Scan) is not as stealthy as one might think. As it doesn't completes the TCP handshake, it has a unique pattern which can be detected by any sufficiently capable security solution.
- Only TCP scan is performed by default. An UDP scan of limited number of ports should be performed to check for services like DNS.
- Nmap scan output in XML format can be imported into Metasploit console using *db_import* command.

NMAP Alternatives:

- Masscan
- Rustscan

3.2 Wireshark

Wireshark is a very powerful network packet analyzer which allows the user to capture and analyze packets in great detail. It is very helpful during a penetration test as it can give insights on what exactly is happening within a network. Due to this feature it can easily detect if any sensitive information is sent via plaintext. Additionally it provides in depth working of network protocols along with advanced statistics.

In the above image, there are 6 sections as follows:-

1. **Main Menu:** Main menu appears on very top of Wireshark and consists of following options as follows:-
   - **File:** Allows user to open and save capture files, export captured data, and perform other file-related tasks.
   - **Edit:** Provides options for editing capture files, such as marking or ignoring packets, copying packets, and adjusting time values.
   - **View:** Allows user to customize the appearance of the Wireshark interface, including the display of packet details, columns, and time values.
   - **Go:** Provides options for navigating the captured data, including moving to the next or previous packet, jumping to a specific packet, and following a TCP stream.
   - **Capture:** Allows user to start and stop packet captures, configure capture options, and manage capture interfaces.
   - **Analyze:** Provides access to various analysis tools, such as decryption, statistics, and graphs.
   - **Statistics:** Allows user to view summary information about the captured data, including protocol statistics, I/O graphs, and endpoint statistics.
   - **Tools:** Provides access to various tools, including firewall rule generation, the export object feature, and the follow stream dialog.
   - **Help:** Provides access to the Wireshark user manual, FAQ, and other help resources.
2. **Main Toolbar:** Main Toolbar appears below Main Menu and consists of following options:-
   - **Start/Stop Capture/Capture Options:** Allows user to start or stop a packet capture and make configuration changes.
   - **Open/Save/Close/Reload:** Allows user to open/close/reload an existing capture file ans save a new one.

- ○ **Find:** Allows user to search for specific packets within the capture file.
- ○ **Zoom In/Out:** Allows user to zoom in or out on the packet details.
- ○ **Go to Packet:** Allows user to jump to a specific packet within the capture file.
- ○ **Previous/Next Packet:** Allows user to navigate to the previous or next packet in the capture file.
- ○ **First/Last Packet:** Allows user to jump to the first or last packet in the capture file.
- ○ **Coloring Rules:** Allows user to apply and manage coloring rules to the captured data.
- ○ **Mark/Ignore:** Allows user to mark or ignore specific packets in the capture file.

3. **Filter Toolbar:** Filter Toolbar lies below Main Toolbar and consists of following options:-
   - ○ **Filter Expression:** Allows user to enter and apply a display filter to the captured data.
   - ○ **History:** Allows user to access a list of recently used display filters.
   - ○ **Apply/Prepare:** Allows user to apply the current filter expression to the captured data or to prepare a filter expression for use.
   - ○ **Clear:** Allows user to clear the current filter expression.
   - ○ **Boolean Operators:** Allows user to use boolean operators (such as "and", "or", and "not") to create more complex filter expressions.
   - ○ **Field Names:** Allows user to access a list of field names that can be used in filter expressions.
   - ○ **Predefined Filters:** Allows user to apply predefined filters to the captured data, such as filters for common protocols (such as HTTP, DNS, and FTP) or for specific network scenarios (such as wireless or VoIP).

4. **Packet List Pane:** Main Toolbar appears below Main Menu and consists of following options:-
   - ○ **Packet Number:** Displays the number of each packet in the capture file.
   - ○ **Time:** Displays the timestamp of each packet in the capture file.
   - ○ **Source/Destination:** Displays the source and destination address for each packet in the capture file.
   - ○ **Protocol:** Displays the protocol used for each packet in the capture file.
   - ○ **Info:** Displays a summary of the information contained in each packet in the capture file.
   - ○ **Packet Color:** Allows user to view and apply color coding to packets in the capture file.

- **Packet Marking:** Allows user to mark or ignore specific packets in the capture file.
- **Packet Searching:** Allows user to search for specific packets within the capture file.

5. **Packet Details Pane:** Main Toolbar appears below Main Menu and consists of following options:-
   - **Packet Dissection:** Displays a detailed breakdown of the contents of each packet in the capture file.
   - **Protocol Tree:** Displays a hierarchical representation of the protocols used in each packet in the capture file.
   - **Byte View:** Displays the raw hexadecimal and ASCII representation of each packet in the capture file.
   - **Field Searching:** Allows user to search for specific fields within the packet details.
   - **Field Filtering:** Allows user to apply filters to the packet details based on the contents of specific fields.
   - **Data Export:** Allows user to export the contents of specific fields in the packet details for further analysis.
   - **Field Detail:** Displays a detailed description of each field in the packet details, including information about the data format and meaning of the field.

6. **Packet Bytes Pane:** Main Toolbar appears below Main Menu and consists of following options:-
   - **Byte-Level Dissection:** Displays a detailed, byte-level representation of the contents of each packet in the capture file.
   - **Data Representation:** Allows user to switch between hexadecimal, ASCII, and EBCDIC representation of the packet data.
   - **Data Export:** Allows user to export the raw binary data of the packet for further analysis.
   - **Data Decoding:** Allows user to decode specific sections of the packet data based on their format.
   - **Data Editing:** Allows user to manually edit the contents of specific bytes within the packet data.

Wireshark also supports the usage of operators to filter the packet captures. Some of the most commonly used operators are given below:-

- Logical Operators:
  - && (AND)
  - || (OR)

- o ^^ (XOR)
- o ! (NOT)
- Filtering Operators:
  - o == (eq)
  - o =! (ne)
  - o > (gt)
  - o < (lt)
  - o >= (ge)
  - o <= (le)

Here are some examples of these operators in action:

§ **Filter by IP:** ip.addr == 192.168.1.1
§ **Filter by destination IP:** ip.dest == 192.168.1.1
§ **Filter by source IP:** ip.src == 192.168.1.1
§ **Filter by IP range:** ip.addr >= 192.168.1.1 and ip.addr <=192.168.1.254
§ **Filter by port:** tcp.port == 80
§ **Filter by MAC address:** eth.addr == 00:80:a2:c4:12:df

Wireshark alternatives:

- Tshark
- Tcpdump

3.3 Nessus

Nessus is a comprehensive and widely-used vulnerability scanner for security professionals. It is designed to help organizations identify and assess potential security threats and vulnerabilities in their networks and systems. Nessus uses a database of known vulnerabilities and security issues to scan target systems and provide a report of potential security risks.

Nessus provides a range of features and capabilities, including multi-platform support, the ability to write custom plugins, integration with other security tools, a comprehensive user interface, and reporting. Additionally, Nessus supports compliance checking and can be used to monitor systems in real-time.

For security professionals, Nessus provides a powerful tool for performing both internal & external vulnerability assessments and identifying potential security risks. The detailed reporting capabilities, including executive summaries and customizable reporting templates, allow security professionals to communicate the results of their assessments to management and other stakeholders.

Some of the most powerful features of Nessus are as follows:-

- **Vulnerability Scanning:** Nessus uses a database of known vulnerabilities and security issues to scan target systems and provide a report of potential security risks.
- **Multi-Platform Support:** Nessus supports a wide range of platforms, including Windows, Linux, macOS, and others.
- **Custom Plugins:** Nessus provides the ability to write custom plugins to extend its functionality and tailor it to specific needs.
- **Integration with Other Tools:** Nessus provides integration with other security tools and platforms, making it easy to integrate into an organization's overall security posture.
- **Intuitive User Interface:** The Nessus user interface provides a comprehensive and intuitive interface for managing scans, viewing results, and generating reports.
- **Compliance Checking:** Nessus provides the ability to check systems for compliance with various regulations and standards, including PCI DSS, HIPAA, and others.
- **Remote Scanning:** Nessus supports remote scanning, making it possible to perform vulnerability assessments on systems that are not directly accessible.
- **Scheduled Scans:** Nessus provides the ability to schedule scans, allowing organizations to automate regular vulnerability assessments.
- **Reporting:** Nessus provides detailed reporting, including executive summaries, detailed technical reports, and customizable reporting templates.
- **Continuous Monitoring:** Nessus provides the ability to monitor systems in real-time, providing alerts and notifications when new vulnerabilities are detected.

Nessus Alternatives:

- Metasploit Pro
- OpenVAS
- Qualys
- Nexpose

3.4 Burp Suite

Burp Suite is a powerful web application security testing platform for security professionals. It is designed to provide a comprehensive set of tools and resources for testing the security of web applications and APIs.

The main components of Burp Suite are:

- **Proxy:** This component allows security professionals to intercept and inspect network traffic, making it easy to identify and assess potential security risks.
- **Scanner:** This component automates and performs manual vulnerability scans on web applications to identify potential security threats.
- **Intruder:** This component provides a platform for testing the security of web applications through targeted attacks.
- **Repeater:** This component allows security professionals to repeat a request with modified parameters.
- **Sequencer:** This component allows the user to determine whether the web tokens are random or similar in some manner.
- **Decoder:** This component allows the encoding and decoding of data between various formats.
- **Comparer:** This component allows the user to compare two files/messages/ payloads in detail.
- **Logger:** This component records the HTTP traffic between Burp Suite and the target applications.
- **Extensions:** This component allows the user to extend the Burp's functionality by installing available extensions or writing custom ones.

In addition to these components, Burp Suite offers several other key features, including:

- **Custom plugin support:** Burp Suite provides a wide range of plugins that can be customized to meet the specific needs of a security professional.
- **Manual testing and analysis:** The platform provides the ability to perform manual testing and analysis to identify vulnerabilities that may not be detected through automated scans.
- **Comprehensive reporting:** Burp Suite offers comprehensive reporting capabilities, including executive summaries and customizable reporting templates, which makes it easy to communicate the results of security assessments to management and stakeholders.

Burp Suite when properly utilized can find a multitude of security vulnerabilities. Some of the most impactful ones are given below:-

- Injection (OS command, code, javascript, XML, CSS, and SQL)
- Cross Site Scripting (XSS)
- Cross Site Request Forgery (CSRF)

- Vulnerable libraries

However it's worth mentioning that Burp Suite can do a lot more than finding vulnerabilities. It has a module named Intruder which can perform password spraying attacks on Web Applications. It should be noted that the community version comes with a rate-limited version of Intruder which will take forever while working with a significant number of payloads. For this purpose the use of other tools is recommended.

One of the things that make Burp Suite very powerful is the BApp Store. It provided a large number of extensions to further enhance the capabilities of the Burp Suite. However not all of these extensions are available for the community version. Also these extensions could increase significant load on the CPU.

Few of the useful extensions are listed below:-

- Logger++
- Authorize
- Active Scan++
- JWT Editor

Burp Suite Alternatives:

- OWASP Zed Attack Proxy
- Acunetix

3.5 Metasploit Framework

Metasploit Framework provides a comprehensive collection of tools and resources for penetration testing, vulnerability scanning, and exploitation. This powerful framework enables security researchers and penetration testers to identify and exploit vulnerabilities in network systems and applications. It includes an extensive library of pre-built exploits and payloads, which can be easily adapted and customized for specific testing scenarios. Additionally, the framework provides a comprehensive set of tools for post-exploitation activities, such as data collection and privilege escalation.

The Metasploit Framework consists of the following components:

- **Auxiliary Modules:** Additional tools and modules that can be used to perform various tasks, such as reconnaissance and information gathering.
- **Encoder Modules:** Tools that can be used to encode payloads to evade detection by security systems.

Figure 3. Payload generation with Msfvenom

- **Exploit Database:** A collection of exploits and payloads that can be used to target specific vulnerabilities.
- **MSFconsole:** A command-line interface for interacting with the Metasploit Framework.
- **MSFRPC:** A remote procedure call interface that allows other programs and scripts to interact with the Metasploit Framework remotely.
- **NOP Generator:** A tool that can be used to generate No-operation (NOP) sleds for buffer overflow exploits.
- **Payload Generator:** A tool that allows users to create custom payloads for specific exploitation scenarios, such as MSFvenom.
- **Post-Exploitation Modules:** Tools and modules that can be used to perform actions after a successful exploitation, such as data collection and privilege escalation.

Out of these components, perhaps Msfvenom is one of the most powerful. With the ability to generate custom payloads as well as the encode them to prevent detection makes it very handy tool.

Now that we got some idea about Metasploit in general, let's see how we can actually interact with it. For this purpose users can either use a GUI like Armitage or CLI like Msfconsole. Both of these have some advantages and disadvantages and their usage differ from case to case basis. In this chapter, author will only discuss about Msfconsole.

Here is a list of some of the most commonly used commands in MSFconsole:

- **search:** Searches the Metasploit Framework's module database for a specific module.
- **use:** Loads a specific module into the framework.
- **show:** Displays information about the currently loaded module.
- **info:** Shows detailed information about a specific module.
- **options:** Displays and sets options for the currently loaded module.
- **set:** Changes the value of a specific option for the currently loaded module.
- **run/exploit:** Executes the currently loaded module.

- **sessions:** Shows information about active sessions.
- **sessions -i:** Interacts with an active session.
- **sessions -k:** Kills an active session.
- **back:** Exits the current context and returns to the previous context.
- **exit:** Exits MSFconsole.

4. PENETRATION TESTING TECHNIQUES

In this section, readers will be introduced to Network Penetration Testing and Web Application Penetration Testing. Some good places to learn new techniques and payloads are HackTricks [1] and PayloadOfAllThings [2].

4.1 Network Penetration Testing

Network penetration testing, also known as network penetration testing, is a process of evaluating the security of a computer network by simulating an attack. The objective of network penetration testing is to identify vulnerabilities in the network and provide recommendations to improve the security posture of the network.

The network penetration testing process typically involves several steps, including reconnaissance, scanning, enumeration, vulnerability analysis, exploitation, and post-exploitation. During the reconnaissance phase, the tester gathers information about the target network to identify potential vulnerabilities and attack vectors. In the scanning phase, the tester scans the network to identify open ports, services, and operating systems. In the enumeration phase, the tester gathers more detailed information about the network, such as user accounts and network topology.

In the vulnerability analysis phase, the tester identifies potential vulnerabilities and exploits them to gain access to the network. This is done in a controlled manner to avoid causing damage to the network. During the exploitation phase, the tester attempts to gain access to the network using various methods, including exploiting software vulnerabilities or using social engineering techniques.

Finally, in the post-exploitation phase, the tester assesses the level of access gained and identifies any further vulnerabilities that may exist on the network. The tester then provides a report detailing the vulnerabilities found and recommendations for how to address them.

Overall, network penetration testing is a critical component of an organization's cybersecurity strategy and helps to identify vulnerabilities and weaknesses in the network before they can be exploited by attackers. A good mindmap of network attacks is available at Networknightmare [3].

Network Penetration Testing can be broadly classified in two major categories as follows:-

- **External Network Penetration Testing**
- **Internal Network Penetration Testing**

Now let's see both of these in a bit more detail.

- **External Network Penetration Testing**:

   § External network penetration testing is a process of evaluating the security of a company's computer network from an external perspective, i.e., from the Internet-facing side. The objective of external network penetration testing is to identify vulnerabilities in the network that could be exploited by external attackers to gain unauthorized access to the company's systems and data.

   § The external network penetration testing process typically involves several steps. During the reconnaissance phase, the tester gathers information about the target company's network from publicly available sources, such as the company's website, social media accounts, and domain name registration records. This information is used to identify potential vulnerabilities and attack vectors.

   § In the scanning phase, the tester uses various tools to scan the target network for open ports, services, and operating systems. This information is used to identify potential vulnerabilities that can be exploited.

   § During the enumeration phase, the tester gathers more detailed information about the target network, such as user accounts, network topology, and other network assets.

   § In the vulnerability analysis phase, the tester identifies potential vulnerabilities and attempts to exploit them to gain access to the network. This is done in a controlled manner to avoid causing damage to the network.

   § Finally, in the post-exploitation phase, the tester assesses the level of access gained and identifies any further vulnerabilities that may exist on the network. The tester then provides a report detailing the vulnerabilities found and recommendations for how to address them.

   § Overall, external network penetration testing is an essential component of an organization's cybersecurity strategy, as it helps to identify vulnerabilities and weaknesses in the network that could be exploited by external attackers. By conducting external network penetration testing, companies can proactively identify and address vulnerabilities before they can be exploited.

Here are some common techniques with examples that can be used for external network penetration testing:

- **Port Scanning:** This technique involves scanning for open ports on a target system, which can then be used to determine the services and applications running on the system. For example, an external penetration tester may use a tool like Nmap to scan a target network for open ports and identify vulnerable services.
- **Vulnerability Scanning:** This technique involves using an automated tool to scan a target system for known vulnerabilities. For example, a penetration tester may use a tool like Nessus or OpenVAS to scan a target network for known vulnerabilities in web applications, servers, and other software.
- **Social Engineering:** This technique involves attempting to trick users into providing sensitive information or access to a system. For example, a penetration tester may use phishing emails or phone calls to try and gain access to the network.
- **Web Application Testing:** This technique involves assessing the security of web applications hosted on a target network. For example, a penetration tester may use tools like Burp Suite or OWASP ZAP to test for vulnerabilities like SQL injection or cross-site scripting (XSS).
- **Password Attacks:** This technique involves attempting to crack passwords or gain access to systems using default or weak credentials. For example, a penetration tester may use tools like John the Ripper or Hashcat to crack passwords or brute-force login credentials.
- **Network Traffic Analysis:** This technique involves analyzing network traffic to identify vulnerabilities or sensitive information. For example, a penetration tester may use tools like Wireshark to capture and analyze network traffic to identify sensitive data being transmitted in clear text.
  - **Internal Network Penetration Testing:**
    - § Internal network penetration testing is the process of assessing the security of an organization's internal network infrastructure, including servers, workstations, and other network-connected devices. The main objective of internal network penetration testing is to identify vulnerabilities that could be exploited by attackers to gain unauthorized access to sensitive data, systems, or applications.
    - § Internal network penetration testing can be conducted using a variety of methods, including vulnerability scanning, manual testing, and

social engineering. The testing can be conducted by an internal security team or a third-party security consultant.

§ The testing process typically begins with information gathering, which involves identifying the network infrastructure, mapping out the network, and identifying potential entry points for attackers. This is followed by vulnerability scanning and manual testing, which is done to identify security weaknesses and vulnerabilities that could be exploited.

§ Once vulnerabilities have been identified, the penetration testers will attempt to exploit them to gain access to the network. This can include attempting to gain access to systems and data, escalate privileges, or pivot to other systems on the network. The penetration testers will then document their findings and provide recommendations for remediation to the organization.

Here are some common techniques with examples that can be used for internal network penetration testing:

- **Port scanning:** This technique is used to identify open ports on a target system. For example, nmap can be used to scan the internal network for open ports and services.

- **Vulnerability scanning:** This technique is used to identify known vulnerabilities in a system or network. For example, tools like OpenVAS or Nessus can be used to scan the internal network for known vulnerabilities.

- **Password cracking:** This technique is used to identify weak passwords used by employees. For example, tools like John the Ripper can be used to crack passwords on the internal network.

- **Privilege escalation:** This technique is used to gain higher levels of access on a target system. For example, an attacker can exploit a vulnerability in a service or application to gain root access to a system.

- **Social engineering:** This technique is used to exploit the human element in security. For example, an attacker can use phishing emails or phone calls to trick employees into providing sensitive information or giving the attacker access to the internal network.

- **Lateral movement:** This technique is used to move from one system to another within the internal network. For example, an attacker can use a compromised system to gain access to other systems on the internal network.

- **Backdoors and Trojans:** This technique is used to create a persistent backdoor or Trojan on a target system. For example, an attacker can use a

RAT (Remote Access Trojan) to gain access to the internal network and maintain access even if the original entry point is discovered and closed.

4.2 Web Application Penetration Testing

Web application penetration testing is a process of identifying and exploiting vulnerabilities in web applications by simulating attacks from potential attackers. It is a proactive approach to identify security issues that can potentially cause harm to the application, its data, and the users.

The process typically begins with information gathering and reconnaissance, where the tester tries to gather as much information as possible about the target application. This includes identifying the technologies and frameworks used, understanding the application architecture, and mapping the attack surface. Next, the tester attempts to identify and exploit vulnerabilities in the application. This can include injection attacks, such as SQL injection and cross-site scripting (XSS), authentication and authorization issues, and other common vulnerabilities identified in the OWASP Top Ten list.

Once vulnerabilities have been identified, the tester attempts to exploit them to gain access to sensitive data or control of the application. This can include accessing restricted areas of the application, manipulating data or functionality, and even taking control of the underlying system. The final stage is to provide a comprehensive report of the findings, which includes a description of the vulnerabilities discovered, the potential impact of each vulnerability, and recommendations for remediation. The report is then used to prioritize the necessary fixes to ensure the application is secure and protected against future attacks.

Overall, web application penetration testing is an important aspect of any comprehensive security program, as it helps identify potential vulnerabilities and provides a roadmap for remediation.

Here are some types of web application penetration testing:-

- **Black box testing:** This type of testing is carried out without any prior knowledge of the system under test. Testers will perform a range of attacks in order to identify vulnerabilities.
- **White box testing:** This type of testing is carried out with full knowledge of the system under test. Testers will have access to source code and system architecture in order to identify vulnerabilities.
- **Gray box testing:** This type of testing is carried out with partial knowledge of the system under test. Testers will have some access to system information in order to identify vulnerabilities.

- **Automated testing:** This type of testing uses automated tools to identify vulnerabilities in web applications. This is an efficient way to test large applications, but may not be as thorough as manual testing.
- **Manual testing:** This type of testing is performed by testers who manually identify vulnerabilities. Manual testing can be time-consuming, but is more thorough than automated testing.
- **Network testing:** This type of testing is focused on testing the network infrastructure that supports the web application.
- **Client-side testing:** This type of testing is focused on testing the client-side of the web application, which includes HTML, JavaScript, and other front-end technologies.

Some of the widely used Web Application attacks are discussed below:-

- **Injection Attacks:**

  - **SQL Injection:** An attacker can inject malicious SQL commands through input fields, such as login fields, that can result in stealing, modifying, or deleting sensitive information from a database.

Example: *' OR 1=1; --*

- **Command Injection:** An attacker can execute arbitrary system commands through input fields, such as search bars, that can result in gaining remote access to the system or stealing sensitive information.

Example: *;ls -la*

- **XML Injection:** An attacker can inject malicious XML input that can result in Denial of Service (DoS) attacks, or steal sensitive information.

Example: *<!DOCTYPE foo [<!ENTITY xxe SYSTEM "file:///etc/passwd">]>*

- **Cross-Site Scripting (XSS) Attacks:**

  - **Reflected XSS:** An attacker can inject malicious code, such as a script, that is reflected from the server to the client.

Example: *<script>alert('XSS')</script>*

- **Stored XSS:** An attacker can inject malicious code, such as a script, that is stored on the server and is displayed to all users who access the vulnerable page.

  Example: *<script>alert('XSS')</script>*

- **DOM-based XSS:** An attacker can inject malicious code, such as a script, that is executed by modifying the Document Object Model (DOM) of the web page.

  Example: *javascript:alert('XSS')*

- **Cross-Site Request Forgery (CSRF) Attacks:** An attacker can create a malicious website that sends unauthorized requests to the web application server on behalf of an authenticated user.

  Example: *<img src="https://vulnerable-site.com/delete-account?id=123" width="1" height="1">*

- **File Inclusion Attacks:**

  ○ **Local File Inclusion (LFI):** An attacker can access sensitive files on the server by including them in the web application through input fields, such as a file path.

  Example: *../../../../etc/passwd*

- **Remote File Inclusion (RFI):** An attacker can execute arbitrary code on the server by including a remote file in the web application through input fields, such as a file URL.

  Example: http://example.com/shell.php

- **Authentication and Session Attacks:**

  ○ **Brute Force Attack:** An attacker can try different password combinations until the correct one is found, and gain unauthorized access to the system. Example: Using a password list to try different passwords for a user account.

○ **Session Hijacking:** An attacker can steal a valid session cookie and use it to impersonate the authenticated user.

Example: Intercepting and using the session cookie of an authenticated user.

- **Other Common Attacks:**

○ **Server-Side Request Forgery (SSRF):** Exploiting the ability of a web application to make HTTP requests to other systems to access resources or services that should not be available.

Example: http://example.com/?url=http://attacker.com/private

- **Insecure Direct Object Reference (IDOR):** Accessing a sensitive resource on a web application by manipulating a parameter or URL.

Example: http://example.com/user?id=123

- **Remote Code Execution (RCE):** Running malicious code on a target server through a vulnerability in a web application.

Example: *;system('id');*

4.3 Password Cracking

Password cracking or Password hacking is the process of guessing or recovering a password from stored data or transmitted communication. It is often used to test the strength of passwords used in a system. The process usually involves using automated tools or scripts to attempt to guess the password by trying various combinations of letters, numbers, and symbols.

There are different types of password cracking techniques, including brute force attacks, dictionary attacks, and hybrid attacks. Brute force attacks involve trying every possible combination of characters until the correct password is found, while dictionary attacks use a pre-computed list of common passwords and commonly used words. Hybrid attacks combine both techniques by adding numbers and symbols to common words.

A very comprehensive collection of tutorials is available at Password Cracking 101+1 [4]. These tutorials cover a large number of tools and techniques for cracking passwords.

4.4 Report Writing

At the very end after finding and exploiting vulnerabilities, comes the part of providing the client with a well documented report of all the findings. This report should be written with utmost care and precision listing all the vulnerabilities detected during the assessment and how the tester approached it. Readers please note that client will only see this report so the emphasis should be on quality.

A typical penetration testing report has following sections:-

- Confidentiality statement
- Disclaimer
- Contact Information
- Assessment overview and components
- Severity ratings
- Risk factors
- Scope
- Executive summary
- Vulnerability summary
- Technical findings
- Additional information

Please note that the above mentioned list is not an exhaustive and it differ from company to company. However that being said, some things are common in all reports like executive summary, assessment overview, vulnerability summary, and technical findings.

Here are some tips to enhance a report:

- Report should be tailored to the client's organization and risks should be matched accordingly.
- Severity should be calculated with proper models like CVSS/CVSS v3.
- Risk factors should map to the client's infrastructure.
- Include findings that are relevant to the client and their operations.
- Executive summary should include the impact (financial, reputation, data, etc) and risk if the vulnerabilities are exploited.
- To protect the confidentiality always redact sensitive information (passwords, hashes, etc) from the report.
- Include proper screenshots, exploitation methodology, and warnings in technical findings.

5. PENETRATION TESTING KNOWLEDGE BASE

In this section, readers will be made familiar with some of the existing knowledge bases which are being used by security professionals throughout the world.

5.1 MITRE ATT&CK®

MITRE ATT&CK (Adversarial Tactics, Techniques, and Common Knowledge) [5] is a globally accessible knowledge base of adversary tactics and techniques based on real-world observations. The framework provides a comprehensive and structured understanding of the tactics and techniques used by attackers and is used by organizations to assess and improve their security posture.

ATT&CK is divided into several matrices that cover various stages of an attack, such as initial access, execution, persistence, privilege escalation, and lateral movement. Each technique is described in detail, including its purpose, prerequisites, and security implications.

The information in ATT&CK is based on real-world observations and is continuously updated by the security community. The framework is widely adopted by organizations, security vendors, and governments, and is used as a reference in threat intelligence, incident response, and security product evaluations.

The use of ATT&CK enables organizations to better understand the tactics and techniques used by attackers, prioritize their defense efforts, and measure the effectiveness of their security controls.

5.2 OWASP Top Ten

The OWASP Top 10 Project is an open-source, community-driven project that aims to identify and raise awareness of the most critical security risks faced by web applications. The project is maintained by the Open Web Application Security Project (OWASP), a non-profit organization focused on improving the security of software. This project provides a comprehensive list of the top ten most critical security risks faced by web applications. This list is updated every three years to reflect the current threat landscape and keep up with the evolving nature of cyber attacks.

The OWASP Top 10 Project provides a wealth of information and resources for developers, security professionals, and organizations, including detailed information on each of the top ten risks, best practices for mitigating those risks, and practical guidance for building secure web applications. It is widely used as a standard for evaluating the security of web applications and is referenced by security professionals, organizations, and government agencies around the world. In recent years, OWASP has also published Top 10 list for Application Programming Interface (API) security which provides strategies and solutions for managing them securely.

Apart from these Top 10 lists, OWASP also has a great checklist for Web Application Audit named Web Security Testing Guide (WSTG) [6]. This checklist provides a long list of steps to test a Web Application for security vulnerabilities.

The current list of OWASP Top 10 (2021) features the following vulnerabilities:-

1. Broken Access Control
2. Cryptographic Failures
3. Injection
4. Insecure Design
5. Security Misconfiguration
6. Vulnerable and Outdated Components
7. Identification and Authentication Failures
8. Software and Data Integrity Failures
9. Security Logging and Monitoring Failures
10. Server-Side Request Forgery

REFERENCES

Caster. (2023, February 8). *C4S73R/Networknightmare: Network pentesting mindmap by caster*. GitHub. https://github.com/c4s73r/NetworkNightmare

OWASP Foundation. (2022, December 03). *OWASP Web Security Testing Guide*. OWASP. https://owasp.org/www-project-web-security-testing-guide/v42/

Polop, C. (n.d). *HackTricks*. HackTricks. https://book.hacktricks.xyz/welcome/readme

Security, In. (n.d.). *Password cracking 101+1*. In Security. https://in.security/technical-training/password-cracking

Swissky. (n.d.). *PayloadOfAllThings*. Github. https://github.com/swisskyrepo/PayloadsAllTheThings

The MITRE Corporation. (2022, October 25). *MITRE ATT&CK*. MITRE. https://attack.mitre.org/versions/v12/

Chapter 13
Social Engineering Attacks and Countermeasures

Kshyamasagar Mahanta

iD https://orcid.org/0000-0002-5385-9463

Maharaja Sriram Chandra Bhanja Deo University, India

Hima Bindu Maringanti

Maharaja Sriram Chandra Bhanja Deo University, India

ABSTRACT

This book chapter examines the increasing danger of social engineering attacks in cybersecurity. These attacks focus on exploiting human vulnerabilities instead of technical weaknesses and target the human element of organizations. The chapter outlines the various types of social engineering attacks such as phishing, pretexting, baiting, and quid pro quo, and explores the strategies employed by social engineers including the creation of urgency, trust, and fear. It also covers countermeasures that can be employed to guard against social engineering attacks, such as education and awareness programs for employees and technical solutions like spam filters and multi-factor authentication. By understanding the threat of social engineering attacks and taking proactive steps to mitigate this risk, individuals and organizations can protect themselves against this growing cybersecurity menace.

1. INTRODUCTION

Social engineering attacks have become an increasingly common and dangerous threat in the realm of cybersecurity. These attacks differ from traditional hacking and cyberattacks in that they target the human element of an organization rather than

DOI: 10.4018/978-1-6684-8218-6.ch013

exploiting technical vulnerabilities. Instead of exploiting system weaknesses, social engineers rely on human vulnerabilities to manipulate individuals into providing access to sensitive information or systems.

Social engineering attacks can have devastating consequences, from financial loss and identity theft to data breaches and reputational damage. Cybercriminals are constantly developing new and sophisticated social engineering techniques to exploit human weaknesses and bypass security defenses. Therefore, it is essential for individuals and organizations to understand the nature of social engineering attacks and implement effective countermeasures to mitigate the risks.

This chapter provides a comprehensive guide to social engineering attacks and strategies for preventing them. This chapter is divided into three parts: the first part covers the fundamentals of social engineering, including its definition, history, psychology, and types of attacks. The second part delves into specific social engineering tactics, such as phishing, pretexting, baiting, and quid pro quo, providing real-world examples and case studies. The final part of the book focuses on countermeasures against social engineering attacks. It provides practical tips and techniques for individuals and organizations to protect themselves from social engineering attacks, such as implementing strong passwords, using multi-factor authentication, conducting security awareness training, and building a culture of security. This chapter also discusses the role of technology in preventing social engineering attacks, such as intrusion detection systems, firewalls, and antivirus software.

Overall, "Social Engineering Attacks and Countermeasures" chapter is a must-read for anyone who wants to understand the threat of social engineering attacks and take proactive steps to protect themselves and their organizations. This chapter provides a wealth of information, insights, and practical advice that will help readers stay safe in an increasingly dangerous digital world.

2. FUNDAMENTALS OF SOCIAL ENGINEERING

Social engineering is the art of manipulating individuals to divulge sensitive information, access confidential data, or perform an action that is not in their best interest. Social engineering attacks are typically carried out by cybercriminals who use a variety of tactics to exploit human weaknesses and bypass security defenses. The fundamental principles of social engineering are rooted in human psychology and behavior.

The history of social engineering attacks dates back even further than the early days of computing. In fact, social engineering tactics have been used by humans for centuries to manipulate and deceive others.

One of the earliest examples of social engineering can be traced back to the Trojan War, where the Greeks used a wooden horse as a pretext to enter the city of Troy and defeat the Trojans. The wooden horse was presented as a gift, but it actually contained Greek soldiers who were able to open the city gates and attack from within.

During World War II, social engineering tactics were used extensively by both Allied and Axis powers to gain intelligence and deceive their enemies. For example, British intelligence agents used double agents to feed false information to the Germans, while the Germans used false radio signals and fake military units to deceive the Allies.

In the 1950s and 1960s, social engineering tactics were used extensively by Soviet intelligence agencies to recruit spies and infiltrate Western governments. The KGB and other Soviet intelligence agencies used a variety of tactics, including blackmail, bribery, and false flag operations, to recruit individuals and gather intelligence.

One of the earliest recorded social engineering attacks was the "Morris Worm" in 1988. The Morris Worm was a computer virus that infected thousands of computers across the internet, causing widespread disruption. The creator of the Morris Worm, Robert Tappan Morris, was a graduate student at MIT who had intended the virus to be a harmless experiment to measure the size of the internet. However, the worm quickly spread out of control, causing significant damage to computer systems(Jajoo, 2021).

In the 1990s and early 2000s, social engineering attacks became more sophisticated, with attackers using a variety of tactics to manipulate individuals and gain access to sensitive information. One notable example was the "Love Bug" virus, which spread through email in 2000 and caused billions of dollars in damages. The Love Bug virus was a classic example of a phishing attack, as the email appeared to be a love letter, but actually contained a malicious attachment that infected the victim's computer.

With the rise of the internet and digital technologies, social engineering attacks have become more prevalent and sophisticated. Today, cybercriminals use a wide range of tactics to deceive and manipulate individuals, including phishing, pretexting, baiting, and quid pro quo.

Despite the increased sophistication of social engineering attacks, the basic principles remain the same. Attackers seek to exploit human weaknesses and psychological traits to gain access to sensitive information or carry out malicious activities. As such, the most effective defense against social engineering attacks is often education and awareness, as individuals who are aware of the risks are better equipped to recognize and respond to these types of attacks.

3. PSYCHOLOGY OF SOCIAL ENGINEERING

The psychology of social engineering refers to the techniques and principles that social engineers use to manipulate their victims into divulging sensitive information or performing actions that may compromise security. Social engineers are skilled at exploiting cognitive biases, social norms, and emotional triggers to gain the trust and cooperation of their targets(Del Pozo et al., 2018).

One common psychological principle that social engineers exploit is the principle of reciprocity. This principle states that people tend to feel obligated to return a favor when one is done for them. Social engineers may exploit this principle by offering something to their targets, such as a small gift or a free service, in exchange for access to sensitive information or systems.

Another psychological principle that social engineers may exploit is the principle of authority. This principle states that people are more likely to comply with requests from authority figures or those who appear to have expertise in a particular area. Social engineers may pose as authority figures, such as IT technicians or security personnel, to gain the trust and cooperation of their targets.

Social engineers may also exploit cognitive biases, such as the confirmation bias and the optimism bias. The confirmation bias is the tendency for people to seek out information that confirms their preexisting beliefs and to ignore information that contradicts those beliefs. Social engineers may exploit this bias by tailoring their messages to match their targets' existing beliefs or by presenting information in a way that reinforces those beliefs. The optimism bias is the tendency for people to overestimate the likelihood of positive outcomes and to underestimate the likelihood of negative outcomes. Social engineers may exploit this bias by presenting their requests or offers in a way that emphasizes the potential benefits while downplaying the potential risks.

To defend against social engineering attacks, it is important to understand the psychology behind these attacks and to train employees to recognize the signs of such attacks. Security awareness training can help employees identify common social engineering techniques and avoid falling victim to these attacks. Additionally, access controls, strong passwords policies, and incident response plans can help mitigate the risks of social engineering attacks.

4. TYPES OF SOCIAL ENGINEERING ATTACKS

Social engineering attacks are created to take advantage of human weaknesses and trick people into disclosing private information or granting access to networks or systems. Attackers utilise a variety of social engineering techniques to accomplish their objectives. We will talk about the most typical forms of social engineering assaults in this part.

Phishing

Phishing is one of the most well-known and widely used social engineering attacks. Phishing attacks typically involve sending an email or message that appears to be from a reputable source, such as a bank, social media platform, or online retailer. The message may contain a link that, when clicked, takes the victim to a fake website that looks legitimate but is designed to steal their login credentials or personal information(Surbhi Gupta et al., 2016).

For example, A victim gets a fake email from their bank informing them that their account has been hijacked and that they need to click a link to change their password. The victim clicks on the link to change their password since the email appears authentic, has the bank's logo, and contains a persuasive message. The victim is sent to a fake website that impersonates the bank's website but is actually under the attacker's control by clicking the link. The victim submits their login information under the impression that they are changing their password; nonetheless, the attacker now has access to the victim's bank account and is able to steal their money or personal data.

In this scenario, the attacker used a phishing email to trick the victim into providing their login credentials. Phishing emails can be very convincing, and attackers often use social engineering tactics to create a sense of urgency or fear in the victim, prompting them to click on the link and divulge their information. Receiving emails that request personal information or include dubious links should be treated with caution.

Pretexting

Pretexting involves an attacker posing as a trustworthy individual or authority figure to gain access to sensitive information. This type of attack often involves the attacker impersonating a government official, law enforcement officer, or IT support technician. The attacker may use various techniques to build trust with the victim, such as providing convincing details and information or using social engineering tactics to create a sense of urgency(Girinoto et al., 2022).

For example, An attacker poses as an IT support technician and calls an employee at a company, claiming that there has been a security breach and that they need the employee's login credentials to verify their account. The attacker provides a convincing story and uses social engineering tactics to create a sense of urgency, urging the employee to provide their credentials quickly to prevent further damage.

The employee, believing that they are helping to prevent a security breach, provides their login credentials to the attacker. The attacker now has access to the company's network and sensitive information, which they can use for their own purposes or sell on the dark web.

In this scenario, the attacker used pretexting to gain the employee's trust and obtain their login credentials. Pretexting often involves impersonating a trusted individual or authority figure to gain access to sensitive information. It is important for employees to be wary of unsolicited requests for personal information and to verify the identity of individuals who ask for sensitive information.

Baiting

Baiting involves the use of enticing offers or promises to lure victims into divulging sensitive information or granting access to systems or networks. This type of attack may involve offering free downloads or prizes in exchange for the victim's personal information or login credentials(Shubham Gupta et al., 2021). Baiting attacks may also involve physical media, such as leaving infected USB drives in public places or in a company's parking lot.

For example, An attacker leaves a USB drive in a public place, such as a coffee shop or a library, with a label that says "Confidential Company Information - HR." The attacker then waits for someone to pick up the USB drive and plug it into their computer.

Once the victim plugs the USB drive into their computer, it automatically installs malware that gives the attacker access to the victim's computer and sensitive information. The attacker can now steal passwords, financial information, and other sensitive data.

The victim in this instance was tricked into connecting the USB drive to their computer by the attacker, who performed a baiting technique. Baiting attacks frequently entail making alluring claims or promises to entice targets into disclosing private information or allowing access to networks or systems. When locating physical media, such as USB drives, in public areas, one should use caution and refrain from putting them into computers or other electronic equipment.

Quid Pro Quo

In a quid pro quo social engineering assault, something is provided in exchange for access to systems or networks or sensitive information. The attacker typically poses as a legitimate individual, such as an IT technician or customer service representative, and offers to help the victim with a problem in exchange for their login credentials or other sensitive information.

Here's an example of a quid pro quo attack:

An attacker poses as an IT technician and calls an employee at a company, claiming that they are conducting a security audit and need the employee's login credentials to verify their account. The attacker offers to provide the employee with a free antivirus software in exchange for their login credentials.

The employee, believing that they are getting a free antivirus software and helping with a security audit, provides their login credentials to the attacker. The attacker now has access to the company's network and sensitive information, which they can use for their own purposes or sell on the dark web.

In this scenario, the attacker used quid pro quo to gain the employee's trust and obtain their login credentials. Quid pro quo attacks often involve offering something in exchange for sensitive information or access to systems or networks, and it is important for employees to be cautious when receiving unsolicited offers from individuals they do not know or trust.

Spear Phishing

Spear phishing is a type of social engineering attack that targets specific individuals or organizations, rather than casting a wide net with generic phishing emails. Spear phishing attacks are often highly personalized and customized to appear more legitimate, making them more difficult to detect(Atmojo et al., 2021).

Here's an example of a spear phishing attack:

An attacker researches a specific employee at a company, including their job role, responsibilities, and personal interests, by reviewing their social media profiles and other online sources. The attacker then sends an email to the employee that appears to be from their supervisor, requesting sensitive information or asking the employee to download a file or click on a link.

The email may reference specific projects or events that the employee is working on, making it appear more legitimate. The employee, believing that the email is from their supervisor, provides the requested information or downloads the file, unknowingly installing malware on their computer.

In this scenario, the attacker used spear phishing to target a specific individual at a company, using personalized information to make the email appear more legitimate. Spear phishing attacks can be highly effective, as they are often difficult to detect and appear more trustworthy than generic phishing emails. It is important for individuals and organizations to be aware of spear phishing attacks and to implement security measures, such as employee training and email filters, to detect and prevent these types of attacks.

Impersonation

Impersonation is a sort of social engineering attack in which an attacker pretends to be a reliable person or high-ranking official in order to access confidential data or computer systems. If the perpetrator is adept at imitating the person or organization they are impersonating, impersonation assaults can take many different forms, including emails, phone calls, or in-person contacts, and they can be challenging to spot.

Here's an example of an impersonation attack:

An attacker poses as a bank representative and calls an individual, claiming that there has been fraudulent activity on their account and that they need to verify their identity by providing their login credentials or other sensitive information. The attacker provides a convincing story and uses social engineering tactics to create a sense of urgency, urging the individual to provide their information quickly to prevent further damage.

The individual, believing that they are working with a legitimate bank representative, provides their login credentials or other sensitive information to the attacker. The attacker can now access the individual's bank account and withdraw funds or use the information for other fraudulent purposes.

In this scenario, the attacker used impersonation to gain the individual's trust and obtain their sensitive information. Impersonation attacks often involve posing as a trusted individual or authority figure, such as a bank representative, government official, or company executive. It is important to be wary of unsolicited requests for personal information and to verify the identity of individuals who ask for sensitive information.

Tailgating

Tailgating is a type of social engineering attack that involves an attacker following a legitimate individual into a restricted area or building, without proper authorization or identification. The attacker may use social engineering tactics, such as pretending to be lost or in need of assistance, to gain the victim's trust and access to the restricted area.

Here's an example of a tailgating attack:

An attacker approaches an employee outside of a secure building and asks for directions to a nearby office. The employee, wanting to be helpful, allows the attacker to follow them through the entrance and into the building.

Once inside the building, the attacker may continue to follow the employee, gaining access to restricted areas or systems. The attacker may use this access to steal sensitive information or install malware on the organization's network.

In this scenario, the attacker used tailgating to gain access to a secure building and potentially steal sensitive information or install malware. Tailgating attacks can be difficult to detect, as the attacker may appear to be a legitimate visitor or employee. Organizations can prevent tailgating attacks by implementing security measures, such as requiring identification badges, conducting security awareness training, and monitoring security cameras.

Watering Hole

Watering Hole is a type of social engineering attack that targets a specific group of individuals or organizations by infecting a website or online resource that is frequently visited by the victims. The attackers target a website that is known to be popular among the targeted group and infect it with malware or other malicious software. When the victims visit the infected website, their computers become infected with the malware, allowing the attackers to gain access to sensitive information or control over the victim's computer.

Here's an example of a watering hole attack:

An attacker targets a website that is popular among employees of a specific organization. The attacker infects the website with malware and waits for employees to visit the site. When an employee visits the infected website, their computer becomes infected with the malware.

Once the attacker has gained control over the employee's computer, they can steal sensitive information or use the computer to launch further attacks against the organization's network. The attacker can also use the compromised computer to launch attacks against other organizations or individuals.

In this scenario, the attacker used a watering hole attack to target a specific organization and gain access to sensitive information. Watering hole attacks are difficult to detect, as the victims may not realize that they have been infected with malware. Organizations can protect against watering hole attacks by implementing security measures such as keeping software up to date, using antivirus software, and conducting security awareness training for employees.

Reverse Social Engineering

Reverse social engineering is a type of social engineering attack where the victim is the one who initiates contact with the attacker, usually seeking help or support. The attacker then uses social engineering tactics to gain the victim's trust and obtain sensitive information or access to systems or networks(Bishnoi et al., 2023).

Here's an example of a reverse social engineering attack:

An attacker creates a fake technical support company and posts ads online offering free computer support services. When the victim contacts the attacker, the attacker pretends to be a legitimate technical support provider and convinces the victim to download and install remote access software.

Once the remote access software is installed, the attacker gains control over the victim's computer and can steal sensitive information or install malware on the victim's computer. The attacker may also use the victim's computer to launch further attacks against other systems or networks.

In this scenario, the attacker used reverse social engineering to gain access to the victim's computer and potentially steal sensitive information or install malware. Reverse social engineering attacks can be difficult to detect, as the victim initiates contact with the attacker, and the attacker appears to be a legitimate service provider. Organizations can prevent reverse social engineering attacks by implementing security measures, such as employee training, multi-factor authentication, and network monitoring.

Dumpster Diving

Dumpster diving is a sort of social engineering attack in which the attacker looks through the recyclables or garbage of a company to locate private papers, passwords or other sensitive data. The attacker may use this knowledge to access the company's systems or take confidential data.

Here's an example of a dumpster diving attack:

An attacker enters an organization's parking lot after hours and searches through their dumpsters. The attacker finds several documents containing confidential information, including employee names, addresses, and social security numbers.

Once the attacker has obtained this information, they can use it to impersonate employees, steal their identities, or launch further attacks against the organization's systems.

In this scenario, the attacker used dumpster diving to obtain sensitive information and potentially steal identities or launch further attacks. Organizations can prevent dumpster diving attacks by implementing security measures such as shredding sensitive documents before discarding them, training employees on proper document disposal procedures, and using secure disposal containers for sensitive information.

In summary, social engineering attacks come in many forms, and attackers are constantly developing new tactics to exploit human vulnerabilities. By understanding the different types of social engineering attacks, individuals and organizations can better protect themselves against these threats. Chapter 1: Introduction to Social Engineering Attacks

Cybersecurity threats include social engineering assaults, which target an organization's human resources. The purpose of a social engineering assault is to persuade people to provide confidential information or take activities that compromise the organization's security. There are several ways that these assaults may be carried out, including phishing, baiting, pretexting, and quid pro quo.

The fact that social engineering assaults depend more on preying on human nature than on technology flaws is one of their most cunning features. Social engineering attacks take use of the fact that people are frequently the weakest link in any security system to obtain access to confidential data or systems. This chapter will examine the many forms of social engineering assaults as well as the strategies employed by attackers to carry them out.

5. TACTICS USED BY SOCIAL ENGINEERS

Social engineers use a variety of tactics to gain the trust of their targets and convince them to divulge sensitive information or perform specific actions. Here are some common tactics used by social engineers for attack:

Authority: Social engineers may impersonate someone in a position of authority, such as a senior executive or IT administrator, to gain the trust of their target.

Urgency: Social engineers may create a sense of urgency or panic to pressure their target into taking immediate action, such as resetting a password or downloading a file.

Scarcity: Social engineers may create a sense of scarcity or exclusivity to convince their target to act quickly, such as offering access to a limited-time offer or a restricted resource.

Familiarity: Social engineers may create a false sense of familiarity or friendship to gain the trust of their target, such as pretending to be a friend or family member.

Flattery: Social engineers may use flattery or compliments to make their target feel valued and important, increasing the likelihood that they will comply with their requests.

Intimidation: Social engineers may use intimidation or threats to coerce their target into compliance, such as threatening to disclose embarrassing information or report the target to their supervisor.

Pretexting: Social engineers may create a false pretext, such as pretending to be conducting a survey or performing a security audit, to gain the trust of their target and obtain sensitive information.

By using these tactics, social engineers can manipulate their targets into divulging sensitive information or performing actions that put an organization's security at risk. To prevent social engineering attacks, organizations should train their employees to recognize and report suspicious requests or behavior, implement strong authentication and access controls, and conduct regular security awareness training .

6. SOCIAL ENGINEERING TESTING

The purpose of social engineering testing is to simulate actual social engineering assaults in order to determine how vulnerable a company's security system is. Attackers can persuade people into disclosing private information or carrying out tasks that compromise the organization's security through social engineering tactics (Li et al., 2019).

Social engineering testing can involve a variety of methods, including email phishing, pretexting, baiting, and tailgating. During the testing process, a security team will attempt to breach the organization's security system by using these techniques to gain access to sensitive information or physical locations.

The goal of social engineering testing is to find security system flaws in an organization and to educate staff members about potential social engineering assaults. Additionally, it may assist organizations in creating and putting into place efficient security policies and processes to guard against social engineering assaults.

An organization's comprehensive security testing approach should include social engineering testing. Social engineering testing focuses on human weaknesses rather than technical ones, which are the focus of more conventional security testing techniques like vulnerability scanning and penetration testing. This is crucial since a company's security system frequently has humans as its weakest link.

Attacks using social engineering can take many different shapes. An attacker may, for instance, send a convincing-looking email purporting to be from a reliable source, such a senior executive or a third-party vendor. The email may request sensitive data or login credentials from the receiver. Alternately, an attacker can try to enter the organization's premises by pretending to be a delivery or maintenance person.

During a social engineering test, a security team will use similar techniques to those used by real-world attackers to attempt to gain access to sensitive information or physical locations. For example, the team might send a phishing email to employees or attempt to tailgate behind an employee who is entering a secure area.

The results of a social engineering test can provide valuable insights into an organization's security posture. By identifying weaknesses in employee training or security policies and procedures, the organization can take steps to improve its overall security. In addition, social engineering testing can help raise employee awareness of the importance of security and help them to identify potential social engineering attacks in the future.

It's important to note that social engineering testing should be conducted ethically and legally. Testing should only be conducted with the permission of the organization being tested, and appropriate safeguards should be put in place to prevent any actual harm to the organization or its employees. In addition, any information obtained during the testing should be kept confidential and should only be shared with authorized personnel.

7. SOCIAL ENGINEERING AND THE DARK WEB

Social engineering and the dark web are two distinct concepts that are often associated with cybersecurity threats.

Social engineering is the practise of using psychological manipulation tactics to trick people into disclosing confidential information or doing activities that might compromise an organization's security. Techniques like phishing emails, pretexting, baiting, and tailgating are examples of this.

The dark web, on the other hand, refers to a portion of the internet that is not accessible through traditional search engines. It is often used by cybercriminals to buy and sell illegal goods and services, such as drugs, weapons, and stolen personal information. The dark web can also be used to exchange information and collaborate on cyber attacks.

The dark web is sometimes used by cybercriminals to obtain personal information that can be used in social engineering attacks. For example, an attacker might purchase a database of email addresses and passwords from the dark web and use this information to send phishing emails to unsuspecting individuals.

In addition, the dark web can be a source of tools and resources that can be used to launch social engineering attacks. For example, an attacker might purchase a pre-made phishing kit that includes templates for fake login pages, as well as instructions on how to use them to steal login credentials.

It's important to note that not all activity on the dark web is illegal or malicious. However, the anonymity provided by the dark web can make it an attractive platform for cybercriminals to carry out their activities. Organizations can take steps to protect themselves against social engineering and dark web threats by implementing strong security policies and procedures, training employees to identify and report suspicious activity, and monitoring for unusual network traffic.

8. SOCIAL ENGINEERING AND IOT DEVICES

Social engineering attacks can also target Internet of Things (IoT) devices, which are connected devices that are embedded with sensors and other technologies that allow them to exchange data with other devices and systems over the internet.

IoT devices can be vulnerable to social engineering attacks because they often have default or weak passwords and may not have proper security configurations. An attacker can use social engineering tactics to trick a user into revealing their login credentials or other sensitive information, which can then be used to access the IoT device and potentially compromise the user's privacy or security.

For example, an attacker might send a phishing email to a user, pretending to be from the manufacturer of an IoT device and asking the user to provide their login credentials to update the device's firmware. The attacker can then use these credentials to access the device and potentially control it or steal sensitive information.

Another example is when an attacker poses as a technical support representative and convinces the user to provide access to the IoT device. The attacker can then use this access to install malware or make unauthorized changes to the device.

It's crucial to adhere to recommended practises for protecting IoT devices, such as changing default passwords, keeping devices updated with the latest security patches, and limiting access to the devices, in order to defend against social engineering attacks that target IoT devices. Users should also be taught to spot suspicious emails and phone calls and report them, as well as to be wary of giving out login information or other sensitive information to anyone who asks for it, even if they seem to be a genuine technical support or device manufacturer representative.

9. LEGAL AND ETHICAL CONSIDERATIONS OF SOCIAL ENGINEERING

Social engineering can have significant legal and ethical implications, as it often involves deception, manipulation, and exploitation of human vulnerabilities. In many cases, social engineering tactics can be illegal, such as phishing, pretexting, or impersonation, which can violate privacy laws, intellectual property laws, and fraud laws.

Organizations and people should be aware of a number of legal issues related to social engineering. For instance, it is against the law to access computer systems without authorization or to use them for unauthorized purposes, according to statutes like the Computer Fraud and Abuse Act (CFAA) in the United States or the Computer Misuse Act in the United Kingdom. Similarly, legislation like the California Consumer Privacy Act (CCPA) in the United States and the General Data

Protection Regulation (GDPR) in the European Union mandate that organizations safeguard the privacy and security of personal data and acquire express authorization before using it.

From an ethical standpoint, social engineering can be controversial, as it involves manipulating people's trust and emotions for personal gain. While some forms of social engineering, such as security awareness training, can be used for legitimate purposes, others, such as phishing or pretexting, can be seen as unethical or even malicious. Moreover, social engineering can have unintended consequences, such as damaging relationships, violating trust, or causing psychological harm.

To ensure that social engineering is conducted in a legal and ethical manner, organizations and individuals should follow established ethical principles, such as honesty, integrity, and respect for privacy. They should also obtain informed consent from participants, use non-deceptive methods whenever possible, and minimize the risk of harm or negative impact. Additionally, organizations should have clear policies and procedures in place to govern social engineering activities, and individuals should seek legal advice if they are unsure about the legality or ethics of their actions.

10. FUTURE TRENDS IN SOCIAL ENGINEERING ATTACKS

Attackers are always developing new techniques to use human weaknesses to obtain sensitive data or systems through social engineering assaults. Here are some potential future developments in social engineering attacks:

AI-powered social engineering: As artificial intelligence (AI) technology continues to advance, attackers may use it to automate social engineering attacks, making them more targeted and personalized. AI-powered social engineering attacks could use machine learning algorithms to analyze social media profiles, emails, and other data sources to create more convincing and believable phishing messages.

Deepfake attacks: Deepfake technology can be used to create realistic videos or audio recordings of people saying or doing things that they never actually did. Attackers may use this technology to impersonate executives or other high-ranking individuals to gain access to sensitive information or systems.

Social engineering through IoT devices: As the Internet of Things (IoT) becomes more prevalent, attackers may use IoT devices, such as smart home devices or medical devices, as a way to gain access to other systems or networks. For example, attackers could use compromised IoT devices to launch phishing attacks or to gain access to other devices on the same network.

Social engineering in the context of remote work: With the rise of remote work, attackers may use social engineering tactics to exploit the increased use of online communication tools and platforms. Attackers may impersonate colleagues or clients to gain access to sensitive information or systems.

Psychological manipulation through social media: Social media platforms are increasingly being used as a way to influence people's opinions, beliefs, and behavior. Attackers may use social engineering tactics to create and spread fake news or misinformation, or to create fake social media accounts to manipulate public opinion.

To defend against these future trends in social engineering attacks, individuals and organizations need to be vigilant and proactive in identifying and mitigating social engineering threats. This includes implementing security best practices, such as using multi-factor authentication, keeping software and systems up-to-date, and providing security awareness training to employees. It is also important to stay informed about the latest social engineering tactics and to be aware of the potential risks associated with new technologies and trends.

11. ROLE OF TECHNOLOGY IN PREVENTING SOCIAL ENGINEERING ATTACKS

Technology plays a critical role in preventing social engineering attacks, as it provides a range of tools and solutions to help detect and prevent these types of attacks.

Multi-factor authentication (MFA), which requires users to give two or more forms of authentication before getting access to a system or application, is one of the most successful solutions for defeating social engineering attacks. Even if an attacker is successful in obtaining a user's password, they will not be able to access the system without the extra form of authentication, which can assist avoid attacks like phishing and pretexting.

Intrusion detection and prevention systems (IDPS), which monitor network traffic for suspicious behavior and can automatically block or notify administrators of possible attacks, are another technology that can be useful in preventing social engineering attacks. Attacks like baiting and quid pro quo, which include attackers seeking to acquire physical access to a system or facility, can be detected by IDPS.

Endpoint security solutions, such as anti-virus and anti-malware software, can also be effective in preventing social engineering attacks. These solutions can help detect and remove malicious software that may be installed on a user's device as a result of a social engineering attack.

Additionally, social engineering attempts are being detected and prevented using machine learning (ML) and artificial intelligence (AI). In addition to using AI and ML algorithms to identify and block questionable emails, texts, and other communications, they may also be used to analyze enormous volumes of data for patterns and abnormalities that may suggest an attack.

However, while technology can be effective in preventing social engineering attacks, it is not a silver bullet. Cybercriminals are constantly evolving their tactics and techniques, and there is no single technology or solution that can prevent all types of social engineering attacks. As such, education and awareness remain critical components of any effective defense against social engineering attacks, as individuals who are aware of the risks and able to recognize and respond to these types of attacks can be the first line of defense against them.

12. COUNTERMEASURES FOR SOCIAL ENGINEERING ATTACKS

There are several countermeasures that organizations can implement to prevent and mitigate the risks of social engineering attacks. Here are some effective countermeasures:

Security Awareness Training: Security Awareness Training refers to the process of educating individuals about the importance of cybersecurity, the potential risks of cyber threats, and best practices to prevent or mitigate them. The goal of such training is to improve the overall security posture of an organization by ensuring that employees are aware of the threats they face and have the knowledge and skills to identify and respond appropriately to those threats.

Security Awareness Training typically includes topics such as password hygiene, phishing attacks, malware, social engineering, and data protection. The training can be delivered in various formats, including online courses, webinars, videos, interactive simulations, and in-person workshops.

Organizations should conduct Security Awareness Training regularly, as cyber threats continue to evolve, and new attack techniques are developed. Additionally, it is essential to tailor the training to the specific roles and responsibilities of different employees and to reinforce the training with periodic assessments to measure its effectiveness.

Effective Security Awareness Training can help organizations reduce the risk of cyber attacks, improve their security posture, and safeguard their sensitive information and assets.

Access Controls: Access Controls refer to the measures put in place to restrict or regulate access to resources or information in a system or organization. These controls are intended to prevent unauthorized access, modification, disclosure, or destruction of sensitive or critical data, applications, or systems.

Access Controls can be classified into three main categories: physical, technical, and administrative controls.

1. **Physical Access Controls**: These controls restrict physical access to facilities, equipment, and resources. Examples of physical controls include security guards, locks, biometric scanners, access cards, and surveillance cameras.
2. **Technical Access Controls:** These controls are implemented through software or hardware to restrict access to digital resources. Examples of technical controls include passwords, authentication mechanisms, firewalls, intrusion detection systems, encryption, and virtual private networks.
3. **Administrative Access Controls:** These controls are policies and procedures that govern access to resources. Examples of administrative controls include role-based access control, access approval processes, separation of duties, and periodic access reviews.

Access Controls are crucial to maintaining the confidentiality, integrity, and availability of resources and information. Implementing strong access controls requires a risk-based approach that considers the sensitivity of the data or resources being protected, the potential threats, and the likelihood and impact of a security breach. Access Controls should be regularly reviewed, tested, and updated to ensure their effectiveness in mitigating risks and meeting compliance requirements.

Encryption: To prevent unauthorized access or interception, clear, readable data or communications are transformed into a coded or incomprehensible form through the process of encryption. Sensitive data's confidentiality and integrity are ensured through encryption, which makes it impossible for unauthorized users to access and read the data. An algorithm and a key are often used in the encryption process to convert plaintext data into ciphertext.

There are two primary types of encryption: symmetric encryption and asymmetric encryption.

1. **Symmetric Encryption:** The same key is utilized in symmetric encryption for both encryption and decryption. To decrypt and read the message, both the sender and the recipient must have access to the same secret key. Advanced Encryption Standard (AES), Data Encryption Standard (DES), and Triple DES (3DES) are a few examples of symmetric encryption methods (Srivastava et al., 2022).
2. **Asymmetric Encryption:** A pair of keys—a public key and a private key—is used in asymmetric encryption. Anyone who needs to transmit an encrypted communication is given access to the public key, while the owner keeps the private key a secret. Only the appropriate private key may be used to decode

messages that have been encrypted with the public key. Digital signatures, secure email, and safe online surfing all employ asymmetric encryption. The RSA and Elliptic Curve Cryptography (ECC) encryption methods are examples of asymmetric encryption algorithms (Srivastava et al., 2022).

Encryption is an essential component of data security and is widely used in various applications, including online banking, e-commerce, messaging apps, and file sharing. However, encryption is not foolproof, and it can be compromised by attackers with sufficient resources and skills. Therefore, encryption must be complemented with other security measures, such as access controls, firewalls, intrusion detection, and response systems, to provide comprehensive protection against cyber threats.

Physical Security Measures: Physical Security Measures refer to the measures put in place to protect physical assets, facilities, and personnel from unauthorized access, theft, damage, or harm. Physical security is a critical component of a comprehensive security program and is essential to safeguarding an organization's operations and assets.

Physical security measures can be broadly classified into three main categories: deterrent, preventive, and detective measures.

1. **Deterrent Measures:** These measures are intended to discourage unauthorized access or criminal activity by creating a perception of risk or increasing the difficulty of gaining access. Examples of deterrent measures include security cameras, warning signs, perimeter fencing, and security patrols.
2. **Preventive Measures:** These measures aim to physically prevent unauthorized access or mitigate the impact of an incident. Examples of preventive measures include access controls, biometric identification systems, intrusion detection systems, fire suppression systems, and environmental controls.
3. **Detective Measures**: These measures are designed to detect and alert security personnel of an intrusion or suspicious activity. Examples of detective measures include motion sensors, alarm systems, video analytics, and security personnel.

Physical security measures should be designed based on a risk-based approach that considers the likelihood and impact of potential security incidents, the criticality of the assets or facilities, and the budget and resource constraints. Physical security measures should be periodically tested, reviewed, and updated to ensure their effectiveness and compliance with applicable regulations and standards.

Effective physical security measures can help organizations prevent or minimize security incidents, reduce losses, and maintain the safety and well-being of their personnel and customers.

Security Audits: Security Audits refer to the process of evaluating an organization's security controls, policies, and procedures to identify vulnerabilities, gaps, and non-compliance with applicable regulations or standards. Security audits can be conducted by internal or external auditors and may cover various aspects of security, including physical, technical, and administrative controls.

Security audits typically involve several steps, including planning, conducting fieldwork, analyzing findings, and issuing a report. During the audit, auditors may review policies and procedures, conduct interviews with personnel, assess the effectiveness of security controls, and perform vulnerability assessments and penetration testing.

Security audits can help organizations identify weaknesses in their security posture and prioritize remediation efforts to mitigate risks. Security audits can also provide assurance to stakeholders, such as customers, partners, and regulators, that the organization is taking appropriate measures to safeguard its assets and data.

However, security audits are not a one-time event but rather an ongoing process that requires continuous monitoring and improvement. Security audits should be conducted periodically or when significant changes occur, such as new technology implementations, business expansion, or regulatory changes.

Security Audits are critical for organizations to ensure that their security controls are effective, meet regulatory requirements, and align with industry best practices. Security audits help organizations maintain a robust security posture, minimize risks, and build trust with stakeholders.

Incident Response Plan: An Incident Response Plan (IRP) is a documented set of procedures and guidelines that outlines the steps an organization must follow to detect, contain, and recover from a security incident. The purpose of an IRP is to minimize the impact of a security incident on an organization's operations, reputation, and financial stability.

An effective IRP should include the following key components:

1. **Preparation:** This phase involves identifying potential security incidents, defining roles and responsibilities, establishing communication protocols, and documenting procedures and guidelines. The preparation phase should also include regular training and testing of the IRP to ensure its effectiveness and compliance with applicable regulations and standards.
2. **Detection and Analysis:** This phase involves monitoring and detecting security incidents, assessing the impact and severity of the incident, and initiating a response. The detection and analysis phase should also involve collecting and preserving evidence for forensic analysis and reporting.
3. **Containment, Eradication, and Recovery**: This phase involves containing the incident to prevent further damage, eradicating the root cause of the incident,

and recovering affected systems and data. The containment, eradication, and recovery phase should also include validating the effectiveness of the response and restoring normal operations.

4. **Post-Incident Activities:** This phase involves analyzing the incident and response, identifying lessons learned, and implementing corrective actions and improvements to prevent future incidents. The post-incident activities should also include reporting the incident to relevant stakeholders, such as customers, partners, and regulators, and maintaining a record of the incident and response.

An effective IRP should be customized to the organization's specific needs, risk profile, and regulatory requirements. The IRP should be regularly reviewed and updated to reflect changes in the organization's operations, technology, and threat landscape.

An Incident Response Plan is a critical component of a comprehensive security program and can help organizations minimize the impact of security incidents, maintain business continuity, and build trust with stakeholders.

Employee Verification: Employee Verification is the process of verifying the identity and credentials of a job applicant or an existing employee. Employee verification is an essential component of the hiring process and is critical to ensuring that the organization hires qualified and trustworthy employees who do not pose a security or safety risk.

Employee verification can involve several steps, including:

1. **Background Checks:** This involves verifying an applicant's employment history, education credentials, criminal history, credit history, and other relevant information. Background checks can be conducted by the organization's HR department or by a third-party vendor.

2. **Reference Checks:** This involves contacting an applicant's previous employers, colleagues, or personal references to obtain information about the applicant's work ethic, job performance, and character.

3. **Identity Verification: This** involves verifying an applicant's identity using government-issued identification documents, such as a passport, driver's license, or national ID card.

4. **Drug Screening**: This involves testing an applicant for drug use, typically using urine or blood samples. Drug screening is especially important for positions that involve safety or security risks.

5. **Security Clearance**: This involves conducting a thorough background check and investigation to grant an employee access to sensitive or classified information. Security clearance is typically required for government positions or positions

in industries that deal with sensitive information, such as defense, intelligence, or finance.

Employee verification should be conducted consistently and fairly for all job applicants and employees. The organization should also comply with applicable laws and regulations, such as the Fair Credit Reporting Act (FCRA) and the Equal Employment Opportunity Commission (EEOC) guidelines.

Employee Verification is a critical component of the hiring process and can help organizations hire qualified and trustworthy employees, minimize security and safety risks, and comply with applicable laws and regulations.

Email Filters: Email filters are software programs that automatically sort and organize incoming email messages based on predefined criteria. Email filters are designed to help users manage their email inbox, reduce clutter, and prioritize important messages.

Email filters can be set up in various ways, including:

1. **Whitelisting**: This involves adding trusted email addresses or domains to a whitelist to ensure that emails from those sources are always delivered to the inbox and not marked as spam.
2. **Blacklisting:** This involves adding email addresses or domains to a blacklist to ensure that emails from those sources are always marked as spam and not delivered to the inbox.
3. **Content Filtering:** This involves analyzing the content of the email message, such as keywords, attachments, and links, to determine if it meets predefined criteria. Content filtering can be used to identify and block spam, phishing emails, and malware.
4. **Sender Reputation Filtering:** This involves checking the sender's reputation based on their email history and behavior, such as the frequency of sending emails, the volume of emails sent, and the use of spammy tactics. Sender reputation filtering can be used to block emails from known spammers and prevent phishing attacks.

Email filters can be set up in email clients, such as Microsoft Outlook or Gmail, or in email server software, such as Microsoft Exchange or Postfix. Email filters can also be configured by IT administrators to apply to an entire organization's email system.

Email filters can help users manage their email inbox efficiently, reduce the risk of phishing attacks and malware infections, and improve productivity. However, users should be aware that email filters are not perfect and may occasionally mark legitimate emails as spam or fail to catch malicious emails.

Strict Information Sharing Policies: Strict information sharing policies refer to a set of rules and procedures that govern how information is shared and accessed within an organization or between organizations. These policies are designed to protect sensitive information from unauthorized access, theft, or misuse.

Information sharing policies typically include the following elements:

- **Data Classification**: This involves categorizing information based on its sensitivity, value, and confidentiality. Different levels of access are then granted based on the data classification.
- **Access Controls:** This involves implementing measures to restrict access to sensitive information to authorized personnel only. Access controls can include passwords, biometric authentication, and physical security measures.
- **Encryption:** This involves using encryption techniques to protect sensitive data while in transit or at rest. Encryption can prevent unauthorized access to sensitive data by encrypting it before it is transmitted or stored.
- **Need-to-know Basis:** This involves limiting access to sensitive information to only those individuals who have a legitimate need to know. This can help reduce the risk of data breaches and minimize the potential impact of a security incident.
- **Non-Disclosure Agreements:** This involves requiring employees, contractors, and partners to sign non-disclosure agreements that prohibit them from sharing sensitive information with unauthorized individuals.
- **Monitoring and Auditing:** This involves monitoring and auditing access to sensitive information to ensure that it is only accessed by authorized personnel and to detect any unauthorized access or misuse.

Strict information sharing policies are important for protecting sensitive information and minimizing the risk of security incidents. Organizations should regularly review and update their information sharing policies to ensure that they are effective and up-to-date with the latest security threats and best practices. Additionally, organizations should provide regular training and awareness programs for their employees to ensure that they understand the policies and procedures for handling sensitive information.

Vulnerability Management: Identification, prioritization, and mitigation of security vulnerabilities in computer systems, networks, and software applications is known as vulnerability management. By locating and patching vulnerabilities before they may be used by attackers, vulnerability management seeks to lower the risk of a security breach.

The vulnerability management process typically includes the following steps:

- **Vulnerability Scanning:** This involves using software tools to scan networks and systems for known vulnerabilities. The results of the scan are typically presented in a report that includes a list of vulnerabilities and their severity levels.
- **Vulnerability Assessment:** This involves analyzing the results of the vulnerability scan to determine which vulnerabilities pose the greatest risk to the organization. The assessment takes into account factors such as the likelihood of the vulnerability being exploited, the potential impact of an attack, and the difficulty of remediation.
- **Risk Prioritization:** This involves prioritizing the identified vulnerabilities based on their severity and the potential impact they could have on the organization. High-severity vulnerabilities are typically addressed first, followed by medium- and low-severity vulnerabilities.
- **Remediation:** This involves fixing or mitigating the identified vulnerabilities. Remediation can include applying software patches, configuring settings, or upgrading software applications or hardware.
- **Verification:** This involves verifying that the remediation efforts have been successful in addressing the identified vulnerabilities. Verification can include conducting follow-up scans, testing systems and applications, and monitoring for any signs of exploitation.

Vulnerability management is an important part of an organization's overall security strategy. It helps to identify and prioritize vulnerabilities, which can reduce the risk of a security breach and minimize the potential impact of an attack. Regular vulnerability scanning and assessment, combined with effective remediation efforts, can help organizations stay ahead of emerging security threats and maintain the confidentiality, integrity, and availability of their systems and data.

Penetration Testing: A simulated cyberattack on the systems, applications, or network infrastructure of an organization is called penetration testing, or pen testing. A pen test's objectives are to find weaknesses that a hostile attacker may exploit and to evaluate how well the organization's security procedures are working (Jayasuryapal et al., 2021).

A typical penetration testing process involves the following steps:

- Planning and Scoping: This involves defining the scope and objectives of the pen test, determining the systems and applications to be tested, and obtaining permission from the organization to conduct the test.
- Reconnaissance: This involves gathering information about the target systems, applications, and network infrastructure that could be used to launch an attack.

- Vulnerability Scanning: This involves using software tools to scan for known vulnerabilities in the target systems and applications.
- Exploitation: This involves attempting to exploit the identified vulnerabilities to gain unauthorized access to the target systems and applications.
- Post-Exploitation: This involves testing the effectiveness of the organization's security controls by attempting to escalate privileges, move laterally within the network, or exfiltrate data.
- Reporting: This involves documenting the results of the pen test, including any vulnerabilities that were identified and recommendations for improving the organization's security posture.

Penetration testing is an important part of an organization's overall security strategy. It helps to identify vulnerabilities and weaknesses that could be exploited by attackers, and provides insight into the effectiveness of the organization's security controls. Regular penetration testing can help organizations stay ahead of emerging threats and minimize the risk of a security breach. However, it's important to conduct pen tests in a controlled environment and to obtain permission from the organization before conducting any testing.

Third-Party Risk Management: Third-party risk management (TPRM) is the process of assessing and managing the risks associated with engaging with third-party vendors, suppliers, or service providers. The goal of TPRM is to identify and mitigate potential risks that could impact an organization's operations, reputation, or security.

The TPRM process typically includes the following steps:

- **Risk Assessment:** This involves identifying the third-party vendors, suppliers, or service providers that have access to the organization's sensitive data or systems, and assessing the level of risk associated with each one. This assessment may include reviewing the third party's security controls, policies, and procedures, and evaluating their track record in terms of security incidents.
- **Due Diligence:** This involves conducting a deeper investigation of the third party's security controls, policies, and procedures, to determine their effectiveness and ensure that they meet the organization's standards and requirements.
- **Contract Negotiation**: This involves negotiating contractual terms that address security requirements, responsibilities, and liabilities, as well as outlining the process for incident reporting and remediation.
- **Ongoing Monitoring:** This involves continuously monitoring the third-party vendor's security posture and performance, to ensure that they are complying

with the contractual terms and meeting the organization's security standards. This may include conducting regular audits and assessments, and reviewing incident reports and security updates.

- **Incident Response**: This involves establishing procedures for responding to security incidents involving third-party vendors, suppliers, or service providers, and communicating with stakeholders, including customers, partners, and regulators.

Effective TPRM helps organizations to identify and manage the risks associated with engaging with third-party vendors, suppliers, or service providers. By conducting due diligence and ongoing monitoring, organizations can ensure that their third-party partners are implementing adequate security controls and complying with contractual terms. This can help to minimize the risk of a security breach, protect the organization's reputation, and maintain the trust of customers and partners.

By implementing these countermeasures, organizations can reduce the risk of social engineering attacks and protect their sensitive information and systems from being compromised. However, it's important to note that social engineering attacks are constantly evolving, and organizations must remain vigilant and adapt their security strategies accordingly.

13. CASE STUDIES OF SOCIAL ENGINEERING ATTACKS

Case studies of social engineering attacks provide insight into how attackers have used various techniques to exploit human vulnerabilities and gain access to sensitive information or systems. Here are a few examples:

The MySpace Worm: In 2005, a hacker named Samy Kamkar created a worm that was able to spread across the social networking site MySpace. The worm exploited a vulnerability in MySpace's code and was able to add thousands of friends to Kamkar's profile. The attack was an example of a type of social engineering known as "clickjacking," in which attackers trick users into clicking on a link or button that performs a hidden action(Grossman, 2007).

The Target Data Breach: In 2013, the retail giant Target suffered a massive data breach in which hackers stole the personal and financial information of over 40 million customers. The breach was the result of a phishing attack in which attackers sent an email to employees of an HVAC company that was a vendor for Target. The email contained a malware-infected attachment that, when opened, gave the attackers access to Target's network(Manworren et al., 2016).

The Ubiquiti Networks Phishing Scam: In 2015, the networking equipment manufacturer Ubiquiti Networks suffered a phishing attack in which attackers stole

$46.7 million from the company. The attackers used social engineering tactics to trick employees into transferring funds to bank accounts controlled by the attackers. The attackers used fake emails and websites that appeared to be from legitimate vendors and employees to carry out the attack(Bakarich & Baranek, 2020).

The Carbanak Cybercrime Group: In 2015, a group of cybercriminals known as Carbanak carried out a series of attacks on banks and financial institutions around the world. The attacks involved a combination of social engineering tactics and malware to steal hundreds of millions of dollars. The attackers used phishing emails to gain access to the banks' networks and then used malware to steal credentials and carry out fraudulent transactions(Park et al., 2015).

The Spear-Phishing Attack on the Democratic National Committee: In 2016, hackers believed to be affiliated with the Russian government carried out a spear-phishing attack on the Democratic National Committee (DNC). The attackers sent phishing emails to DNC employees that appeared to be from Google, asking them to change their passwords. When employees clicked on the link in the email, they were taken to a fake website where they entered their login credentials, which the attackers then used to gain access to the DNC's systems(NCCIC, 2016).

The Bangladesh Bank Heist: In February 2016, cybercriminals attempted to steal $1 billion from the Bangladesh Bank by transferring funds from the bank's account at the Federal Reserve Bank of New York. The attackers used a combination of social engineering tactics and malware to gain access to the bank's systems and carry out the attack. They used stolen credentials to access the bank's SWIFT messaging system, which they used to make the fraudulent transfers(Part et al., 2017).

The Social Engineering Attack on Equifax: In 2017, credit reporting agency Equifax suffered a massive data breach in which hackers stole the personal and financial information of over 143 million customers. The breach was the result of a social engineering attack in which the attackers were able to exploit a vulnerability in Equifax's website to gain access to sensitive data. The attackers were able to trick Equifax employees into clicking on a malicious link, which allowed them to install malware on the company's servers and steal the data(Drenick, 2017).

The SolarWinds Supply Chain Attack: In December 2020, it was discovered that a state-sponsored hacking group had carried out a sophisticated supply chain attack against SolarWinds, a company that provides software used by numerous government agencies and Fortune 500 companies. The attackers were able to gain access to SolarWinds' systems by compromising the company's software build process, and then used that access to distribute malware to SolarWinds' customers. The attack involved a combination of social engineering and technical tactics, and has been described as one of the most significant cyber attacks in recent years(Report, 2021).

The Twitter Bitcoin Scam: In July 2020, hackers gained access to the Twitter accounts of several high-profile individuals, including Elon Musk, Joe Biden, and Barack Obama, and used them to promote a bitcoin scam. The attackers used social engineering tactics to trick Twitter employees into giving them access to the company's internal tools, which they used to take control of the accounts(Witman & Mackelprang, 2022).

The Colonial Pipeline Ransomware Attack: In May 2021, a ransomware attack against the Colonial Pipeline, which supplies nearly half of the fuel to the East Coast of the United States, caused widespread disruptions and panic buying. The attack was carried out by a cybercriminal group known as DarkSide, which used a combination of technical and social engineering tactics to gain access to Colonial Pipeline's systems and encrypt their data. The group then demanded a ransom payment in exchange for the decryption key, which Colonial Pipeline ultimately paid(Yorke, 2010).

COVID-19 Scams: During the COVID-19 pandemic, there have been numerous reports of social engineering attacks that exploit people's fears and concerns about the virus. Scammers have used a variety of tactics, such as phishing emails, text messages, and phone calls, to trick people into revealing personal information, downloading malware, or making fraudulent payments.

For example, scammers have sent emails or text messages claiming to be from health authorities, offering information about the pandemic or links to fake COVID-19 testing sites. When people click on these links, they are redirected to a website that looks like a legitimate health authority but is actually controlled by the scammers. The website may ask people to enter their personal information, such as their name, date of birth, and credit card details, which the scammers can then use for identity theft or financial fraud.

Other COVID-19 scams have involved fake job offers, fake charities seeking donations for COVID-19 relief efforts, and fake government stimulus payments. These scams can be very convincing and can prey on people's vulnerability during a difficult time. It is important to be cautious and skeptical of unsolicited messages, and to verify the authenticity of any information or offers related to COVID-19 before taking action.

14. CONCLUSION

Social engineering attacks represent a serious threat to the security of organizations and individuals alike. These attacks rely on exploiting human weaknesses rather than technical vulnerabilities, making them difficult to detect and prevent. By understanding the different types of social engineering attacks and the tactics used by attackers, individuals and organizations can take steps to protect themselves against these threats. Education and awareness are key components of any effective defense strategy, and technical solutions can also be implemented to help mitigate the risk of social engineering attacks.

Ultimately, preventing social engineering attacks requires a multi-pronged approach that includes both education and technical solutions. Organizations must invest in training programs that teach employees to recognize and respond appropriately to social engineering attacks, as well as implementing technical measures such as spam filters and multi-factor authentication to make it more difficult for attackers to succeed.

It is also important for individuals to remain vigilant and skeptical when it comes to requests for sensitive information or access. By adopting a cautious approach to such requests, individuals can help to protect themselves and their organizations against the threat of social engineering attacks.

In conclusion, social engineering attacks are a significant and growing threat to organizations and individuals. However, by understanding the different types of attacks and the tactics used by attackers, and by implementing a combination of education and technical solutions, it is possible to reduce the risk of falling victim to these attacks. With continued vigilance and awareness, we can all work together to stay one step ahead of the social engineers and protect our sensitive information and systems from harm.

REFERENCES

Atmojo, Y. P., Susila, I. M. D., Hilmi, M. R., Rini, E. S., Yuningsih, L., & Hostiadi, D. P. (2021). A New Approach for Spear phishing Detection. *3rd 2021 East Indonesia Conference on Computer and Information Technology, EIConCIT 2021,* (pp. 49–54). ACM. 10.1109/EIConCIT50028.2021.9431890

Bakarich, K. M., & Baranek, D. (2020). Something phish-y is going on here: A teaching case on business email compromise. *Current Issues in Auditing, 14*(1), A1–A9. doi:10.2308/ciia-52706

Bishnoi, A. & Gupta, N. (2023). Comprehensive Assessment of Reverse Social Engineering to Understand Social Engineering Attacks. Icssit. doi:10.1109/ICSSIT55814.2023.10061054

Del Pozo, I., Iturralde, M., & Restrepo, F. (2018). Social engineering: Application of psychology to information security. *Proceedings - 2018 IEEE 6th International Conference on Future Internet of Things and Cloud Workshops, W-FiCloud 2018,* (pp. 108–114). IEEE. 10.1109/W-FiCloud.2018.00023

Drenick, A. H. (2017). *The 2017 Equifax Hack: What We Can Learn.*

Girinoto, P. D. F., Yulita, T., Zulkham, R. K., Rifqi, A., & Putri, A. (2022). OmeTV Pretexting Phishing Attacks: A Case Study of Social Engineering. *IWBIS 2022 - 7th International Workshop on Big Data and Information Security, Proceedings*, (pp. 119–124). IEEE. 10.1109/IWBIS56557.2022.9924801

Grossman, J. (2007). *Cross-Site Scripting Worms & Viruses - The Impending Threat & thee Best Defense.* White Hat Sec. https://www.whitehatsec.com/assets/WP5CSS0607.pdf

Gupta, S., Singhal, A., & Kapoor, A. (2016). *A Literature Survey on Social Engineering Attacks : Phishing Attack.* Semantic Scholar.

Gupta, S., & Isha, B. A., & Gupta, H. (2021). Analysis of Social Engineering Attack on Cryptographic Algorithm. *2021 9th International Conference on Reliability, Infocom Technologies and Optimization (Trends and Future Directions), ICRITO 2021*, (pp. 1–5). IEEE. 10.1109/ICRITO51393.2021.9596568

JajooA. (2021). *A study on the Morris Worm.* https://arxiv.org/abs/2112.07647

Jayasuryapal, G., Pranay, P. M., & Kaur, H., & Swati. (2021). A Survey on Network Penetration Testing. *Proceedings of 2021 2nd International Conference on Intelligent Engineering and Management, ICIEM 2021*, (pp. 373–378). IEEE. 10.1109/ICIEM51511.2021.9445321

Li, T., Wang, K., & Horkoff, J. (2019). Towards effective assessment for social engineering attacks. *Proceedings of the IEEE International Conference on Requirements Engineering, 2019*, (pp. 392–397). IEEE. 10.1109/RE.2019.00051

Manworren, N., Letwat, J., & Daily, O. (2016). Why you should care about the Target data breach. *Business Horizons, 59*(3), 257–266. doi:10.1016/j.bushor.2016.01.002

NCCIC. (2016). *Grizzly Steppe - E – Russian Malicious Cyber Activity Summary.* NCCIC.

Park, Y., Mccoy, D., Shi, E., Tapia, M. G., Texas, A., Tyagi, A. K., Aghila, G., Hendriksen, H., Andriesse, D., Rossow, C., Stone-Gross, B., Plohmann, D., Bos, H., December, I., Vidros, S., Kolias, C., Kambourakis, G., Coletta, A., Van Der Veen, V., ... Brinkmann, U. (2015). Carbanak APT The Great Bank Robbery. *Computer Fraud & Security, 2015*(June), 1–5. http://www.trendmicro.com/vinfo/us/security/special-report/cybercriminal-underground-economy-series/%5Cnhttp://www.tandfonline.com/doi/full/10.1080/17440572.2016.1157480%5Cnhttp://www.sophos.com/threatreport%5Cnhttp://security-sh3ll.blogspot.com/2011/05/what-is-ze

Part, B. H., Quadir, B. Y. S., Aneez, S., Bergin, T. O. M., Layne, N., Das, K. N., & Spicer, J. (2017). THE The Bangladesh Bank Heist typo helped. *Journal of Consumer and Commercial Law.*

Srivastava, S., Tiwari, A., & Srivastava, P. K. (2022). Review on quantum safe algorithms based on Symmetric Key and Asymmetric Key Encryption methods. *2022 2nd International Conference on Advance Computing and Innovative Technologies in Engineering, ICACITE 2022*, (pp. 905–908). IEEE. 10.1109/ICACITE53722.2022.9823437

Witman, P. D., & Mackelprang, S. (2022). The 2020 Twitter Hack – So Many Lessons to Be Learned. *Journal of Cybersecurity Education, Research and Practice, 2*(2), 1–13. https://digitalcommons.kennesaw.edu/jcerpAvailableat:https://digitalcommons.kennesaw.edu/jcerp/vol2021/iss2/2

Yorke, C. (2010). CYBERSECURITY AND SOCIETY: Bigsociety.com. *The World Today, 66*(12), 19–21. https://www.jstor.org/stable/41963033

Chapter 14
Technological Trends and Recent Statistics of Dark Web

Kamna Solanki
Maharshi Dayanand University, Rohtak, India

Sandeep Dalal
Maharshi Dayanand University, Rohtak, India

ABSTRACT

The depth of the Internet extends well beyond the surface information that many people may quickly access in their routine searches. Some people may think of the web as only being made up of webpages that can be found using conventional search engines like Google. This information, referred to as the "Surface web," represents a very small percentage of the entire internet. The part of the internet that search engines and web crawlers do not index is known as the deep web. On the other hand, a subset of the deep web known as the "dark web" is only accessible using specialized software like Tor (The Onion Router). The surface web is primarily used for acceptable daily online activity, while the dark web is purely anonymous and is known for carrying out illicit transactions. The dark web is a small part of the deep web which can be accessed through the Tor browser. This chapter aims to examine current technology developments and some intriguing recent dark web statistics to evaluate the dark web's present state, technologies, usage, and current trends and data breaches.

DOI: 10.4018/978-1-6684-8218-6.ch014

INTRODUCTION

A portion of the World Wide Web, referred to as the "Surface Web," is easily accessible to the general public and can be explored using popular search engines. It is also known as the "Surface/Visible Web" and represents the part of the Internet that search engines currently index. Surface web or open websites are accessible by regular users without requiring specialized browsers or software like Tor. Search engines play a vital role in indexing surface web pages, making it convenient to locate them. Many widely recognized websites with .com, .net, and .org domains can be found on the surface web. It is worth noting that while the surface web accounts for only around 4% of the information available on the Internet (as depicted in Figure 1), the deep Web encompasses the rest. The deep Web is a segment of the World Wide Web whose contents are not indexed by popular search engines, often due to various reasons. Within the deep Web, you can find commonly used applications such as webmail, online banking services, pay-per-use platforms like video on demand, select online journals, newspapers, and numerous others. The content within the deep Web remains concealed behind HTTP forms. It can be accessed through direct URLs or I.P. addresses, albeit sometimes requiring passwords or other security measures to access specific pages on public websites. Nearly 90% of websites are on the deep Web, and many are used by businesses, governmental organizations, and nonprofitable agencies, as shown below in Figure 2.

The phrase "dark web" denotes a stratum of information and web pages exclusively reachable via overlay networks, which are networks built upon the standard internet and obscure accessibility. Due to its encryption and anonymous hosting of most

Figure 1. Difference between various types of webs

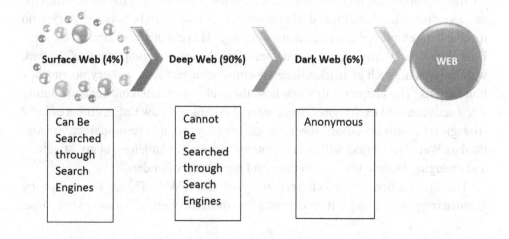

Figure 2. Three layers of the internet
(Bahri, A., 2020)

pages, accessing the Dark Web necessitates specialized software. The dark Web takes up 6% of all web space. Most of the terms used on the dark Web relate to illicit activities, although it is entirely legal.

The chapter's objective is to introduce readers to the concept of the dark Web, explaining its fundamental characteristics and distinguishing it from the surface web. It will set the stage for a deeper exploration of the topic. Another objective is to delve into the technologies and protocols that enable the functioning of the dark Web. This includes discussing Tor, I2P, and other anonymization tools and exploring encryption mechanisms employed to maintain user privacy. This chapter also aims to examine Dark Web Ecosystem by focusing on unraveling the various components of the dark web ecosystem, such as marketplaces, forums, chat services, and cryptocurrency transactions. The chapter will show how these elements interconnect, facilitating illicit activities and exchanges. The chapter also explores Law Enforcement efforts/ strategies and tools law enforcement agencies employ to combat criminal activities on the dark Web. The chapter will discuss notable successes, challenges of investigators, and emerging techniques used to trace and apprehend offenders.

The last section of the chapter analyzes Dark Web Threat Landscape by summarizing recent major threats present on the dark Web, including cybercrime, hacking, identity theft, drug trafficking, and extremist content dissemination. It aims to understand the risks associated with the dark Web comprehensively.

1. WEB ACCESS METHODS OF DARK WEB

- **Tor (The Onion Router)**

The dark Web is a small part of the Deep Web which can be accessed through the Tor browser. Statistics show that only 6.7% of users use Tor for unlawful purposes, such as spreading malware, sharing child exploitation content, or purchasing and selling illegal products. Although Tor is entirely legal, most typical internet users will never need to view information on the dark Web. Freenet, an innovative project initiated by University of Edinburgh student Ian Clarke, revolutionized the concept of information storage and retrieval with its "Distributed Decentralized Information Storage and Retrieval System." This groundbreaking endeavor, initiated in 2000, is widely regarded as the catalyst for the emergence of the dark Web—Clarke's visionary pursuit aimed at creating a novel platform for secure file sharing and anonymous online communication. In 2002, the fruition of Clarke's efforts led to the launch of "The Tor Project," a pivotal milestone in the dark Web's evolution. Building upon Freenet's principles, this project introduced the Tor browser in 2008, facilitating anonymous internet access on an unprecedented scale. As a result, users gained the ability to explore the depths of the dark Web while ensuring their online activities remained concealed. The dark Web has developed into a center for individuals who want to maintain their anonymity worldwide. The U.S. Department of Defense was the first to use it for anonymous communication. Internet users use the dark Web for both legitimate and illegal purposes. You require specialized software, setups, or authorization to access the dark Web. It uses hidden I.P. addresses, and only one web browser may access them. Tor, I2P, Freenet, etc., as examples, as shown in Figure 3 below.

Figure 3. Web access methods of dark web

- **I2P**

The Invisible Internet Project (I2P) is a dark web that prioritizes its function as an anonymity network rather than routing traffic through the open Internet. It differs from Tor in this way, but that isn't the only way. Tor is based on Onion Routing, whereas I2P is based on Garlic Routing, as shown in Figure 4 and Figure 5. The legitimate application of "The dark web" is present in nations where political dissidents may be monitored by the government and subjected to repression. Several dark web sites, including Silk Road, AlphaBay, and Hansa that carried illegal content have been taken down by government authorities lately. Due to the dark Web's anonymity, several data breaches and cybersecurity dangers have occurred during the past two decades. Like any technological advancement, Tor is a tool that can be employed for good or bad. Considering the success of Tor's adoption, studies on its use are challenging. Users who access these services also have their anonymity firmly safeguarded, and onion services are well-hidden unless made public someplace on the open Web.

- **Freenet**

Both TOR and I2P contain hidden servers, enabling connections in which neither the client nor the server is aware of the identity of the other. This indicates that the server cannot be located, making removing or destroying the server's data more challenging. These frequently—but not consistently—contain unlawful material, such as forums containing details on weapons or child pornography. Freenet, on the other hand, is not an anonymous network. The Freenet cannot be used to access the Internet. Freenet users can view the files submitted to it, nevertheless. The Freenet has no servers; all files are hosted (in an encrypted state by users. This indicates that the server cannot be located, making removing or destroying the server's data more challenging. As a result, files cannot be deleted. There is no delete button. Therefore, access to those files cannot be removed by shutting down a server. Although the Freenet frequently contains illicit content, many other files are also kept there by users who take advantage of this capability. Since no one can delete their data, they utilize the Freenet, in essence, as a free cloud drive.

- **Things to remember for maintaining anonymity on Dark Web**

    i.   Never reveal your identity (real name and photos).
    ii. Always utilize anonymous accounts, unique passwords, and fictitious names to prevent being discovered.
    iii. When transacting, always utilize a personal Bitcoin dark wallet.

Figure 4. Working of Tor
(*Temel M.H., 2019*)

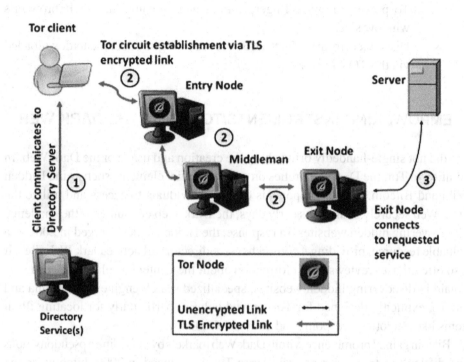

Figure 5. Difference between onion and garlic routing

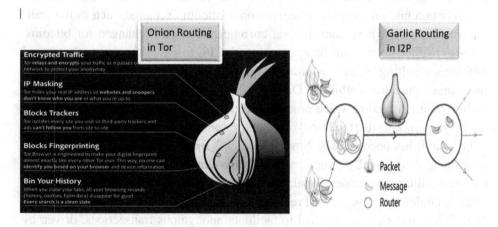

iv. Avoid adding plugins to your browser, and turn off JavaScript and cookies to prevent the exposure of your I.P. address.

v. To prevent browser fingerprinting, never change the TOR browser's window size.

vi. Disconnect from the Internet before accessing any documents downloaded via the TOR browser.

2. EMPOWERING INSTRUMENTS/TOOLS OF THE DARK WEB

Tor did not single-handedly bring about the creation and usage of the Dark Web. In addition to Tor, the Dark Web relies on other essential elements such as the Hidden Wiki and Bitcoin. These components enable individuals to access and utilize the Dark Web securely. During its early days, the Dark Web encountered the challenge of uncovering hidden websites. In response, the Hidden Wiki emerged in 2004 as a valuable resource, providing a comprehensive directory of active Dark Web sites. It also offered user reviews and information about the content available on each site. To aid in discovering hidden websites, specialized search engines like Ahmia and Grams, explicitly designed for Tor, proved helpful, particularly for locating illicit items like narcotics, weapons, and counterfeit money.

Bitcoin gained prominence within Dark Web markets by providing a pseudonymous and difficult-to-trace currency similar to Tor. Introduced in 2009, bitcoins are an unregulated digital currency stored in encrypted wallets. The intricacy of their design makes it highly challenging to trace the identities of individuals involved in Bitcoin transactions. Although every transaction is recorded in a public log, only wallet I.D.s are logged, ensuring the buyer's and seller's identities remain undisclosed.

To obtain bitcoins, users can register on a Bitcoin exchange, such as the well-known Mt. Gox, where conventional currencies can be exchanged for bitcoins. Another method involves utilizing a computer's CPU to solve complex mathematical problems, resulting in the acquisition of bitcoins. While Bitcoin currently serves as the primary currency within the Dark Web, there is anticipation for the upcoming money called Zerocoin, which promises even greater anonymity in transactions.

It is worth noting that the Dark Web, in its present state, was not originally intended to be what it has become. The Naval Research Laboratory (NRL), responsible for developing Tor, initially aimed to establish secure communication channels for overseas military personnel. Similarly, the founders of the Hidden Wiki sought to create an index that would assist regular users in navigating and understanding the Dark Web. Bitcoin was designed to facilitate anonymous transactions, driven by the intentions of its creators to prioritize privacy rather than malevolent purposes.

Table 1. Most popular dark websites

WEBSITE NAME	PURPOSE
The Hidden Wiki	Directory for dark web links
DuckDuckGo	Go to the search engine on the Tor browser
Facebook	Access to Facebook in restricted places
Hidden Answers	Share insights, ask questions, and more
ProPublica	Investigative journalism platform
SecureDrop	The exchange between news outlets and their anonymous sources
Keybase	Secure file sharing on the dark web
ZeroBin	Share encrypted text on the dark web
Riseup	Secure email service primarily for activist groups
The Hidden Wallet	Digital wallet for cryptocurrencies

Despite the good intentions behind these technologies, illegal activities persist within the shadows of the Dark Web.

- **Social Media and Popular Dark Websites**

The Dark Web Social Network is an emerging social media network on the dark Web that resembles those on the World Wide Web (DWSN). The DWSN allows users to make individual pages, add friends, like postings, and blog in forums, just like a conventional social networking website. Facebook and other traditional social media businesses have begun to develop dark web versions of their websites to address concerns with the traditional platforms and sustain their service across the entire World Wide Web. In contrast to Facebook, the DWSN's privacy policy requires that members preserve their anonymity and share no personal information. There are many popular websites on the dark Web. Table 1 below shows the list of the ten most popular shady websites.

3. RECENT STATISTICS OF DARK WEB

This section attempts to delve into the fascinating statistics and information about the dark Web and deep-learn all the necessary information, as shown in Table 2 below.

It is evident from the statistics that the dark Web is full of illicit activities apart from some fair usage. Despite the "Silk Road" closure in 2013, illegal drugs accounted for a staggering 52% of active listings on darknet marketplaces in 2017. This shows that the market for illicit narcotics is still very much alive. Figure 6 below depicts various crimes/illegal activities performed on the dark Web and the distribution of

Table 2. Recent statistics about dark web

Source	Fact/Statistics	Discussion/Explanation
Tor Project, 2022	The most widely used dark web browser right now is Tor.	This browser offers its users privacy and "multi-layer encryption" to ensure that users have the highest safety feeling on the Deep Web, and it has maintained its popularity for years. It's crucial to remember that these activities could make the browser lag.
Tor Project, 2022	In 2022, there are a staggering two million active Tor users.	The count of active users using Tor has been roughly constant since the start of the year, never falling below two million users, according to data that is made accessible to the public on the website of Tor.
LinkedIn, 2020	In 2020, the U.S. had the most significant number of users on the Dark Web.	Dark Web statistics show that 34.81% of daily users were from the United States, which is 831,911 users. Germany was in the third position with 7.16% of all active users, followed by Russia in second place with 11.46%.
Arxiv	Anonymity is the biggest concern of most Tor users on the Dark web	Up to 70.79% of poll participants 2018 stated they used Tor to maintain their privacy. In addition, 62.28% claimed to utilize it for more security, while 27% of users use it only out of interest in the dark Web.
Arxiv, 2018	Most Tor users are highly educated.	Around 59% of Tor users possess a postgraduate degree—approximately 18% of users of the dark Web own Graduates. Only 5.9% of users on the dark Web needed a degree.
Arxiv, 2018	The dark web users are predominantly male.	A 2018 online poll shows that men comprise 84.7% of dark web visitors. Only 9.4% of users were women, a relatively small percentage.
Forbes, 2022	The fact that money moves through cryptocurrencies that preserve anonymity is one of the most notable factors driving the growth of the dark web sector.	In the last ten years, the dark web market has expanded significantly. The number of Bitcoin transactions on the dark web market has climbed by almost $622 million (from $250 million in 2012 to $872 million), according to some dark web statistics, and the trend is still increasing.
PNAS, 2021	Tor is only used for evil reasons by a tiny percentage of its daily users.	According to dark web data, just 6.7% of users globally use Tor for illegal activities, including propagating malware, sharing child exploitation content, or buying and selling illicit goods.
Norton	The dark Web itself isn't illegal.	On the dark Web, numerous websites do not break any regulations. There are, however, many pages with illegal content (such as selling illegal substances and weapons or child pornography), and visiting those pages would be against the law.
DarkNetOne, 2022	In 2022, The Hidden Wiki is among the most widely used deep and dark web search engines.	For the majority of users, Tor is the preferred dark web browser. Other dark web search engines, such as The "Hidden Wiki," "Ahmia," "DarkSearch," "Haystack," and "Torch," have lately grown in popularity among users of the dark Web.
Avast, 2022	Popular dark web services include Facebook and bank account hacking.	Black hat hackers sell the stolen credentials of banks for around $2,000- $120 and breach Facebook accounts for as little as $65. Additionally, depending on the time frame selected, you can order a DDoS assault to be launched against an unprotected website for the specified duration at the correct cost. The attack cost can be as low as $15 per hour or as high as $1,000 monthly.
DIGIT, 2018	There is an exponential growth in demand for malware on the dark Web.	The most renowned location for both purchasing and accidentally getting malware is, without a doubt, the dark Web. Dark web statistics from 2018 revealed a sharp rise in demand for malware, and supply is well on time.
MSSP Alerts, 2021	Ransomware assaults broke all previous records in 2021.	Many willing hackers on the dark Web produce and distribute ransomware and plan various assaults, putting many people's cybersecurity at risk. In 2021, ransomware's popularity skyrocketed, but not in a good way. More than 190.4 million ransomware attack attempts were made in Q3 of that year, almost as many as in the first three quarters of 2020 combined.
The SUN	Who developed the Dark web?	The U.S. government developed the dark Web to enable covert communications and the exchange of sensitive information. Then, this technology was made available to the general public to protect the spies' privacy. Separating the government spies from other users became more challenging as the number of users increased due to the thousands of additional users being granted access.
Statistica	8% of Telegram Users use it for accessing darknet channels	Dark web statistics demonstrate the relationship between the finest privacy-focused chat apps and the darknet. The fact that 8% of Telegram's users also engage in the app's darknet channels is unsurprising, given that the tight encryption encourages privacy-conscious internet users to interact on services like Telegram.

Figure 6. Crime distribution on dark web

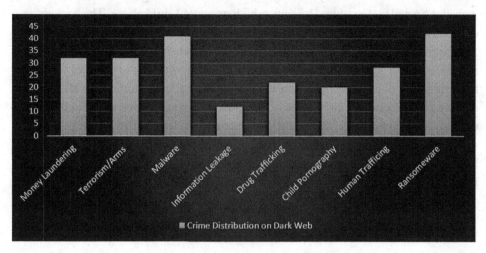

countrywide usage of the dark Web, respectively. The calculation is based on the average percentage of data collected from various sources and agencies available on the Internet. Ransomware and Malware related crimes are increasing at a faster pace on the dark Web, followed by FireArms and Money Laundering.

4. BENEFITS/DRAWBACKS AND GREY AREAS OF DARK WEB

Whistleblowing and hacktivism are two distinct phenomena within the realm of the Dark Web, each possessing certain qualities that defy simple categorization. While neither can be unequivocally labeled as wholly positive, they exhibit commendable aspects worth acknowledging. Whistleblowing, for instance, serves as a vital mechanism in upholding democratic systems, ensuring transparency and accountability. However, it is crucial to emphasize the potential risks associated with bypassing official channels, as it may inadvertently expose sensitive government practices and sources. Prominent figures such as Edward Snowden, Julian Assange, and Chelsea Manning used the Dark Web to unveil classified information. However, alternative avenues, such as filing complaints through legitimate congressional methods, could have been pursued.

Similarly, hacktivism represents another complex topic that defies simplistic categorization. While the goals of certain hacktivists may raise questions, their chosen tactics often trespass into offensive and illegal territory. In October 2011, for instance, a hacktivist collective launched a Distributed Denial of Service (DDoS) attack, resulting in the shutdown of the Freedom Hosting internet hosting service.

This action was motivated by identifying server-to-server signatures associated with websites hosted by Freedom Hosting, which promoted child abuse. The hackers also exposed login details of 1500 visitors to the child abuse website known as Lolita City. Although the objective of dismantling child abuse websites is undoubtedly laudable, it is essential to acknowledge the inherent flaws in vigilante-style retribution due to the absence of proper online abuser accountability. In summary, whistleblowing and hacktivism encompass shades of complexity that cannot be reduced to notions of absolute good or bad. While they may possess praiseworthy elements such as upholding democratic principles or targeting reprehensible activities, caution must be exercised to ensure their methods align with legal frameworks and respect due process.

5. DARK WEB MONITORING

Dark Web Monitoring is a procedure businesses use to keep an eye on their secret information on the dark Web and to alert them if it turns up there. This will enable you to mitigate the effects of a data breach and take the required precautions to safeguard your company, employees, clients, and other assets against a prospective attack.

- **Advantages of Dark Web Monitoring**
  - Finding Data Breach - A sophisticated piece of software is essential to identify the stolen credentials and other personal information distributed on the dark web networks. You can configure your queries on the software to find any pertinent data or information. They constantly scan the open, dark, and deep Web, then pass these searches via A.I. algorithms to determine relevant ones. We later sent someone to conduct additional information screening for a better outcome.
  - Finding Physical Threats to People and Property - The fact that all users must use an encrypted browser to access the dark Web, which is fully anonymous as to their presence, is a significant draw for lawbreakers. This means that criminals can and do brag about their actions or use them as a component of their plans. You may continuously watch the dark Web with dark web monitoring, and you'll be alerted as soon as criminal talks or potentially compromises one of your employees or resources.
  - Predicting Potential Terrorist Attacks: Businesses monitor online threats to them for physical threats; in addition, the dark Web is where terrorists go to coordinate, plan, and launch attacks. The corporation can join

its discussion and use the information gathered by scanning the dark Web to potentially anticipate and identify terrorist threats that have been made against the company.

- **Best Practices to Follow for Protecting Organization's Data**
  - Encrypted cloud services must be made available by the businesses; these services must encrypt and authorize access to the file for repository use. Access to the file is restricted to the authorized user. An additional security measure safeguards passwords and data.
  - Remind your team that any URLs they click might be infected with malware. Users should only open attachments from reputable and well-known sites, no matter how "official" they may be.
  - Organizations should consistently back up their data weekly to prevent loss and swiftly fix issues.
  - The business should conduct an appropriate cloud security assessment regularly to safeguard the security of employees and the organization's data.
  - Create password policies, such as using password managers instead of sticky notes to record them.
  - Employers should implement a safer function and turn off the USB/HDD plug. It is essential because rogue software or unknown dangers could target the corporate computer.
  - Businesses should monitor data exfiltration. They knew whether any data leaks would be helpful.
  - Inform staff members of the risks of accessing unprotected public networks.

6. ROLE OF LAW ENFORCEMENT AGENCIES

The Dark Web, characterized by its anonymity, poses a challenge in distinguishing between criminals and legitimate users. To address this, law enforcement authorities need to employ strategies that safeguard the privacy of regular users while identifying offenders. A practical approach is to target unlawful websites rather than individual users, ensuring that confidentiality is maintained for the majority. By deploying authorized software to deanonymize users accessing these sites, authorities can deter potential visitors and discourage illicit activities without merely shutting down websites, as new ones often emerge in their place.

Attempting to break Tor or tracking down every Tor user would likely result in the emergence of a more resilient version of the service, undermining government efforts and curtailing the freedoms that genuine users, such as dissidents, rely upon.

Recognizing the importance of upholding online freedom of expression, unlike certain nations, the United States has a constitutional commitment to protect it. Given the ambiguous nature of enforcement jurisdiction, governments must collaborate to establish mutually accepted regulations for the Dark Web. While the Dark Web provides dissidents with a means to communicate freely, it also presents a dual-edged sword of online anonymity that necessitates cautious use. Policymakers must continually monitor its evolution and ensure law enforcement agencies receive sufficient funding and legal support to regulate it effectively. Striking a delicate balance between safeguarding user privacy and the state's responsibility to combat criminal behavior calls for comprehensive and well-considered laws that govern the Dark Web.

Cybercriminals can victimize individuals and organizations; geographic boundaries do not constrain them. Law enforcement has difficulty figuring out how criminals use borders, especially given how boundaries and borders have changed.

Physical Limits. Jurisdictional lines have been drawn between nations, states, and other localities for law enforcement purposes. Law enforcement entities have the power to dispense justice within specific territories. When crimes cross jurisdictional lines, one organization's control over criminal prosecution may no longer be absolute, and laws may not be uniformly applied. These phenomena have long been understood and used by criminals.

Physical–Cyber Borders. In the virtual world, the boundaries are not always as distinct as in the physical world. In addition to helping legal businesses expand, high-speed Internet connectivity has also improved the ability of criminals to operate in a setting where they can quickly exploit their victims and expand their pool of possible targets. Face-to-face frauds and schemes can now be carried out remotely from other parts of the country or even the world. For instance, hackers can use botnets to target people worldwide without ever crossing a single border.

Cyber Borders. Although there are no physical borders in cyberspace, there are still legal and technological barriers. For example, some web domains are country-specific, and those countries are in charge of running those websites. The distinctions between the Surface Web and the Deep Web represent another barrier in cyberspace. Subscriptions or paying a price to access a particular website's content may be necessary to cross these limits. Some companies may charge for access, including news websites, periodicals, and file-sharing websites. Other websites require an invitation to view them.

7. RECENT DATA BREACHES IN INDIA ON DARK WEB

Recent research by Lithuania's Nord Security for Nord-VPN places India at the center of a significant data breach. In the Russian bot markets, thousands of Indians' stolen data were discovered. Data on more than 50 lakh persons were stolen worldwide and sold on both markets for 490 Indian rupees, according to a recent analysis by Lithuania's Nord Security and Nord-VPN. According to the study, India was the source of the data breach because hackers were found to have stolen and sold 6 lakh Indian people's personal information online. Cybercriminals sell data they have stolen from their victims' devices using bot malware, which is internet marketplaces. The information offered in packages contains the complete digital identity of the victim, including logins, cookies, digital fingerprints, and other details.

According to the investigation, hackers have obtained user camera images, screenshots, current login credentials, cookies, and digital fingerprints. The study uncovered 26.6 million compromised logins. Approximately 720,000 Google logins, 654,000 Microsoft logins, 647,000 Facebook logins, 223,000 Netflix logins, and other user data from numerous big firms were discovered to have been compromised.

- 2easy Marketplace

The 2easy platform debuted in 2018, and since then, it has expanded quickly. 2easy was considered a minor participant in this dark Web and information-stealing market. Analysis since then suggests that the website's "high-quality" goods stoked the curiosity of online criminals. Hackers aim to find the following network they can break into. The data logs are archives of data stolen from computers or web browsers infected with malware. Cookies saved credit card data and account credentials are frequently found in catalogs. Due to the 2easy platform's complete automation, users may register accounts, send funds to wallets, and make purchases without dealing with merchants directly. Logs are available to hackers for as little as $5 per item. This is almost five times less than what a typical rival provides.

- Data Breach of India Railway-December 2022

On Tuesday, December 27, the Indian Railways had a data breach that resulted in the theft of the personal information of about 30 million people. On the dark Web, user data is rumored to have been advertised for sale by a hacker. The hacker claims that "it's one of the biggest railroads databases in India" without providing the information's source. User information and invoices, some of which have due dates of December 31, 2022, are included in the stolen data collection.

- Tweeter Data Breach- Dec 2022

There has allegedly been a data breach at the social media site. A "threat actor," according to a report, is said to have listed the data of almost 400 million Twitter users for sale on the dark Web. According to a report by Hudson Rock, the 'threat actor' or hacker under investigation allegedly sold the personal data of several accounts. These include well-known accounts' phone numbers, email addresses, and other contact details. "Hudson Rock revealed a credible threat actor is selling 400,000,000 Twitter user's data," states a tweet from Hudson Rock. Devastating amounts of information, including phone numbers and emails, are contained in the private database.

- Data Breach of Clear-trip-July 2022

One of India's most well-known sites for booking travel has confirmed a data breach after hackers allegedly posted the stolen information on the dark Web. Taking legal action against the hackers was launched by Clear Trip. Whether immediately known are the exacts of the stolen material, including whether or not it contains sensitive information. Hackers offered the data for sale on a closed, invite-only forum on the dark Web.

- Data Breach of Tata Power Server-Oct 2022

Hive, a ransomware organization, allegedly released a packet of crucial data stolen from Tata Power servers on the dark Web. On October 14, Tata Power announced that a cyberattack had struck its I.T. infrastructure, and as a result, some of its systems had been impacted. With its headquarters in Mumbai, the company stated in a Bombay Stock Exchange statement that all crucial operational systems were operating and that it has "taken steps to retrieve and restore its systems." On Tuesday, Hive took ownership of the cyberattack and started disclosing the stolen information on its dark web forum.

Less than a year old, cybersecurity professionals rank Hive among the top three ransomware threats. Together with other ransomware affiliates, it is known to target industries like energy, healthcare, financial services, media, and education. The company's bank accounts, bank statements, and information about its personnel, including their salaries and passport numbers, were all among the material that was released. Tata Power's battery usage information and schematics for some of its grids were also included in the hacked material.

- Data Breach of Banks by Biden Cash -Oct 2022

Security experts at Cyble have found a significant data leak involving the financial information of credit card clients of major institutions like the State Bank of India (SBI) and American Express. Security experts have found a massive data leak involving the financial information of credit card clients of significant institutions like the State Bank of India (SBI), American Express, and Fiserv Solutions LLC. According to threat intelligence firm Cyble, the financial information of a whopping nine million credit card users was compromised. American Express (U.S.) was the corporation most impacted by the credit card data leak, according to Cyble's analysis. The US, India, Brazil, the U.K., Mexico, Turkey, Spain, Italy, and France were the top 50 nations with affected consumers.

" Some leading financial firms impacted included American Express, Fiserv Solutions LLC, and State Bank of India. According to security researchers, over 508,000 debit cards were compromised, with 414,000 records belonging to the Visa payment network coming in second place. Threat actors frequently establish markets like Biden Cash, exchanging sensitive card information for carding and cloning services. Threat actors often check and implement new methods to bypass them, even when contemporary security systems can reduce the impact. Early in February 2022, the Biden Cash forum started operating. After that, the threat actor turned to other strategies, such as leaving spammy comments on other websites, to drive traffic to his website.

- Data Breach of Domino's -April 2021

According to Alon Gal, CTO of cyber intelligence company Hudson Rock, 180 million Domino's India pizza orders were for sale on the dark Web. Gal discovered a person offering ten bitcoin (approximately \$535,000 or 4 crore) for 13 T.B. of data that purportedly contained 1 million credit card records and information on 180 million Dominos India pizza purchases, along with the names, phone numbers, and email addresses of the purchasers. Gal uploaded a screenshot that revealed the hacker also claimed to have the personal information of the 250 employees of Domino's India.

- Data Breach of 1.90 Lac CAT Applicants Leaked -May 2021

For the 2020 Common Admission Test, used to choose candidates for admission to the Indian Institutes of Management (IIMs), 190,000 candidates' personally identifiable information (PII) and test results were leaked and sold on a cybercrime website. The hacked database contained information about candidates' names, dates of birth, email addresses, mobile numbers, addresses, 10th and 12th-grade test results, bachelor's degrees, and CAT percentile scores. The information was taken from the CAT exam administered on November 29, 2020, but security intelligence firm Cloud-SEK claims that the same thread actor also exposed the 2019 CAT exam database.

- Trading Platform Upstox Data Breach- April 2021

Know-your-customer (KYC) data of Upstox was compromised, which Upstox has openly admitted. Financial services organizations gather KYC data to verify the identity of their clients and stop fraud or money laundering, but hackers can also exploit it to steal identities. Upstox informed clients on April 11 that it would reset their passwords and take other security measures. Customers were apologized to by Upstox for the inconvenience and reassured that the event had been reported to the appropriate authorities, that security had been strengthened, and that the company had increased its bug bounty program to entice ethical hackers to stress-test its systems.

- Juspay Data Breach -January 2021

Juspay disclosed in early January that information about around 35 million customer accounts, including disguised card data and fingerprints, were obtained from a server using an unrecycled access key. It said the incident happened in August of last year. Independent cybersecurity analysts claim that the price of user data on the dark Web is around $5000.

- Big-Basket Data Breach -October 2020

According to Atlanta-based cyber intelligence company Cyble, Big-Basket's online grocery platform sells user information on the dark Web. Cyble announced on November 7 that a portion of a database containing the private data of over 20 million customers was offered for sale for 3 million rupees ($40,000). Names, email addresses, password hashes, PINs, cellphone numbers, addresses, dates of birth, localities, and I.P. addresses were all included in the data. Cyble claims to have discovered the data on October 30 and, after validating it by comparing it to user data from Big-Basket, notified Big-Basket of the alleged breach on November 1.

- Un-Academy Data Breach -May 2020

Un-Academy, a start-up in edtech, announced a data breach that affected 22 million user accounts. Usernames, email addresses, and passwords were listed for sale on the dark Web, according to cybersecurity company Cyble. Un-Academy was founded in 2015 with funding from companies like Facebook, Sequoia India, and Blume Ventures.

8. FUTURE RESEARCH ASPECTS OF THE DARK WEB

The dark Web, a part of the Internet not indexed by search engines and requires specific software to access, presents several future research aspects and needs. Here are some areas that may require attention in the future:

- Security and Anonymity: As law enforcement agencies and security measures evolve, individuals on the dark Web will likely continue to seek improved methods to remain anonymous and secure. Research is needed to understand emerging technologies, encryption methods, and potential vulnerabilities to enhance security and anonymity.
- Criminal Activities and Detection: The dark Web is associated with various illicit activities, such as drug trafficking, hacking, human trafficking, and more. There is a need for ongoing research to understand and combat these criminal activities. This includes developing advanced techniques for identifying and tracking individuals involved in illegal transactions and networks.
- Cybersecurity and Threat Intelligence: Dark web marketplaces and forums often serve as platforms for trading stolen data, malware, and hacking tools. Research is crucial to avoiding emerging cyber threats, understanding new attack vectors, and developing effective defense mechanisms.
- Cryptocurrencies and Financial Transactions: Cryptocurrencies like Bitcoin are frequently used for anonymous dark web transactions. Future research may focus on analyzing the role of cryptocurrencies, identifying patterns in financial transactions, and developing tools for tracking and tracing these transactions.
- Online Extremism and Radicalization: The dark Web can provide a breeding ground for extremist ideologies and radicalization efforts. Research is essential to monitor and counteract extremist content spread, study recruitment strategies, and develop effective interventions.
- Dark Web Governance and Policy: The dark Web's decentralized nature challenges regulation and governance. Research is needed to explore potential policy frameworks, legal approaches, and international cooperation to address issues related to the dark Web effectively.
- User Behavior and Social Dynamics: Understanding individuals' motivations, behaviors, and interactions on the dark Web can provide valuable insights. Research could focus on analyzing user communities, social dynamics, and the impact of anonymity on online behavior.

It's important to note that researching on the dark Web presents ethical and legal challenges. Researchers must carefully consider the potential implications, follow ethical guidelines, and adhere to legal boundaries while investigating this hidden part of the Internet.

CONCLUSION

Statistics show that only 6.7% of users use Tor for unlawful purposes, such as spreading malware, sharing child exploitation content, or purchasing and selling illegal products. Although Tor is entirely legal, most typical internet users will never need to view information on the dark Web. Freenet, a University of Edinburgh student Ian Clarke's thesis project that aimed to develop a "Distributed Decentralized Information Storage and Retrieval System," is thought to have marked the start of the dark Web in 2000. Using specialized software like Tor, Freenet, and the Invisible Internet Project (I2P), people can access the Dark Web by routing users' web traffic through several other users' computers. Tor uses a network of volunteer computers to ensure that the originating user cannot be identified from where the traffic is being directed. Using the dark Web might initially seem against the law, but many good reasons exist to utilize Tor and browse anonymously. The dark Web is used for dubious and illegal reasons due to its anonymity. The dark Web's anonymity is like a technology that can be a great enslaved person but a dangerous enslaver. The effectiveness of the dark Web can be a bane for defense services agencies and other legitimate activists, while it has threatened humanity and law enforcement agencies.

Chapter Achievements

- Comprehensive Overview: The chapter provides readers with a thorough introduction to the dark Web, covering its technology, ecosystem, and inherent risks. It offers a foundational understanding of this hidden part of the Internet.
- Insightful Analysis: By examining the dark web threat landscape, the chapter enhances readers' awareness of the various criminal activities and potential dangers associated with the dark Web. It sheds light on the scale and nature of illicit activities, contributing to a broader understanding of the risks involved.
- Case Studies and Examples: Including relevant case studies and real-world examples illustrates the practical implications of the dark Web. It allows readers to grasp the complexities of investigations, the impact on individuals and society, and the challenges law enforcement agencies face.

- Policy and Legal Implications: The chapter explores the governance and policy challenges surrounding the dark Web, offering insights into legal frameworks, international cooperation, and potential avenues for regulation. It encourages critical thinking and debate on strategies to address dark web-related issues.
- Ethical Considerations: Recognizing the ethical dilemmas associated with researching the dark Web, the chapter addresses the importance of conducting research within ethical boundaries and complying with legal requirements. It emphasizes the need for responsible exploration and responsible dissemination of knowledge. This chapter attempts to address every aspect of comprehending the dark Web.

REFERENCES

Al Nabki, M. W., Fidalgo, E., Alegre, E., & De Paz, I. (2017, April). They classify illegal Tor network activities based on web textual content. In *Proceedings of the 15th Conference of the European Chapter of the Association for Computational Linguistics,* (pp. 35-43). ACL.

AlSabah, M., & Goldberg, I. (2016). Performance and security improvements for Tor: A survey. [CSUR]. *ACM Computing Surveys, 49*(2), 1–36. doi:10.1145/2946802

Aminuddin, M. A. I. M., Zaaba, Z. F., Samsudin, A., Juma'at, N. B. A., & Sukardi, S. (2020, August). Analysis of the paradigm on tor attack studies. In *2020 8th International Conference on Information Technology and Multimedia (ICIMU)* (pp. 126-131). IEEE.

Bahri, A. (2020). The Dark and Deep Web. *Medium.* https://ahedbahri.medium.com/the-dark-and-deep-web-6d6629923968b

Cambiaso, E., Vaccari, I., Patti, L., & Aiello, M. (2019, February). Darknet Security: A Categorization of Attacks to the Tor Network. In ITASEC (pp. 1-12).

Cascavilla, G., Tamburri, D. A., & Van Den Heuvel, W. J. (2021). Cybercrime threat intelligence: A systematic multi-vocal literature review. *Computers & Security, 105,* 102258. doi:10.1016/j.cose.2021.102258

Catakoglu, O., Balduzzi, M., & Balzarotti, D. (2017, April). Attacks landscape in the dark side of the Web. In *Proceedings of the Symposium on Applied Computing* (pp. 1739-1746). ACM. 10.1145/3019612.3019796

Chen, H. (2011). *Dark Web: Exploring and data mining the dark side of the Web.* Springer Science & Business Media.

Erdin, E., Zachor, C., & Gunes, M. H. (2015). How to find hidden users: A survey of attacks on anonymity networks. *IEEE Communications Surveys and Tutorials, 17*(4), 2296–2316. doi:10.1109/COMST.2015.2453434

Evers, B., Hols, J., Kula, E., Schouten, J., Den Toom, M., van der Laan, R. M., & Pouwelse, J. A. (2016). *Thirteen Years of Tor Attacks.*

Faizan, M., Khan, R. A., & Agrawal, A. (2022). Ranking potentially harmful Tor hidden services: Illicit drugs perspective. *Applied Computing and Informatics, 18*(3/4), 267–278. doi:10.1016/j.aci.2020.02.003

FinkleaK. (2015). https://digital.library.unt.edu/ark:/67531/metadc700882/

Gupta, A. (2018). *The dark Web as a phenomenon: a review and research agenda* [Doctoral dissertation, University of Melbourne].

Jain, K., Gupta, M., & Abraham, A. (2021). A Review on Privacy and Security Assessment of Cloud Computing. *Journal of Information Assurance and Security., 16*(5), 161–168.

Karunanayake, I., Ahmed, N., Malaney, R., Islam, R., & Jha, S. (2020). Anonymity with Tor: A survey on Tor attacks. *arXiv preprint arXiv:2009.13018.*

Kaur, S., & Randhawa, S. (2020). Dark Web: A web of crimes. *Wireless Personal Communications, 112*(4), 2131–2158. doi:10.100711277-020-07143-2

Kaur, S., & Randhawa, S. (2020). Dark Web: A web of crimes. *Wireless Personal Communications, 112*(4), 2131–2158. doi:10.100711277-020-07143-2

Nazah, S., Huda, S., Abawajy, J., & Hassan, M. M. (2020). Evolution of dark web threat analysis and detection: A systematic approach. *IEEE Access : Practical Innovations, Open Solutions, 8*, 171796–171819. doi:10.1109/ACCESS.2020.3024198

Patel, P. B., Thakor, H. P., & Iyer, S. (2019, March). A comparative study on cybercrime mitigation models. In *2019 6th International Conference on Computing for Sustainable Global Development (INDIACom)* (pp. 466-470). IEEE.

Rawat, R., Mahor, V., Chouhan, M., Pachlasiya, K., Telang, S., & Garg, B. (2022). Systematic Literature Review (SLR) on social media and the Digital Transformation of Drug Trafficking on Darkweb. In *International Conference on Network Security and Blockchain Technology* (pp. 181-205). Springer, Singapore. 10.1007/978-981-19-3182-6_15

Saleem, J., Islam, R., & Kabir, M. A. (2022). The Anonymity of the Dark Web: A Survey. *IEEE Access : Practical Innovations, Open Solutions, 10,* 33628–33660. doi:10.1109/ACCESS.2022.3161547

Saleh, S., Qadir, J., & Ilyas, M. U. (2018). Shedding light on the Internet's dark corners: A survey of tor research. *Journal of Network and Computer Applications, 114,* 1–28. doi:10.1016/j.jnca.2018.04.002

Samtani, S., Chai, Y., & Chen, H. (2022). Linking exploits from the dark Web to known vulnerabilities for proactive cyber threat intelligence: An attention-based deep structured semantic model. *Management Information Systems Quarterly, 46*(2), 911–946. doi:10.25300/MISQ/2022/15392

Singh, A., & Jain, K. (2022). An Automated Lightweight Key Establishment Method for Secure Communication in WSN. *Wireless Personal Communications, 124*(4), 2831–2851. doi:10.100711277-022-09492-6

Singh, A., & Jain, K. (2022). An efficient, secure key establishment method in a cluster-based sensor network. *Telecommunication Systems, 79*(1), 1–14. doi:10.100711235-021-00844-4

Siuda, P., Nowak, J., & Gehl, R. W. (2023). Darknet imaginaries in Internet memes: the discursive malleability of the cultural status of digital technologies. *Journal of Computer-Mediated Communication, 28*(1), zmac023.

Sulaiman, M. A., & Zhioua, S. (2013, July). Attacking Tor through unpopular ports. In *2013 IEEE 33rd International Conference on Distributed Computing Systems Workshops* (pp. 33-38). IEEE. 10.1109/ICDCSW.2013.29

Temel, M. H. (2019). Research Review: Traffic Analysis Attack Against Anonymity in TOR and Countermeasures. School of Informatics. The Eindhoven University of Technology.

Weimann, G. (2016). Going dark: Terrorism on the dark Web. *Studies in Conflict and Terrorism, 39*(3), 195–206. doi:10.1080/1057610X.2015.1119546

Yang, M., Luo, J., Ling, Z., Fu, X., & Yu, W. (2015). Deanonymizing and countermeasures in anonymous communication networks. *IEEE Communications Magazine, 53*(4), 60–66. doi:10.1109/MCOM.2015.7081076

Chapter 15
Unraveling the Server:
Mastering Server–Side Attacks for Ethical Hacking

Aviral Srivastava
Amity University, Rajasthan, India

ABSTRACT

This chapter, per the authors, delves into the realm of server-side attacks within the context of ethical hacking and penetration testing. It aims to provide a comprehensive understanding of server-side vulnerabilities, focusing on web, database, and application servers. Additionally, the chapter explores prevalent server-side attack techniques, guiding readers through the process of identifying, exploiting, and mitigating server-side vulnerabilities in a responsible and ethical manner. Legal and ethical considerations surrounding server-side attacks are also discussed, emphasizing the importance of responsible disclosure and collaboration with vendors. Lastly, the chapter concludes by examining the role of server-side attacks in ethical hacking and highlighting future trends and challenges that ethical hackers may encounter in the ever-evolving digital landscape.

1. INTRODUCTION

In the ever-evolving field of cybersecurity, an uncompromising vigilance in the protection of digital assets is required since cybercriminals are becoming more sophisticated, and the malicious strategies they employ are becoming more prevalent. Specifically, server-side assaults have evolved as a formidable threat vector that threatens the basic foundation of information systems. This is a relatively recent

DOI: 10.4018/978-1-6684-8218-6.ch015

development. Ethical hackers and penetration testers can play a crucial role in the protection of an organization's digital infrastructure if they have a full grasp of the vulnerabilities that exist and take measures to eliminate or mitigate them.

1.1. The Importance of Server-Side Attacks

As a result of the fact that server-side assaults have the potential to inflict major damage on an organization's operations, reputation, and financial status, they have risen to prominence in recent years. Malicious actors can steal sensitive data, destroy systems, or gain unauthorised access to crucial resources by focusing their attention on the servers that support websites, applications, and databases. Understanding the complexities of server-side threats has become an essential component of any comprehensive cybersecurity plan as a direct consequence of this development.

In addition, as businesses continue to gradually transition to server-side infrastructures hosted in the cloud, the ramifications of server-side attacks on data security have become even more evident. Because of the complexity and interdependence of these environments, the potential for cascading failures is significantly increased. As a result, it is absolutely necessary for cybersecurity professionals to have an in-depth understanding of server-side vulnerabilities and the ways in which they can be exploited.

1.2. Ethical Hacking and Penetration Testing Context

Ethical hacking and penetration testing have evolved as essential components of a proactive cybersecurity posture in this volatile environment. Ethical hackers are able to conduct a comprehensive review of an organization's defences, locate potential weak spots, and offer advice on how to patch those holes in order to thwart potential intrusions by dishonest adversaries. This is accomplished through the simulation of real-world attack scenarios. This preventative method of approaching security gives organisations the ability to stay one step ahead of new security risks and to foster a culture of robust security.

Exploring server-side attack vectors in a strategic and responsible manner is an essential component of both the profession of ethical hacking and penetration testing. Ethical hackers can shed light on the chinks in an organization's armour by painstakingly studying server-side vulnerabilities and designing techniques to attack those flaws. This provides the business with actionable intelligence that can be used to strengthen its defences. This chapter delves deeper into the complexities of server-side assaults, elucidates the significance of these attacks within the framework of ethical hacking and penetration testing, and provides recommendations for limiting the possible impact of these attacks.

2. UNDERSTANDING SERVER-SIDE VULNERABILITIES

As the digital ecosystem continues to become more complex, server-side vulnerabilities will continue to proliferate, making it necessary to have an in-depth understanding of the risks that are inherent to the situation. In this part of the article, we will outline the major categories of server-side security problems and elaborate on the vulnerabilities that affect application, database, and web servers.

2.1. Common Server-Side Security Flaws

Misconfigurations, out-of-date software, or a lack of proper security controls frequently cause problems on the server's side of the network. The following are examples of common server-side security weaknesses; however, this list is not exhaustive.

Inadequate verification and cleaning of the input data.

Mechanisms of authentication and authorization that are not secure

Unprofessional management of errors and dissemination of information

Transmission and storage of sensitive information are not protected by encryption.

Components of software that are out of date or missing necessary patches

2.2. Web Server Vulnerabilities

Because of their essential function in hosting and delivering web content, web servers are frequently the focus of attacks launched by hostile actors. Examples of notable vulnerabilities in web servers include:

Directory traversal refers to the practise of navigating file systems and gaining access to sensitive data by utilising insufficient access constraints.

Web servers with incorrect configuration may have insecure default settings or access controls that are not setup correctly, which may expose important information or functionality.

Exploiting server-side programmes in order to carry out unauthorised network requests on behalf of the attacker is referred to as server-side request forgery (SSRF).

2.3. Database Server Vulnerabilities

Because they are used to store and manage essential data for a business, database servers are a very desirable target for cybercriminals. Prominent database server vulnerabilities encompass:

SQL injection refers to the modification of SQL queries through the injection of malicious code, which may provide unauthorised access, result in data alteration, or lead to the exfiltration of data.

Weak authentication and authorization refers to the practise of gaining unauthorised access to databases by utilising credentials that are either default or weak, or by engaging in insecure account management activities.

Unprotected data storage and transmission refers to the practise of storing or transferring sensitive data without applying adequate encryption measures. This leaves the data open to the risk of being stolen or intercepted.

2.4. Application Server Vulnerabilities

Application servers are an essential component of today's digital ecosystem because of their ability to assist the execution of business logic as well as the distribution of dynamic content. The following are notable examples of vulnerabilities in application servers:

Exploiting vulnerable code or configurations to execute arbitrary commands on a target server is referred to as remote code execution (RCE). This can lead to unauthorised access or a compromised system.

Unsafe deserialization refers to the manipulation of serialised data in order to accomplish unwanted results, such as the execution of malicious code, the elevation of privileges, or the denial of service.

Injecting malicious scripts into web applications, often known as cross-site scripting (XSS), gives an attacker the ability to hijack user sessions, deface websites, or redirect visitors to dangerous information.

3. SERVER-SIDE ATTACK TECHNIQUES

As malicious actors continue to refine their craft, a multitude of server-side attack strategies have arisen, each of which is aimed to exploit a particular vulnerability. In this section, we will present an explanation of various well-known server-side attack strategies as well as insight into the workings of those tactics.

3.1. Command Injection Attacks

Attacks known as command injection include the execution of arbitrary commands on a target server by manipulating user input. This is often accomplished through inadequate input validation and sanitization. These attacks have the potential to result in unauthorised access, the exfiltration of data, or even a complete compromise of the system.

3.2. SQL Injection Attacks

Attacks known as SQL injection take advantage of security flaws in database-driven web applications by modifying SQL queries using user input. Attackers are able to circumvent authentication, modify data, and remove sensitive information from the database by injecting malicious SQL code.

3.3. Cross-Site Scripting (XSS) Attacks

The injection of malicious scripts into trusted websites is what constitutes a cross-site scripting attack. This gives the attacker the ability to hijack user sessions, deface web pages, or redirect users to harmful information. Attacks using XSS are often divided into the following three categories: stored, reflected, and DOM-based.

3.4. File Inclusion and Path Traversal Attacks

Attacks using file inclusion take use of vulnerabilities in online applications that allow the inclusion of external or local files. These attacks can result in the execution of remote code or the disclosure of sensitive data. On the other side, path traversal attacks take advantage of weak access constraints in order to browse a server's file system and get access to sensitive files or directories.

3.5. Server-Side Request Forgery (SSRF) Attacks

Attacks known as server-side request forgeries involve manipulating apps running on the server itself in order to get them to make unauthorised network requests on behalf of the attacker. Attackers have the ability to acquire access to internal resources, exfiltrate sensitive data, or execute network reconnaissance by exploiting vulnerable functionality on the target network.

3.6. Server-Side Template Injection (SSTI) Attacks

Attacks known as server-side template injection target online applications that make use of server-side templating engines and look for vulnerabilities in those programmes to exploit. Attackers have the ability to execute arbitrary code on the server if they insert malicious code into template expressions. This could result in unauthorised access to the system or a compromise of the data contained inside it.

3.7. XML External Entity (XXE) Attacks

Attacks using XML external entities take use of flaws in XML parsers, which are programmes that interpret user-supplied XML data. An attacker can force the server to divulge sensitive information, execute unwanted network requests, or commit denial-of-service attacks by referencing external entities within the XML document.

3.8. Insecure Deserialization Attacks

Attacks that use insecure deserialization entail manipulating serialised data in order to accomplish undesired results, such as the execution of malicious code, the elevation of privileges, or the denial of service. Attackers can get unauthorised access to system resources or manipulate application logic if they exploit vulnerabilities in the deserialization process and use those vulnerabilities to their advantage.

3.9. Server-Side Cache Poisoning Attacks

Attacks that store harmful content on the server by exploiting flaws in caching methods are referred to as server-side cache poisoning attacks. Attackers can trick the server into serving harmful material to unwary users by tampering with the cache headers or by employing cache poisoning techniques. This could result in data theft or unauthorised access.

4. EXPLOITING SERVER-SIDE VULNERABILITIES

Exploiting server-side vulnerabilities is a complex endeavour that requires a wide variety of knowledge, abilities, and strategies. This art form requires a diverse approach. In this section, we will discuss the process of locating and evaluating vulnerabilities, developing and deploying exploits, and making use of sophisticated exploitation strategies.

4.1. Identifying and Assessing Vulnerabilities

The first thing that needs to be done in order to exploit server-side vulnerabilities is to locate possible targets and determine how vulnerable they are to being attacked. This method includes reconnaissance, scanning for vulnerabilities, and manual analysis, and it ultimately results in a full understanding of the target environment and the dangers that are connected with it.

4.2. Crafting and Executing Exploits

As soon as a vulnerability has been located, the following step is to devise an exploit that takes advantage of the found defect in order to accomplish the objective that has been set. In order to accomplish this, it may be necessary to develop unique payloads, make use of extant exploit frameworks, or modify well-known attack techniques so that they are compatible with the specific requirements of the target environment.

4.3. Advanced Exploitation Techniques

Exploiting server-side vulnerabilities may, in certain circumstances, call for more advanced strategies that go beyond the purview of more conventional methods of exploitation. A privilege escalation, lateral movement, or the chaining together of various vulnerabilities to generate a synergistic impact are some examples of these types of attacks.

4.4. Post-Exploitation Activities

After successfully exploiting a server-side vulnerability, it is frequently essential to engage in operations known as post-exploitation in order to sustain access, exfiltrate data, or accomplish other goals. This can require the development of data exfiltration techniques, the usage of command and control infrastructure, or the deployment of persistence mechanisms.

5. SERVER-SIDE ATTACK MITIGATION STRATEGIES

It is necessary to develop a strong security posture that incorporates a variety of mitigation measures in order to properly prevent server-side assaults. Only then will you be able to effectively thwart these attacks. These tactics will be broken down in further detail in the following sections, presenting a more in-depth comprehension of their significance in terms of protecting server-side settings.

5.1. Input Validation and Sanitization

For the purpose of reducing server-side attacks that make use of data supplied by users, input validation and sanitization are quite helpful. Organisations are able to prevent the introduction of malicious code or data if they enforce severe validation standards for user inputs and require users to comply. Sanitising input data also guarantees that any potentially dangerous elements are neutralised or removed from the data before it is used. The following are important practises for validating and sterilising inputs:

By using allow-lists, you can restrict the values, formats, and ranges that are permissible for input.

Using centralised input validation routines to ensure consistency while reducing the amount of effort required for maintenance.

Sanitising user inputs involves filtering out content that is not permitted, deleting whitespace that is not necessary, and escaping or encoding special characters.

Adding server-side validation to client-side validation as a second line of defence against those trying to get around the system.

5.2. Secure Configuration and Patch Management

When it comes to shielding server-side systems from newly discovered dangers, a secure configuration and careful patch management are two of the most important factors. It is possible to lower the attack surface and the likelihood of successful exploitation by ensuring the secure deployment and management of web, database, and application servers. Secure configuration and patch management practises encompass:

Maintaining regular software updates for the server's components and libraries in order to fix any known vulnerabilities.

Using principles of least privilege and imposing robust authentication and permission restrictions are two important steps to take.

ensuring that the server configuration follows both the best security practises and the standards provided by the vendor.

Auditing the server configurations on a regular basis to search for and resolve any violations from the established security policies and baseline requirements.

5.3. Intrusion Detection and Prevention Systems

When it comes to recognising and responding to server-side threats, intrusion detection and prevention systems, also known as IDPS, serve as critical lines of defence. IDPS is able to identify unusual patterns of behaviour that may be suggestive of possible threats since it monitors network traffic, host activity, and application events. The following are important IDPS features:

detection based on signatures, which can recognise previously seen attack patterns and harmful payloads.

A detection method that is based on anomalies and looks for departures from predetermined baselines and behavioural profiles.

Capabilities for automated responses in real time, such as blocking, quarantining, or mitigating suspected attacks.

Integration with several additional security tools and platforms in order to provide a comprehensive view of the security posture of the organisation.

5.4. Security Incident Response and Recovery

When it comes to managing server-side threats and minimising the impact they have on business operations, having a security incident response and recovery plan that is well defined is absolutely necessary. Organisations are able to reduce the potential damage caused by server-side assaults if they put in place a method that is structured for detecting security problems, responding to such incidents, and recovering from them. A security incident reaction and recovery plan should always include the following essential components:

It is important to have defined duties and responsibilities for each member of the incident response team.

a set of protocols that can be followed in order to recognise, categorise, and report security events.

Communication procedures to ensure that information is sent to important stakeholders in a timely manner and in an accurate manner.

plans and processes for fallback in the event that the primary plan fails to assist the recovery and restoration of the affected systems and services.

Analysis of what happened after the incident and the lessons that were learnt are used to continuously improve the organization's security posture and its ability to respond to incidents.

Organisations are able to fortify their defences and be more proactive in the face of the ever-evolving threats that exist in the digital realm if they put these server-side attack mitigation measures into practise.

6. LEGAL AND ETHICAL CONSIDERATIONS

Ethical hackers have the difficult task of navigating the complicated legal and ethical landscape that surrounds their profession as the prevalence of server-side attacks continues to rise. This section examines the legal frameworks that regulate cybersecurity, digs into the details that differentiate ethical hacking from crimes, and covers responsible disclosure practises.

6.1. The Fine Line Between Ethical Hacking and Cybercrime

At first look, unethical hacking and criminal hacking may appear to be the same thing because they both require exploiting flaws and finding ways around security features. The distinction between the two, however, comes down to the persons involved and what they had in mind when they gave their consent. Hackers that act within the confines of the law and ethical norms approach their targets with a request for permission and a plan to improve their security measures. On the other hand, cybercriminals take use of vulnerabilities for nefarious reasons, such as the theft of data, the acquisition of financial gain, or the disruption of service. In order to guarantee that ethical hacking stays within the confines of the law, practitioners are required to do the following:

Before beginning any testing activities, you must first obtain the explicit authorisation of the organisation that will be tested.

Maintain adherence to the conditions of engagement, which include both the scope and the constraints of your work.

Always maintain the organization's and its data's privacy, as well as their secrecy.

Avoid causing any damage or causing any interruption to the systems or services that are the target.

6.2. Legal Frameworks and Compliance Requirements

The fields of ethical hacking and cybersecurity are governed by a number of different legal frameworks and regulations, with the overarching objective of protecting the confidentiality, safety, and authenticity of sensitive data. Ethical hackers and the organisations that they service absolutely need to ensure that they are in compliance with these regulatory frameworks. The following is a list of important legal frameworks and compliance requirements:

The Computer Fraud and Abuse Act (CFAA) is a law that was passed in the United States that makes it illegal to gain unauthorised access to computer systems that are protected.

The General Data Protection Regulation (GDPR), which is found in the European Union, is a regulation that places rigorous standards on organisations that handle personal data in regards to data protection and privacy.

The Payment Card Industry Data Security Standard, also known as PCI-DSS, is a standard that outlines the security standards that must be met by any organisation that processes, stores, or transmits payment card data.

regulations that are unique to a particular industry, such as the Health Insurance Portability and Accountability Act (HIPAA) in the medical field or the Federal Information Security Management Act (FISMA) in the public sector.

6.3. Responsible Disclosure and Collaboration With Vendors

It is considered an ethical practise to notify newly identified vulnerabilities to the affected vendor or organisation. This gives the impacted party the opportunity to fix the problem before it is made public or exploited by bad actors. The practise of responsible disclosure encourages collaboration and trust between ethical hackers and providers, which in turn helps to develop a digital ecosystem that is more secure. The following are important components of responsible disclosure:

disclosure of found vulnerabilities either directly to the organisation that was impacted or through a specific channel created for the reporting of vulnerabilities.

Providing in-depth information regarding the vulnerability, its possible effects, and the steps that should be taken to remediate it.

coordinating with the organisation in order to develop a schedule for public disclosure that is mutually acceptable to all parties, taking into account the complexity of the problem as well as the possible harm to consumers.

until the vulnerability has been sufficiently resolved, refraining from revealing sensitive information such as proof-of-concept code or exploitation techniques.

Ethical hackers are able to effectively differentiate themselves from cybercriminals and contribute to the continuous mission of securing digital assets and information when they adhere to legal and ethical principles in their hacking practises.

7. WHAT WE UNDERSTOOD

Server-side attacks continue to be a tough obstacle for businesses and other organisations all over the world, notwithstanding the ongoing transformation of the digital ecosystem. Contributing to the development of a digital environment that is both more secure and more resistant to disruption are ethical hackers, who play an essential part in the process of locating and neutralising potential risks.

7.1. The Role of Server-Side Attacks in Ethical Hacking

When it comes to ethical hacking and penetration testing, attacks on the server side are an essential component. Ethical hackers are able to do an accurate assessment of an organization's security posture, locate vulnerabilities, and make recommendations for corrective actions since they have mastered the tactics and methodologies connected with the assaults. Ethical hackers offer businesses the benefit of their experience in the form of the construction of strong defences, a reduction in the likelihood of successful assaults, and ultimately the protection of sensitive data and systems.

7.2. Future Trends and Challenges

Ethical hackers need to maintain a state of vigilance and adaptability in order to stay ahead of the curve as technology continues to evolve and new attack avenues appear. The following are examples of potential future trends and problems that may have an impact on server-side attacks and ethical hacking:

The increasing use of cloud-based infrastructure and serverless computing, both of which present new opportunities for cyberattacks and raise new security issues.

The proliferation of technologies like artificial intelligence (AI) and machine learning (ML), which have the potential to be utilised by both attackers and defenders alike.

The growing number of devices connected to the Internet of Things (IoT), which frequently have inadequate security measures and are therefore vulnerable to being used as entry points for assaults on servers.

The ongoing development of quantum computing, which has the potential to displace more conventional techniques to encryption and call for the adoption of novel security strategies.

Ethical hackers can continue to play an important role in defending against server-side assaults and protecting the digital landscape if they remain abreast of evolving trends and adjust their abilities accordingly. In conclusion, the mastery of server-side attacks in ethical hacking is vital for organisations to better their security posture and successfully respond to the ever-evolving threats in the digital environment. This is because server-side attacks are the most common type of attack in ethical hacking.

8. CASE STUDIES: REAL-WORLD SERVER-SIDE ATTACKS

It is vital to look at examples from the actual world in order to acquire a deeper comprehension of the consequences and ramifications associated with server-side assaults. This section includes a study of four significant case studies, stressing the impact of these attacks and the useful lessons that can be drawn from them. Additionally, the section concludes with some recommendations for future action.

8.1. Equifax Data Breach: Exploiting Unpatched Vulnerabilities

In 2017, a server-side attack on Equifax, one of the main credit reporting agencies in the United States, resulted in the exposing of sensitive personal data belonging to about 147 million individuals. These individuals' data included social security numbers, birth dates, addresses, and more. The attackers took use of a known flaw in the Apache Struts web application framework, which Equifax had not patched in a timely manner despite the company being aware of the vulnerability.

The following are some important lessons that can be drawn from the Equifax data breach:

The significance of maintaining timely and thorough patch management in order to minimise the impact of known vulnerabilities.

The need of doing a risk analysis and ranking potential vulnerabilities before beginning any remedial work.

The need of enhancing incident response and communication procedures in order to lessen the damage caused by a security breach.

8.2. Apache Struts Vulnerability: Impact and Lessons Learned

Because of its extensive adoption and the presence of various serious vulnerabilities, the Apache Struts web application framework is a common target of server-side assaults. This is the case despite the fact that it is used by numerous organisations across the globe. As a result of these vulnerabilities, attackers have been granted the ability to run arbitrary code, which has led to the compromise of underlying servers and the exfiltration of sensitive data.

The following are some important takeaways from the cases involving the Apache Struts vulnerability:

The requirement that, throughout the software development lifecycle, developers give security a high priority and incorporate security testing at each step of development.

The significance of staying up to date with the latest security bulletins and swiftly resolving any vulnerabilities that are discovered.

The importance of implementing many levels of security controls as part of a defense-in-depth strategy for the purpose of protecting against possible threats.

8.3. The Mirai Botnet: IoT Devices as Attack Vectors

The Mirai botnet, which first surfaced in 2016, is an example of an innovative form of server-side assault. It launched devastating Distributed Denial of Service (DDoS) attacks by exploiting weak Internet of Things (IoT) devices. Through the use of default credentials and lax security configurations, the Mirai malware was able to infect millions of Internet of Things devices, such as cameras and routers, and transform them into a massive botnet that was capable of bringing down online services and infrastructure.

Case study highlights from the Mirai botnet include the following:

The urgent requirement for increased security measures in IoT devices, which must include the incorporation of robust authentication systems and consistent security updates.

The scope and potential impact of server-side attacks that make use of botnets and other coordinated methods.

The significance of coordinated efforts between the public and commercial sectors in the fight against widespread cyberattacks.

8.4. SolarWinds Orion Breach: Supply Chain Compromise

The SolarWinds Orion security breach, which was found in 2020, was the result of a complex attack on the supply chain. In this attack, bad actors exploited a vulnerability in the update system of the SolarWinds Orion network management software, which is extensively deployed. As a result of this penetration, the attackers were able to enter the networks of a large number of high-profile organisations, including government agencies and private sector companies, which resulted to a significant security breach with far-reaching implications.

Important takeaways from the hack at SolarWinds Orion include the following:

The imperative for businesses to move towards a zero-trust security architecture, which operates under the presumption that any part of their environment may be vulnerable to attack.

The significance of keeping an eye on and protecting third-party software and supply chains in order to forestall attacks of a similar nature.

The need of sharing threat intelligence and working together across organisations that have been impacted by large-scale incidents in order to quickly respond to incidents and reduce their effects.

9. BEST PRACTICES FOR ETHICAL HACKERS AND ORGANIZATIONS

It is imperative that ethical hackers and organisations implement best practises that build a robust security posture in light of the lessons learnt from real-world server-side attacks. These attacks have provided invaluable insight. In this section, essential advice for both parties are outlined, with the goals of reducing the danger of server-side attacks and improving their overall cybersecurity resilience.

9.1. Establishing a Security-First Mindset

It is imperative that security be given top priority in each and every facet of an organization's operations, from long-term planning to day-to-day operations. What this means is:

Defining well-defined security policies and operating procedures that will serve to direct the actions of the organisation.

In order to foster a culture that values alertness and accountability, it is important to provide security awareness training to all of the personnel.

In order to ensure that there is a unified strategy for security management, it is important to encourage open communication and collaboration between security teams and other departments.

9.2. Continuous Education and Skill Development

Ethical hackers and organisations alike should make investments in ongoing education and the development of their skills in order to keep one step ahead of both new security threats and advances in technology. This necessitates:

By taking part in industry conferences, workshops, and training courses, one may ensure that they are up to date on the most recent developments in methods, instruments, and trends.

Participating in professional forums and networks in order to exchange information and ideas on what works best.

pursuing the appropriate certifications and accreditations in order to demonstrate your level of expertise and dedication to your professional development.

9.3 Taking a Preventative Role With Regards to Safety and Security

Organisations ought to move from a reactive to a proactive security posture by actively finding out vulnerabilities and correcting them before they can be exploited. This would allow the organisations to prevent vulnerabilities from being exploited. These are the following:

carrying out vulnerability assessments and penetration tests on a regular basis in order to locate holes in their environment and take corrective action.

Putting in place methods for continuous monitoring and threat detection in order to identify potential security breaches and formulate a response to them.

Collaborating with third-party security researchers and ethical hackers by means of bug bounty programmes and other responsible disclosure initiatives.

9.4. Embracing Collaboration and Information Sharing

Organisations and ethical hackers need to collaborate and share knowledge in order to be able to effectively tackle server-side assaults and other types of cyber threats. What this means is:

Exchanging threat intelligence and working together on incident response activities while participating in information sharing and analysis centres (ISACs).

Participating in public-private partnerships with the goal of utilising the knowledge and assets of both the public and commercial sectors in the fight against cybercrime.

Establishing solid connections with relevant law enforcement authorities, regulatory bodies, and other industry participants in order to allow cooperation and coordination in the case of a security breach.

Ethical hackers and organisations can better prepare themselves for the challenges posed by server-side attacks and contribute to a more secure and resilient digital environment by adopting these recommended practises and contributing to a more secure and resilient digital ecosystem.

10. DEMONSTRATING SERVER-SIDE ATTACKS: PRACTICAL EXAMPLES

It is helpful to examine real-world examples that demonstrate how server-side assaults are carried out and the potential effects they can have on target systems in order to acquire a more in-depth comprehension of these attacks. The following examples illustrate a variety of server-side attack approaches and serve to demonstrate how important it is to employ robust security measures in order to protect oneself from threats of this nature.

10.1. SQL Injection Attack

Users are able to search for products on an e-commerce website by typing a term into a search bar that is shown on the page. However, the backend of the website does not adequately sanitise user input, which leaves it open to the possibility of a SQL injection attack. In order to take advantage of this vulnerability, an attacker would need to type in the search bar a SQL query that has been carefully constructed, such as:

' OR 1=1; --

Because of this input, the database server would return all product entries, so circumventing any access controls that were intended to be in place. The malicious actor might possibly utilise these tactics to modify or exfiltrate sensitive data from the database, which would cause considerable harm to both the company and its clients.

10.2. Attack Known as "Local File Inclusion" (LFI)

Administrators of a content management system (CMS) have the ability to incorporate files from other locations by giving a file path parameter to the CMS via a URL. The Content Management System (CMS), on the other hand, does not properly check the given file path, which makes it vulnerable to a local file inclusion (LFI) attack. An adversary could take advantage of this vulnerability by constructing a URL that contains sensitive system files like the following:

https://example.com/cms/include.php?file=../../../../etc/pas
swd

This URL would cause the server to return the contents of the "/etc/passwd" file, exposing user account information and potentially granting the attacker further access to the system. In a worst-case scenario, the attacker could leverage this vulnerability to achieve remote code execution and take full control of the server.

10.3. Server-Side Template Injection (SSTI)

A web application uses a popular template engine to dynamically generate HTML content based on user input. However, the application does not properly sanitize user input before passing it to the template engine, making it vulnerable to server-side template injection (SSTI). An attacker could exploit this vulnerability by submitting input that includes malicious template code, such as:

{{7*7}}

If the programme evaluates the input and displays it as "49," it would mean that the attacker was successful in running arbitrary code on the server. After that, the attacker could try to elevate their privileges, carry out other instructions, or compromise other parts of the system.

These real-world examples highlight the potential consequences of server-side attacks and emphasise the significance of rigorous security practises in protecting an organization's digital assets. These practises include input validation, secure coding, and frequent vulnerability assessments.

11. PENETRATION TESTING METHODOLOGIES FOR SERVER-SIDE ATTACKS

Ethical hackers and professionals in the security field should apply penetration testing procedures that are both methodical and comprehensive in order to properly detect and address server-side vulnerabilities. These approaches provide a methodical approach to evaluate an organization's security posture and discovering holes that could be exploited in server-side attacks. This assessment can be used to determine whether or not an organisation is adequately protecting its data.

11.1. Testing Guide for the Open Web Application Security Project (OWASP)

The OWASP Testing Guide is a methodology that focuses on ensuring the safety of web applications. It has gained widespread use. It protects against a wide variety of different types of server-side attacks, such as injection issues, faulty access controls, and unsafe configurations. The following are important components of the OWASP Testing Guide:

The process of mapping the application's attack surface and detecting entry points is known as information gathering.

Evaluation of the safety of the application's infrastructure, including testing of server configurations and patch management. This type of testing falls under the purview of configuration and deployment management.

Testing for authentication and session management involves determining how robust access control measures and user session management are.

Testing for data validation involves locating possible injection vulnerabilities and input validation weak spots.

11.2. The Penetration Testing Execution Standard, Also Known as PTES

The Penetration Testing Execution Standard (PTES) is a comprehensive methodology that offers a standardised framework for the execution of penetration tests across a wide range of system types and situations. The PTES addresses vulnerabilities in web, database, and application servers, and one of its primary focuses is the protection of servers from attacks that originate on the server side. The following are important stages of the PTES:

Interactions that take place before the actual engagement consist of defining the penetration test's scope, objectives, and rules of engagement.

Collecting information on the target environment, such as server configurations, network topology, and potential attack vectors, is an example of intelligence gathering.

Analysing server-side vulnerabilities involves both automated and human testing methodologies. Vulnerability analysis refers to the process of identifying and analysing these server-side vulnerabilities.

The term "exploitation" refers to the process of attempting to take advantage of vulnerabilities that have been uncovered in order to evaluate the impact those vulnerabilities could have in the actual world.

Evaluation of the level of compromise and locating other entry points to the system is part of the post-exploitation phase.

Documenting the findings and offering recommendations for corrective action and risk reduction are both aspects of reporting.

Ethical hackers are able to reduce the likelihood of successful attacks and improve an organization's overall security posture by following certain penetration testing procedures and systematically identifying and addressing server-side vulnerabilities.

12. REAL-WORLD SERVER-SIDE DEFENSE SCENARIOS

Let us explore a few hypothetical scenarios that illustrate how organisations can successfully protect themselves against server-side attacks so that we can emphasise how important it is to have strong server-side defences and demonstrate how effective the tactics and approaches that have been described throughout this chapter.

12.1. Preventing SQL Injection Attacks

The online banking programme provided by a financial institution offers a transaction history search tool that enables users to search for transactions based on the dates, quantities, and descriptions of the transactions. The development team performs stringent input validation and sanitization to prevent SQL injection attacks. This ensures that all user-supplied data is correctly validated and cleaned before being sent to the database server.

In addition, the group utilises parameterized queries and stored procedures to partition SQL code from data, which further reduces the likelihood of injection attacks occurring. As a consequence of this, the company is able to protect the honesty and privacy of the information pertaining to its clients while also ensuring that the service it offers is trustworthy and reputable.

12.2. Protecting Against Local File Inclusion (LFI) Attacks

Popular news website employs a content management system (CMS) that enables authors to embed multimedia files in their articles by giving a file path. This feature is available on the CMS used by the website. The development team implements a whitelist strategy in order to defend against LFI attacks. This strategy restricts the types of files that can be included to just certain categories, such as photographs, videos, and audio files, and ensures that all file paths are confined to a single media folder.

Because this organisation has implemented these security controls, it is able to prevent attackers from accessing sensitive system files or running arbitrary code. As a result, the organization's web server is able to keep its security and stability intact.

12.3. Mitigating Server-Side Template Injection (SSTI) Vulnerabilities

A template engine is utilised by a cloud-based project management platform in order to generate dynamic and personalised content for the platform's users. Before the user input is sent to the template engine for processing, the development team makes sure that it has been completely sanitised so that any potential SSTI vulnerabilities can be avoided. This involves eliminating characters that could potentially be dangerous and validating input against a preset set of values that are acceptable.

In addition, the group makes use of sandboxing strategies to separate the execution environment of the template engine, which helps to reduce the potential damage that could be caused by an SSTI assault. Because of this, the company is in a position to offer its customers a service that is both secure and dependable, while also reducing the likelihood that the server would be compromised.

The usefulness of implementing best practises, secure coding techniques, and proactive security measures in warding off server-side threats is illustrated by the real-world defence scenarios that are presented in this section. Organisations can greatly enhance their server-side security posture and minimise the likelihood of successful attacks by applying the tactics mentioned in this chapter, which can significantly improve their server-side security posture.

13. ENHANCING SERVER-SIDE SECURITY WITH EMERGING TECHNOLOGIES

Because the digital landscape is constantly shifting, businesses have a responsibility to remain current on developing technologies that have the potential to improve the server-side security posture of their operations. The following technologies have the potential to have a substantial influence on server-side security and offer new alternatives for businesses to defend themselves against assaults launched from the server side.

Artificial intelligence and machine learning are discussed in this section.

Utilising artificial intelligence (AI) and machine learning (ML), one can examine massive volumes of data and detect patterns that may be suggestive of server-side assaults. This can be accomplished by leveraging both of these technologies. When organisations use security solutions that incorporate AI and ML, they are able to spot anomalies, malicious actions, and potential vulnerabilities in real time. This enables the organisations to respond and remediate problems more quickly.

13.1 Serverless Architectures in Computer Systems

Serverless architectures, such as AWS Lambda and Google Cloud Functions, make it possible for companies to design and deploy applications without requiring the management of the underlying infrastructure. This frees up the company to focus on other priorities. When organisations outsource the operation of their servers to cloud providers, they are able to reap the benefits of built-in security features such as automatic patching and secure configuration defaults. This helps to minimise the risk of server-side attacks and reduces the attack surface available to potential attackers.

Containerization and microservices are covered in the next section.

Containerization and microservices architectures are two approaches that can help businesses improve their server-side security by isolating individual application components and reducing the potential damage that could result from a breach. By running apps within containers, businesses can be certain that each component functions in its own safe and isolated environment. This reduces the damage that can be caused by an attack and makes it easier for businesses to recover quickly.

13.2. Security Models Based on Zero Trust

A zero-trust security model is one that requires continual validation and verification because it operates under the presumption that every user, device, and network connection is at risk of being compromised. Organisations are able to minimise the danger of unauthorised access and lessen the possibility of successful assaults if they implement a zero-trust strategy to the server-side security of their systems. Multi-factor authentication, least-privilege access, and continuous monitoring are three essential elements that make up a zero-trust approach.

Organisations are able to remain ahead of the constantly shifting threat landscape and defend their precious digital assets from server-side attacks if they embrace these developing technologies and incorporate them into their server-side security plans.

Chapter Summary

This chapter explored the critical domain of server-side attacks, highlighting their importance in the context of ethical hacking and penetration testing. By understanding server-side vulnerabilities, including common security flaws and specific weaknesses in web, database, and application servers, ethical hackers and security professionals can better identify and address potential threats.

A range of server-side attack techniques were examined, such as SQL injection, cross-site scripting (XSS), local file inclusion (LFI), remote file inclusion (RFI), and server-side template injection (SSTI), providing insight into how these attacks can be executed and the potential consequences they may have on target systems.

The chapter also discussed various server-side attack mitigation strategies, including input validation and sanitization, secure configuration and patch management, intrusion detection and prevention systems, and security incident response and recovery. These strategies are essential for organizations to maintain a strong security posture and defend against server-side attacks.

Legal and ethical considerations surrounding ethical hacking and server-side attacks were addressed, emphasizing the fine line between ethical hacking and cybercrime, the importance of compliance with legal frameworks, and the need for responsible disclosure and collaboration with vendors.

Real-world server-side attack examples, case studies, and hypothetical defense scenarios were presented to underscore the potential impact of server-side attacks and demonstrate the effectiveness of robust security measures. Penetration testing methodologies, such as the OWASP Testing Guide and the PTES, were explored to provide a structured approach to assessing server-side security.

Finally, the chapter touched upon emerging technologies that can help enhance server-side security, including artificial intelligence, machine learning, serverless architectures, containerization, microservices, and zero-trust security models. By staying informed about these technologies and incorporating them into their server-side security strategies, organizations can better protect their digital assets and minimize the risk of server-side attacks.

Chapter 16
Wireless Hacking

Shubh Gupta
University of Petroleum and Energy Studies, India

Oroos Arshi
University of Petroleum and Energy Studies, India

Ambika Aggarwal
University of Petroleum and Energy Studies, India

ABSTRACT

The network has become portable as a result of digital modulation, adaptive modulation, information compression, wireless access, and multiplexing. Wireless devices connected to the internet can possess a serious risk to the information security. These devices communicate among themselves in a public domain which is very easily susceptible to attacks. These devices only depend upon the encryption and their shared keys to help them mitigate the risk when data is in transit. Also WEP/WPA (wired equivalent privacy/ wireless protected access) cracking tools are taken care to avoid break into attacks. Several wireless networks, their security features, threats, and countermeasures to keep the network secure are all covered in this chapter. It analyses various wireless encryption techniques, highlighting their advantages and disadvantages. The chapter also explores wireless network attack techniques and provides countermeasures to safeguard the information systems and also provide a wireless penetration testing framework for safeguarding the wireless network.

1. INTRODUCTION

With the use of wireless technology (Restuccia et al., 2020), the computer industry

DOI: 10.4018/978-1-6684-8218-6.ch016

is entering a new phase of technical development. The use of wireless networking is changing how people work and play. To make data portable, transportable, and accessible, people can use networks in fresh ways by removing the physical link or cable. A wireless platform includes to devices and gathers data by utilising radio frequency technologies. It is an uncontrolled data transmission system. By using this network, users are released from complex and numerous wired connections. Without creating a physical link between them, it uses electromagnetic waves to connect to various sites. In wireless networks, electromagnetic waves are employed to send data and signals through the connection line. In addition to being quick and easy to set up, wireless networks (Restuccia et al., 2020) also eliminate the need for wiring via ceilings and walls, provide connectivity in locations where it is difficult to run cables, and allow access remotely that is within reach of a base station. You can get a wireless LAN connection to the internet at public locations like airports, libraries, schools, and even coffee shops.

1.1 Wireless Terminology

Electromagnetic waves are used in wireless networks to transport signals through the communication line and transmit data. Below are several terms related to wireless networks:

1. GSM (Global System for Mobile communication) (Restuccia et al., 2020): Worldwide wireless networks employ an all-encompassing system for mobile mobility.
2. Bandwidth: It tells about how much data can be exchanged through a connection. The term "bandwidth" typically refers to the data transfer rate. The bandwidth is measured in bits (amount of data) per second (bps).
3. BSSID: A Basic Service Set Identifier (BSSID) is the MAC (Čisar & Čisar, 2018) (Multiple Access Point) address of an access point (AP) or base station that has configured a Basic Service Set (BSS). Users frequently have no idea which BSS they are a part of. When a user transfers a device from one location to another, the BSS that the device uses may change due to a fluctuation in the range that the AP covers, though this may not have an impact on the wireless device's connectivity.
4. ISM (Industrial, Scientific, and Medical) Band (Čisar & Čisar, 2018): A collection of frequencies for the global scientific, industrial, and medical sectors.
5. Access point: Wireless devices are connected to a wireless/wired network through access points. Through wireless protocols like Bluetooth and Wi-Fi,

it enables wireless communication devices to connect to a wireless network. It acts as a switch or hub for the wireless and wired local area networks.

6. Hotspot: localities with open-access wireless networks. Wi-Fi hotspots are locations where users can activate Wi-Fi on their devices and connect to the Internet via a hotspot.

7. Association: tethering is the process of joining a wireless device to an access point.

8. Set (SSID): The wireless local area network (WLAN) is given a 32-character, unique identification called an SSID, which serves as the network's wireless identifier. Among a list of available networks, the SSID allows connecting to the needed network. The same SSID should be used by devices connected to the same WLAN to establish a connection.

9. Orthogonal Multiplexing (OFDM) (Čisar & Čisar, 2018): A signal at a selected frequency is divided into multiple orthogonal (occurring at right angles to each other) carrier frequencies using the digital modulation technique of OFDM. Sharing bandwidth with some other independent channels, OFDM maps data on variations in carrier phase, frequency, amplitude, or any combination of these. In comparison to parallel channel operation, it creates a transmission architecture that supports larger bit rates. Additionally, it is a technique for digital data encoding on several carrier frequencies.

10. Multiple input, multiple output-orthogonal frequency-division multiplexing (MIMO-OFDM) (Čisar & Čisar, 2018): The spectral efficiency of 4G and 5G wireless communication services is impacted by MIMO-OFDM. Using the MIMO-OFDM technology improves the channel's robustness while reducing interference.

11. Direct-sequence Spread Spectrum (DSSS): A pseudo-random noise spreading code is multiplied by the original data stream in the spread spectrum technique known as DSSS. This technology, which guards against interference and jamming, is also known as a data transmission strategy or modulation scheme.

12. Frequency-hopping Spread Spectrum (FHSS): FHSS, commonly referred to as Frequency-Hopping Code Division Multiple Access (FH-CDMA), is a technique for sending radio signals that involves quickly switching a carrier across several frequency channels. It reduces the effectiveness of unlawful telecommunications interception or jamming. In FHSS, a transmitter uses a predetermined method to hop between available frequencies in a pseudorandom sequence that is known to both the sender and receiver.

2. WIRELESS NETWORKS

Radio wave transmission is used for transmission in wireless networks. The physical layer of the network structure is where this typically happens. With the advent of wireless communication, significant changes are being made to data networking and telecommunication. The term "wireless network" (also known as "Wi-Fi") refers to wireless local area networks (WLAN) (Pimple et al., 2020), which are based on the IEEE standard and enable a device to access the network from any location within the range of an access point. Wi-Fi is a commonly used technology for radio-based wireless communication. DSSS, FHSS (Pimple et al., 2020), Infrared (IR), and OFDM are only a few of the multiple connections that Wi-Fi sets up between the transmitter and the receiver. Wi-Fi allows devices to connect to network resources like the internet, such as personal computers, video game consoles, and smartphones.

The following are a few benefits and drawbacks of wireless networks:

Benefits

1. Wiring through walls and ceilings is avoided during installation since it is quick and simple.
2. In locations where cable laying is challenging, connectivity is more readily available.
3. Anywhere that is within an access point's range can access the network.
4. Using Wireless LAN, public locations such as airports, libraries, schools, or even coffee shops provide you with ongoing Internet connectivity.

Drawbacks

1. Security is a major concern that may not live up to expectations.
2. The bandwidth decreases as the number of machines on the network rise.
3. For Wi-Fi improvements, new wireless cards and/or gateways may be necessary.
4. Some technological devices can obstruct Wi-Fi networks.

2.1 Wi-Fi Networks in Public and Residential Areas

*   Wi-Fi in the home (Pimple et al., 2020): You may use a laptop or other portable device wherever you want at home thanks to Wi-Fi networks, which eliminate the need to bury or dig ditches for Ethernet wires.
*   Wi-Fi in Public Zone (Pimple et al., 2020): Wi-Fi connection is available for free or at a cost in public places such as coffee shops, libraries, airport terminals, and hotels.

Figure 1. Types of wireless networks

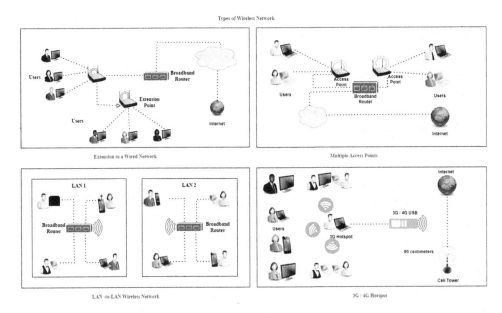

2.2 Types of Wireless Networks

The various kinds of wireless networks are as follows:

1. Extension of a Network (Reddy & Srikanth, 2019): By installing APs between a wired network and wireless devices, a user can extend a wired network. Using an AP, a wireless network may also be built.

    i) Various AP types include:
        a. Software APs (SAPs): They can run on a computer with a wireless NIC and be linked to a wired network.
        b. Hardware APs (HAPs): It supports the majority of wireless functionality.

    The AP functions as a switch in this kind of network, connecting PCs with wireless network interface cards (NIC). The AP can link wireless clients to a wired LAN, enabling access to LAN resources like file servers and internet connections for wireless computers.

2. Multiple Access Points: Multiple APs are used in this form of network to wirelessly connect computers. Multiple Access points can be constructed if

Figure 2. Open system authentication process

a single AP is unable to cover a given area. The wireless range of each AP must overlap that of its neighbor. The use of a function known as roaming, enables users to move about without any disruption. Some manufacturers create wireless relay extension points that increase the range of a single AP. To extend wireless coverage to areas remote from the main AP, several extension points can be connected.

3. LAN-to-LAN (Reddy & Srikanth, 2019) Wireless Network: Local computers have wireless connectivity thanks to APs. and local PCs on several networks can connect. All physical APs can communicate with one another. But linking LANs together through wireless connectivity is a challenging undertaking.

4. 3G/4G Hotspot: It is a type of wireless network that provides Wi-Fi connectivity to a variety of devices that support the technology, such as netbooks, Media players, cameras, PDAs, and tablets.

2.3 Wi-Fi Authentication Modes

Open System Authentication Process (Sen, Maity, & Das, 2020): Any wireless client attempting to connect to a Wi-Fi network sends an authentication request to the wireless AP. The station transmits an authentication administration frame with the authenticity of the sending station in it for recognition and linkage with the other mobile station throughout this process. The AP then returns a verification frame to the requesting station to finish the authentication process and confirm access.

Shared Key Authentication Process: It is obtained by each wireless station taking part in this process across a secure channel that is different from the 802.11 wireless connection communication channels.

The AP may reject the station and prohibit it from connecting to either the 802.11 or Ethernet networks if the decrypted text does not meet the initial challenge text.

Figure 3. Shared key authentication key

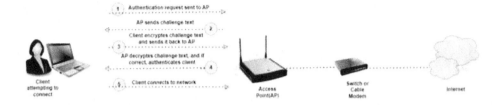

3. WIRELESS THREATS

To secure wireless networks, a network administrator needs to understand the various possible inabilities (weaknesses) of encryption algorithms (Nair et al., 2018) that lure attackers to crack wireless communications. The wireless network can be at risk of various types of attacks, including access control attacks, integrity attacks, confidentiality attacks, availability attacks, authentication attacks, etc. This section will discuss types of security risks, threats, and attacks associated with wireless networks.

3.1 Access Controls Attacks

To enter a network, these attacks attempt to get through wireless LAN access control measures. Access control assaults can take many different shapes, including:

1. War Driving: Either by sending probe requests via a connection or by listening for web beacons, wireless LANs can be found. A breach point can be used by an attacker to launch additional LAN attacks. The attacker may deploy several tools, such as NetStumbler and KisMAC, to carry out wardriving attacks.
2. MAC Spoofing (Khasawneh et al., 2014): A MAC address can be modified by an attacker so that it appears to a host on a trusted network as an authorized AP. To carry out this kind of assault, the attacker may employ tools like SMAC.
3. AP Misconfiguration: Any of the access control settings at any of the APs could be poorly configured by the user, leaving the entire network vulnerable to attacks. Since the system sees the AP as a legal device, most intrusion-detection systems cannot send-off alerts in response to them.
4. Promiscuous Client: A flaw in 802.11 wireless cards allows an attacker to take advantage of their constant search for a stronger signal. An attacker installs an AP close to the target Wi-Fi network, gives it a generic SSID (Baharudin et al., 2015) name, and then offers a signal and speed that is incomparably stronger and faster than the victim's Wi-Fi network. The goal is to persuade the

Client to join the attacker's AP rather than an authorized Wi-Fi network. An attacker can send target network traffic over a fake AP thanks to promiscuous clients. It resembles the wireless network's "evil twin" threat, in which an attacker establishes an AP that seems to be an authorized AP by broadcasting the WLAN's SSID.

3.2 Integrity Attacks

Changing or modifying data while it is being transmitted constitutes an integrity attack. Attackers that want to carry out another kind of assault on wireless devices send falsified control, management, or data frames via a wireless network in wireless integrity attacks (e.g., DOS).

1. Data Frame Injection: A frame being injected into your website by an attacker. The effects of this vulnerability have evolved alongside browser evolution. To carry off the attack, the attacker may use a variety of techniques, including Airpwn, libra date, and wnet dinject/reinject.
2. WEP Injection (Baharudin et al., 2015): A method of changing the content of a web page on the client side and inserting custom content by inserting malware into the URL bar and intercepting all HTTP requests and server responses.
3. Bit-Flipping Attacks: A bit-flipping attack is a method of breaking a cryptographic cypher in which the intruder can alter the ciphertext in a way that alters the plaintext predictably without having access to the plaintext itself. As opposed to cryptanalysis, which would target the cypher directly, this form of attack targets a specific message or set of messages. At its worst, this may develop into a denial-of-service attack against all messages sent using that encryption over a specific channel.
4. Extensible AP Replay: An attacker can actively participate in one of these protocols or actively target them by observing the traffic and trying to gather relevant information. The attacker may attempt to pose as a MitM participant by pretending to be the server, the client, or both.

3.3 Confidentiality Attacks

It makes no difference whether the system sends data in clear text or encrypted form; the goal of these attacks is to intercept sensitive data received over a wireless network. If the system transmits data in encrypted form, a hacker will attempt to circumvent the encryption (such as WEP or WPA). threats to wireless networks' privacy Include:

Table 1. types of attack

Types of Attack	Description	Method and Tools
Eavesdropping	Capturing and decoding unprotected application traffic to obtain potentially sensitive information.	BSD-airtools, Ethereal, Ettercap, Kismet, commercial analyzers
Evil twin AP	Posing as an authorized AP by the WLAN's SSID to lure users.	CqureAP, HostAP, OpenAP
Session Hijacking	Manipulating the network so the attacker's host appears to be the desired destination.	Manipulating
Traffic analysis	Inferring information from the observation of external traffic characteristics.	
MITM Attack	Running traditional MITM attack tools on an evil twin AP to intercept TCP sessions or SSL/SSH tunnels.	sniff, Ettercap

3.4 Availability Attack

By disabling such resources or refusing them direct exposure to WLAN (Baharudin et al., 2015) resources, availability attacks seek to block the provision of wireless systems to authorized users. Wireless network services are inaccessible to authorized users due to this assault. Attackers can carry out these kinds of attacks in a variety of methods, which prevent wireless networks from being accessible.

3.5 Authentication Attacks

To get unauthorized access to network resources, the goal is to steal the identity of Wi-Fi clients as well as their login credentials and other personal information.\

3.6 Rouge Access Point Attack

SSIDs are used by APs to authenticate before connecting to client NICs (Baharudin et al., 2015). Anyone who has an 802.11-equipped device may be able to join the business network using unauthorized (or rogue) APs. An attacker may get access to the network using an unauthorized AP. Vendor name authorized MAC address, and security parameters can all be discovered from APs using wireless sniffing programs. The attacker can then cross-reference the list of MAC addresses discovered through sniffing with a table of MAC addresses of authorized APs on the target LAN. Then, an attacker can build a fake AP and install it close to the business network that is the target. Attackers can control legitimate network users' connections by inserting a malicious AP into an 802.11 network. The malicious AP will offer to connect to the

Table 2. types of attack

Type Of Attacks	Description	Methods and tools
AP theft	Physically removing an AP from its installed location	Stealth and/or speed
EAP- Failure	Observing a valid 802. IX EAP exchange, and then sending the client, a forged EAP-Failure message.	File2air and libradiate
Denial-of-service	Exploiting the CSMA/CA Clear Channel Assessment (CCA) mechanism to make a channel appear busy.	An adapter that supports CW Tx mode, with a low-level utility to invoke continuously transmissions
Routing Attacks	Distributing routing information within the network	RIP protocol
ARP Cache Poisoning Attack	Creating many attack vectors.	
Power saving Attacks	Transmitting a spoofed TIM or DTIM to the client while in power-saving mode, making the client vulnerable to a DoS attack.	
Beacon flood	Generating thousands of counterfeit 802.11 beacons to make it hard for clients to find a legitimate AP.	FakeAP

Table 3. types of attack

Type of Attack	Description	Methods and tools
LEAP Cracking	Recovering user credentials from captured 802.1X Lightweight EAP (LEAP) packets using a dictionary attack tool to crack the NT password hash.	Anwrap, Asleap, THC-LEAPcracker
PSK Cracking	Recovering a WPA PSK from captured key handshake frames using a dictionary attack tool.	coWPAtty, KisMAC, wpa_crack, wpa-psk-bf
Domain Login Cracking	Recovering user credentials (e.g., Windows login and by cracking NetBIOS password hashes, using a brute force or dictionary attack tool.	John the Ripper, L0phtCrack, Cain, & Abel
Shared Key Guessing	Attempting 802.11 Shared Key Authentication with guessed vendor default or cracked WEP keys.	WEP Cracking Tools
Password Speculation	Using a captured identity, repeatedly attempting 2.1X authentication to guess the password.	Password dictionary
Key Reinstallation Attack	Exploiting the 4-way handshake of the WPA2 protocols	Nonce reuse technique

Figure 4. Rouge access point attack

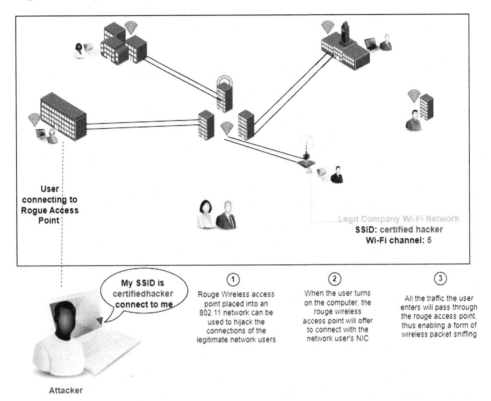

network user's NIC when a user switches on a computer. By transmitting an SSID, the attacker tempts the users to link to the malicious AP. All traffic entered by the user will go through the rogue AP if they establish a connection to it pretending it is a legitimate AP. Consequently, a type of wireless network scanning is enabled. Even usernames and passwords could be found in the analyzed packets.

3.7 Client Mis-Association Attack

A network client connecting to a nearby AP might lead to mis-association, a security issue. Client mis-associations can occur for a variety of reasons, including incorrectly configured clients, inadequate corporate Wi-Fi coverage, a lack of Wi-Fi policies, restrictions on internet use at work, ad-hoc links that administrators seldom manage, tempting SSIDs, etc. The mobile user and the malicious AP may be unaware of this or not.

An attacker deploys a rogue AP outside the organizational perimeter to accomplish client mis-association. The target wireless network's SSID (Sato et al., 2011) is the

Figure 5. Client mis-association attack

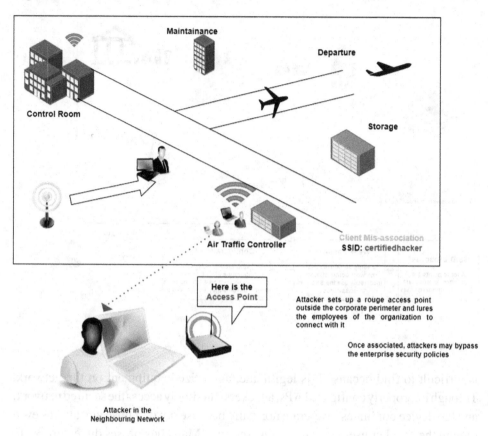

first thing the attacker discovers. The intruder may then emit beacons promoting the fake AP to entice clients to connect using a spoof SSID. This can be a way for the attacker to get around corporate security measures. Once a client connects to the malicious AP, an attacker can use EAP dictionary, MITM, or Metasploit attacks to take advantage of client mis-association to obtain sensitive data such as usernames and passwords.

3.8 Misconfigured Access Point Attack

The majority of businesses invest a lot of time in developing and executing Wi-Fi security rules, however, it is feasible for a wireless network client to mistakenly modify the security configuration on an access point (AP) (Sen, Maity, & Das, 2020). This may then result in AP misconfigurations. An otherwise well-secured gateway can become vulnerable to assaults due to a misconfigured AP. A misconfigured AP

Figure 6. Misconfigured access point attack

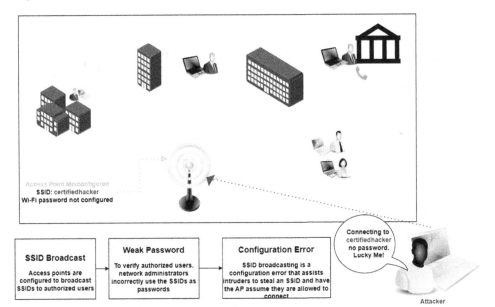

is difficult to find because it is legitimate, authorized equipment on the network. Through improperly configured APs, attackers can simply access the secured network, and the device continues to operate normally because it won't raise any alarms even though the attacker uses it to undermine security. Many businesses disregard Wi-Fi security protocols and do not take the necessary steps to close this security gap.

3.9 Unauthorized Association

A wireless network is seriously threatened by unauthorized affiliation (Jiang et al., 2018). There are two variations of it: malevolent association and unintentional affiliation. Instead of using corporate APs, an intruder performs illegal association with the aid of soft APs. Typically, on a computer, an attacker establishes a soft AP by executing a tool to make the NIC appear to be a real AP. After that, the attacker accesses the victim's wireless network by using the soft AP. On client cards or inbuilt WLAN transmitters in some PDAs and laptops, software APs are accessible, which an attacker can run manually or through a malware program. By infecting the victim's computer and turning on soft APs, the attacker makes it possible for an unauthorized user to connect to the company network. Passwords may be stolen, wired networks may be attacked, or Trojans may be installed by an attacker who acquires network access through unlawful association.

The accidental association is another sort of unlawful association that includes unintentionally connecting to the victim network's AP from a nearby organization's overlapping network.

Step 1: User tokens or integrated WLAN transmitters in some Personal digital assistants and pcs are soft access points that can be activated unintentionally or by virus programs.

Step 2: Attackers activate soft APs on the victim's computer after infecting it, enabling them to connect to the enterprise network without authorization.

Step 3: Attackers use soft access points (APs) rather than real access points to connect to the enterprise network.

3.10 Ad Hoc Connection Attack

Ad hoc mode is used by Wi-Fi clients to communicate without the aid of an access point (AP) (Nam et al., 2012) to relay messages. Ad hoc networks enable clients to easily exchange information. Most Wi-Fi clients use ad hoc networks to exchange music and video material between users. Ad hoc mode can occasionally be forced on a network by an attacker. Only in ad hoc mode are some network resources accessible, but this mode lacks effective authentication and encryption and is inherently unsafe. Therefore, a client running in ad hoc mode is simple for an attacker to connect to and compromise.

The security of the company's cable LAN can be jeopardized by an attacker who breaks into a wireless network using an ad-hoc connection.

3.11 Honeypot Access Point Attack

A user can access any available network if there are numerous WLANs present in the same location. Such a numerous WLAN is more open to intrusions. Typically, a wireless client will scan adjacent wireless networks for a certain SSID when it first turns on. By deploying a rogue AP to create an unauthorized wireless network, an intruder exploits this wireless client behavior. This AP utilizes the same SSID as the target network and has high-power (high gain) antennas. Users may connect to the rogue AP if they often use several WLANs. The attacker mounted what is known as "honeypot" Aps (Nam et al., 2012). They send out a beacon signal that is more powerful than the authentic APs. NICs may communicate to the rogue AP when looking for the strongest signal. An authorized user connecting to a honeypot AP introduces a security flaw and gives the attacker access to sensitive user data including username, identification, and password.

Figure 7. Ad hoc connection attack

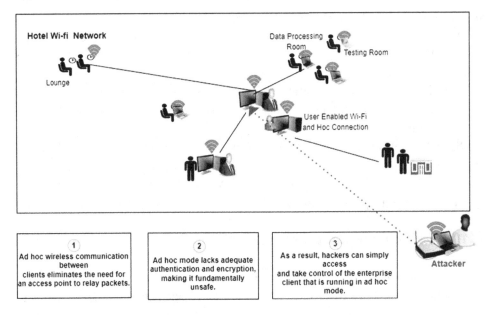

3.12 AP MAC Spoofing

In wireless networks, Aps broadcast probes in response to requests (beacons) to signal their availability and presence. Information on the AP identity (MAC address) and the network it supports is contained in the probe answers (SSID). Based on the MAC address and SSID that these beacons contain, the nearby clients join the network through them. The MAC (Sheldon et al., 2012) addresses and SSIDs of AP devices can often be set using software tools and apps. By creating a malicious AP to broadcast the same identification details as the legal AP, an attacker can fake the MAC address of the AP. An attacker who connects to the AP pretending to be a legitimate client can access the entire network.

3.13 Denial-of-Service Attack

DoS attacks can target wireless networks. Radio signals are used for data transfer on these networks, which operate in unlicensed bands. The MAC protocol's creators tried to keep it straightforward, however, the protocol contains several vulnerabilities that make it susceptible to DoS attacks. Mission-critical applications including database access, internet access, VoIP, and project data files are typically carried by

Figure 8. Honey access point attack

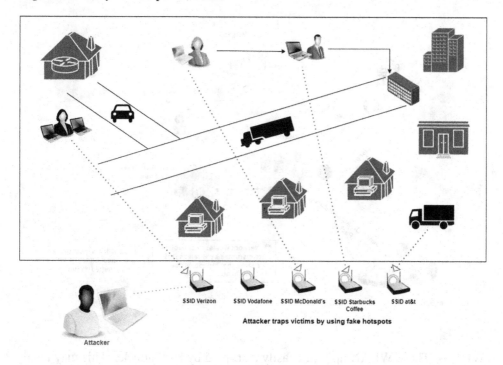

Figure 9. AP MAC spoofing

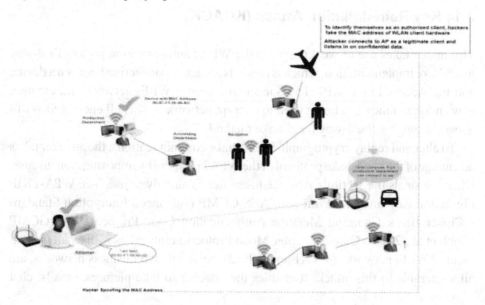

Figure 10. Denial-of-service attack

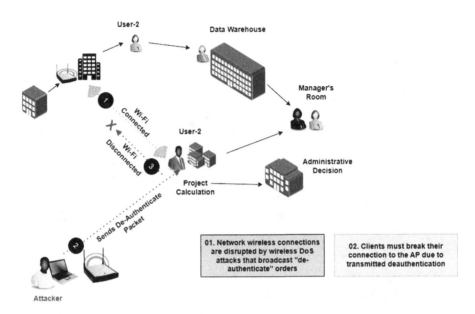

WLANs. These WLAN apps are easily disrupted by DoS attacks. This may result in decreased productivity or network outages. Network wireless connections are disrupted by wireless DoS attacks that transmit de-authenticate orders. The clients must cut off their connection to the AP due to transmitted de-authentication.

3.14 Key Reinstallation Attack (KRACK)

This attack takes use of weaknesses in the WPA2 authentication protocol's 4-way handshake implementation, which is needed to create a connection between a device and the Access Point (AP). To connect to a secure Wi-Fi network and create a new encrypted message being used to encrypt network traffic, all encrypted Wi-Fi networks employ the 4-way handshake method.

To alter and replay cryptographic handshake communications, the attacker takes advantage of the handshake protocol of the WPA2 protocol by imposing Nonce reuse, where he steals the victim's ANonce token that is already in use. The WPA-TKIP (Temporal Key Integrity Protocol), AES-CCMP (Advanced Encryption Standard - Cipher Block Chaining Message Authentication Code Protocol), and GCMP (Cobb et al., 2019) (Galois/Counter Mode Protocol) ciphers, as well as all current secured Wi-Fi networks (WPA1 and WPA2), personal, and business networks, are all vulnerable to this attack. It enables the attacker to take pictures, emails, chat

Figure 11. Key reinstallation attack (KRACK)

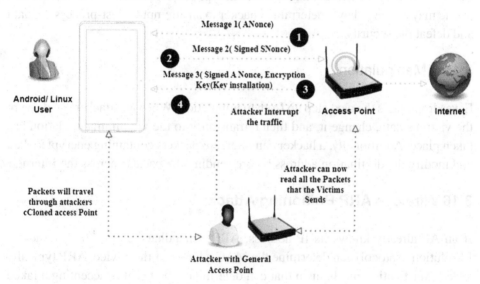

messages, passwords, and sensitive data including credit card numbers. Any device running one of the following operating systems is susceptible to a KRACK attack: Linux, Android, OpenBSD, Windows, Apple, or MediaTek.

3.15 Man-in-the-Middle Attack

An active online attack known as a MITM is one in which the attacker tries to read, intercept, or modify data passing between two computers. Both cable communication systems and 802.11 WLANs are susceptible to MITM attacks.

3.15.1 Eavesdropping

A wireless network makes it simple because there isn't physical media involved in communication. Without much effort or expensive equipment, an attacker who is near the wireless network can pick up radio waves on it. The actual data frame sent over the network can be examined by the attacker, or it can be stored for later analysis.

Put in place multiple encryption levels to stop hackers from obtaining sensitive data. Data-link encryption or WEP can be useful. Use a security method like SSH (Secure Shell Protocol), IPsec (Internet Protocol Security), or SSL (Secure Sockets Layer) to protect transferred data; otherwise, it will be public and subject to hacker assault.

However, using free online tools, an attacker can break WEP. It is dangerous to access email via the POP or IMAP standards since they allow email to be sent over a wi-fi network without any further encryption. Gigabytes of Enabled devices traffic may be logged by a determined hacker in an attempt to post-process the data and defeat the security.

3.15.2 Manipulation

Eavesdropping is the next step up from manipulation. When an attacker can decrypt the victim's data, change it, and then retransmit it to the victim, manipulation has taken place. Additionally, a hacker can intercept packets containing encrypted data and modify the destination address before sending the packets across the Internet.

3.16 Wireless ARP Poisoning Attack

If an AP already knows its IP address, ARP (Chaudhary et al., 2012) (Address Resolution Protocol) can determine the MAC address of the device. ARP typically lacks a verification mechanism that can determine whether it is accepting a faked response or responses from legitimate hosts. An attack method that takes advantage of the absence of verification is ARP poisoning. This method corrupts the OS's ARP cache by using the incorrect MAC addresses. This is done by an attacker delivering an ARP Flashback pack that is built with the incorrect MAC address.

All of the hosts in a subnet are affected by the ARP poisoning attack. Since most APs serve as transparent MAC-layer bridges, all stations connected to a subnet compromised by the ARP poisoning attack are at risk. If the AP is directly linked to a switch or hub, all hosts connected to that switch or hub are vulnerable to ARP poisoning attacks. without a router or firewall in the middle.

The attacker first fakes the victim's Wi-Fi laptop's MAC address before attempting to use the Cain & Abel ARP poisoning tool, a Windows password recovery tool, to

Figure 12. Man-in-the-middle attack

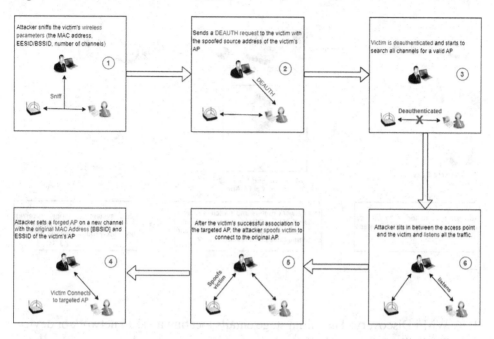

connect to API. The network routers and switches receive the updated MAC address information from API, which they then use to update their routing and switching tables. The system directs traffic that is currently being sent from the core network to the targeted computer via AP1 rather than AP2.

4. WIRELESS HACKING METHODOLOGY

To hack wireless networks, an attacker follows a hacking methodology. This process provides systematic Steps to perform a successful attack on a target wireless network. This section Will explain the steps of the wireless hacking methodology. A wireless hacking methodology helps an attacker to reach the goal of hacking a target wireless network. An attacker who does not follow a methodology may fail to hack a wireless network. An attacker usually follows a hacking methodology to be sure of finding every single-entry point to break into the target network.

Using wireless hacking techniques, it is possible to infiltrate a Wi-Fi network and obtain access to its resources without authorization. Attacks employ a range of wireless hacking techniques, such as:

Figure 13. Wireless ARP poisoning attack

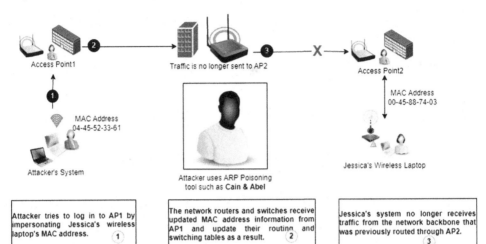

1. Wi-Fi Discovery: The initial stage entails locating a Wi-Fi network or device. Tools like inSSIDer Office, NetSurveyor, etc. are used by an attacker to do Wi-Fi discovery to find a target wireless network. To find the right target network that is within range of the assault, Wi-Fi discovery processes include footprinting the wireless networks.
2. The first steps in an attack on a wireless network are discovery and footprinting. Locating and examining (or comprehending) the network is part of footprinting. The footprinting of a wireless network can be done in two different ways.

3. The BSS offered by the AP must be recognized by the attacker to footprint a wireless network. With the aid of the wireless network's SSID, an attacker may be able to determine the BSS or Independent BSS (IBSS). The attacker must therefore locate the target wireless network's SSID. This SSID can be used by the attacker to connect to the AP and weaken its security.
4. GPS Mapping: This is the second phase in the wireless hacking approach. When an attacker finds a target wireless network, they might go on to wireless hacking by creating a network map. The attacker may employ several automated tools in this step to map the target wireless network.
5. The location, timing, and existence of physical objects on earth are all provided by the Global Positioning System (GPS), a satellite navigation system situated in space. Anyone can locate a specific point on the planet and its surrounding

Figure 14. Types of footprinting methods

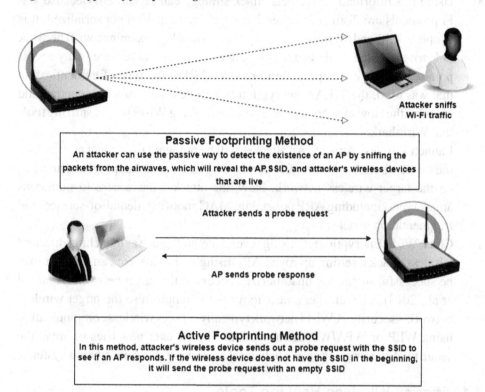

Passive Footprinting Method
An attacker can use the passive way to detect the existence of an AP by sniffing the packets from the airwaves, which will reveal the AP, SSID, and attacker's wireless devices that are live

Attacker sends a probe request

AP sends probe response

Active Footprinting Method
In this method, attacker's wireless device sends out a probe request with the SSID to see if an AP responds. If the wireless device does not have the SSID in the beginning, it will send the probe request with an empty SSID

area using a GPS tool. This GPS tool is used by an attacker to identify and map the target wireless network in a specific geographic location.

6. A GPS receiver processes specially coded satellite signals to determine position, time, and velocity. Attackers are aware that free Wi-Fi is widely available and that some networks can be unsafe. Attackers typically make maps of Wi-Fi networks that they have found and compile statistics using software like NetSurveyor, inSSlDer Office, etc. The coordinates published to websites like WiGLE (Chaudhary et al., 2012) and newly discovered Wi-Fi networks can both be followed using GPS. Attackers may sell or trade this information with other hackers to profit.

7. Wireless Traffic Analysis: Examining the detected wireless network's traffic is the third step in the technique. Before launching an actual attack on the wireless network, an attacker analyses wireless traffic. The attacker can identify the target network's weaknesses and susceptible targets with the use of this wireless traffic analysis. The attacker examines the traffic of the target wireless network using a variety of tools and techniques.

8. Using this information, the best attack strategy can be chosen. Because Wi-Fi protocols are distinct at Layer 2 and airborne traffic is not serialized, it is simple to sniff and examine wireless packets. Attackers examine a wi-fi network to learn about the SSID that was broadcast, whether there were many access points, whether it was possible to recover SSIDs, the type of authentication that was used, the WLAN encryption techniques, etc. Attackers capture and examine the flow of a target wireless network using Wi-Fi packet sniffing tools like Wireshark.

9. Launch Wireless Attacks: An attacker will be prepared to conduct an attack on the target wireless network after performing discovery, mapping, and analysis on the target wireless network. Now, the attacker has access to numerous attack types, including ARP poisoning, MAC spoofing, denial-of-service, and fragmentation attacks.

10. Crack Wi-Fi Encryption: By using several methods, such as launching different wireless attacks, setting up rogue APs, using evil twins, etc., an attacker may be successful in gaining unauthorized access to the target network (Agarwal et al., 2013). The attacker's next move is to compromise the target wireless network's security. A Wi-Fi network typically secures wireless communication using WEP or WPA/WPA2 encryption. The attacker now tries to break the security of the target wireless network by cracking these encryption systems.

4.1 Different Wireless Hacking Tools

4.1.1 WEP/WPA Cracking Tool for Mobile

1. WIBR-WIFI BRUTEFORCE HACK

WPA/WPA2 PSK Wi-Fi networks are tested for security using a program called WIBR+. It can identify passwords that are weak. It allows queuing, brute-force assaults, custom dictionaries, and extensive monitoring.

In WIBR+, two assault types are supported:

A. **Dictionary attack:** WIBR+ tests each password individually from a pre-determined list. Users can import their password lists with WIBR+.

B. **Bruteforce attack:** Custom alphabets and custom masks are supported by WIBR+. You can set the mask to "hacker [x] [x]" and select the alphabet of numbers if you know the password, which is something like "hacker" and two digits. All passwords, including "hacker00," "hacker01," and "hacker99," will be tried by the software.

Additional WEP/WPA cracking tools for smartphones include:

A. AndroDumper (WPS Connect)

B. WIFI WPS WPA TESTER
C. Wi-fi Password WPA-WEP FREE
D. iWep PRO
E. WPS WPA Wi-Fi Tester

4.1.2 Wi-Fi Sniffer

An application or piece of hardware known as a "packet sniffer" blocks network traffic between two computers to capture it. They are sometimes known as protocol analyzers or packet analyzers.

1. Kismet

It is a wireless network sensor, skimmer, and intrusion detection solution for 802.11 Layer 2. By continuously gathering packets and scanning for recognized named networks, it may identify networks. Through data flow, it can identify mysterious networks.
Features:

1. This Wi-Fi sniffer application supports standard PCAP logging.
2. Provides a modular client/server design.
3. Key features can be enhanced using the plugin architecture.
4. Allows for a range of capture sources.
5. Dispersed remote sniffing with straightforward remote capture
6. XML output for use with further tools.

Additional Wi-Fi packet sniffer tools include:

2. **Wireshark:** It is a wireless network sniffer utility that captures packets in real-time and displays them in an understandable way. A variety of file types are supported by the network administration team for reading, writing, and gathering network data.

Features:

1. Captures instantaneously de-compressible gzip-compressed files.
2. XML (Extensible Markup Language), CSV (Comma Seperated values), PostScript, and plain text are available as export options for output.
3. Support for a variety of operating systems, such as Windows, Linux, FreeBSD, and NetBSD.

4. It is possible to read live data from sources such as Ethernet, PPP (Point-to-Point Protocol)/HDLC (High level data control link), Bluetooth, Token Ring, USB, and others.

5. A supported platform is Windows.

Paessler: You may inspect all devices on a network, including Wi-Fi routers, using the Wi-Fi monitoring and analyzer tool PRTG.

Features:

1. You can easily analyze all the data acquired about your Wi-Fi connection networks with PRTG.

2. Makes it possible to analyze wireless traffic.

3. Provides four sensors for capture by a packet sniffer.

4. The LAN sniffer header sensor included with PRTG wireless network monitor allows you to configure the network so that you are notified right away if there is a problem with the Wi-Fi networking.

5. Enables you to customize the alarm system on your Wi-Fi network.

6. Check all of your wireless networks' components.

7. The platforms supported include Windows and the hosted version.

3. Acrylic Wi-Fi Professional: Users of Acrylic Wi-Fi, a wireless internet network sniffer, can scan and evaluate adjacent access points to produce a table containing helpful information. It provides each essential metric, such as vendor, MAC address, channel, and SSID.

Features:

1. It facilitates the investigation and remediation of issues on 802.11 a/b/g/n/ac networks in real-time.

2. It allows Linux systems to record Wi-Fi traffic.

3. It is affordable, widely accessible, and easy to purchase because it is compatible with the majority of hardware and doesn't require any specific components to function.

4. Easily communicates with current versions of Wireshark.

5. Compatible with the most recent and well-liked Wi-Fi USB cards now on the market.

6. A supported platform is Windows.

4. Tcpdump: TCPdump offers a command-line packet analyzer and a portable C/C++ library named libpcap for capturing network traffic. It is a fantastic option for those who want a simple tool and need a rapid scan.

Features:

1. It gives you the ability to do operations like viewing and saving recorded packets to files.

2. The display interface that is available.

3. Supported Platforms: Solaris, FreeBSD, DragonFly BSD, Mac OS, Linux, NetBSD, FreeBSD, etc.

4.1.3 Free Network Analyzer

A free network is a tool for sniffing networks. This can be used to monitor and examine all of the information moving through your network adapter. You may instantly collect network traffic using this application. The best packet sniffer inspection capabilities are also provided by it.

Features:

1. Because it enables real-time protocol analysis, this is one of the best Wi-Fi sniffer tools available.
2. This top packet sniffer tool saves system resources while enabling uninterrupted high data-speed network operation.
3. This free network program supports sophisticated data filtering and layout customization.
4. It offers excellent customer service, enabling you to ask questions and get knowledgeable answers.

4.1.4 WLAN Traffic Analysis Software

It assesses (Tiernan et al., 2022), keeps track of, and maintains data traffic usage, security, and localized network and internet efficiency of the services. They keep track of data transit across a network ethernet port or toll-free connection, analyze it, and then transmit it in a legible way. With the help of this tool, users can get a complete view of the network or wireless LAN connection's traffic. These tools look at network traffic to find security issues or follow specific transactions. However, attackers use them for malicious purposes. The technologies listed below are used to examine the traffic of the target wireless networks.

1. Wi-Fi Analyzer by AirMagnet

Features:

1. Wi-Fi dashboard that displays the WLAN's overall health.
2. Notifications right away about issues with WLAN security and performance.
3. WLAN client roaming analysis.
4. Wi-Fi troubleshooting instruments.
5. Detection and investigation for comprehensive Wi-Fi troubleshooting
6. 802.11ac detection and analysis.

4.2 Other Hacking Tools

- **Wardriving Tools**

Customers can identify all APs in the area that are broadcasting beacon signals by using war driving tools. In addition to helping the user set up new APs, it makes sure that hardly any APs are interfering. These programs examine the network settings, identify WLAN stations with poor coverage, and search for any potentially interfering adjacent networks. This can also reveal unauthorized, rogue APs. Here is a list of some of the combat weapons.

- MacStumbler
- AirFart
- G-MoN
- Airbase-ng
- 802.11 Network Discovery Tools
  - **RF Monitoring Tools**

With the aid of radio frequency (RF) monitoring technologies, Wi-Fi networks can be located and observed. These tools regulate and observe network interfaces, particularly wireless ones. They support the control of network interfaces and the visualization of network activities.

- NetworkManager
- sigX
- Sentry Edge II
- CPRIAdvisor
- WaveNode
- satID
- xosview
  - **Raw Packet Capturing Tools**
- WLAN packet activity is recorded and tracked using raw packet-capturing software. These programs can gather every packet, support Ethernet LAN and 802.11, and display network activity down to the MAC level. The following sentences provide descriptions of a few of these tools:
- RawCap
- Tcpdump
- Airodump-ng

4.3 Countermeasures

The security of the wireless network is improved by an ethical hacker. The correct countermeasures must be adopted and put into action to ensure the security of wireless networks. The measures and suggested practices for wireless network security are described in this section.

Wireless Security Layers

A wireless security method has six layers. This multi-layered approach increases the chance that the offender will be caught and broadens the range of situations in which an attacker can be prevented from compromising a network.

- **Wireless Signal Security:** Continuous network management and RF spectrum monitoring in wireless networks aid in detecting threats and raising awareness levels. The Wireless Intrusion Detection System (WIDS) examines and monitors the RF (Agarwal et al., 2013) spectrum. Setting up alarms makes it easier to find unlicensed wireless devices that violate the network's security guidelines. Higher bandwidth usage, RF interferences, rogue wireless APs that are unknown, etc. may all be signs of a malicious network invader. Continuous network monitoring is the only defense against such attacks and breaches in network security.
- **Connection Security:** Per-frame/packet authentication provides a defense against MITM attacks. The connection is protected while two authorized customers are corresponding with one another since it stops the attacker from sniffing data.
- **Device Security:** Patch management and vulnerability management are essential components of the security infrastructure.
- **Data Protection:** Techniques for data encryption like WPA2 and AES are available.
- **Network Protection:** Only authorized users can access a network thanks to strong authentication.
- **End-user Protection:** The end-user system's firewalls on the WLAN prohibit the attacker from accessing files, even if the attacker has connected to APs.

5. WIRELESS PENETRATION TESTING

To identify design faults, technical issues, and vulnerabilities in a wireless network, a technique known as wireless penetration testing is used to actively assess the

Figure 15. Wireless penetration testing

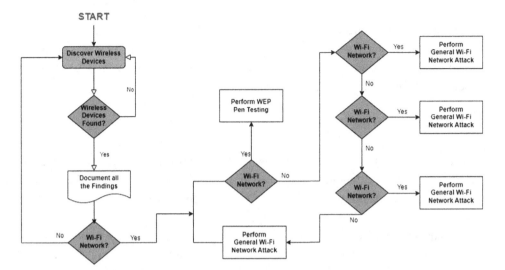

information security measures that have been put in place. Management, e Executive, and professional audiences receive a thorough, complete report on the findings and the range of advised countermeasures (Lu & Yu, 2021).

- Threat Evaluation: Determine the wireless dangers to the information assets of a company.
- Infrastructure upgrade: Modifying or modernizing the software, hardware, or network architecture that is currently in place.
- Risk Prevention and Response: Outline a thorough plan of action for preparations that can be made to thwart exploitation that is unavoidable.
- Auditing security controls: To examine and confirm the effectiveness of wireless security safeguards and controls.
- Data Theft Detection: Sniff the traffic to find streams of sensitive data.
- Information system management: Gather data on connected devices, network stability, and security standards.

REFERENCES

Agarwal, A., Bora, N., & Arora, N. (2013). Goodput enhanced digital image watermarking scheme based on DWT and SVD. *International Journal of Application or Innovation in Engineering & Management*, 2(9), 36–41.

Baharudin, N., Ali, F. H. M., Darus, M. Y., & Awang, N. (2015, August). Wireless intruder detection system (wids) in detecting de-authentication and disassociation attacks in ieee 802.11. In *2015 5th International Conference on IT Convergence and Security (ICITCS)* (pp. 1-5). IEEE.

Chaudhary, R., Singh, P., & Agarwal, A. (2012). A security solution for the transmission of confidential data and efficient file authentication based on DES, AES, DSS and RSA. *International Journal of Innovative Technology and Exploring Engineering, 1*(3), 5–11.

Čisar, P., & Čisar, S. M. (2018). Ethical hacking of wireless networks in kali linux environment. *Annals of the Faculty of Engineering Hunedoara, 16*(3), 181–186.

Cobb, C. L., Meca, A., Branscombe, N. R., Schwartz, S. J., Xie, D., Zea, M. C., Fernandez, C. A., & Sanders, G. L. (2019). Perceived discrimination and well-being among unauthorized Hispanic immigrants: The moderating role of ethnic/racial group identity centrality. *Cultural Diversity & Ethnic Minority Psychology, 25*(2), 280–287. doi:10.1037/cdp0000227 PMID:30284850

Jiang, P., Wu, H., Wang, C., & Xin, C. (2018, May). Virtual MAC spoofing detection through deep learning. In *2018 IEEE International Conference on Communications (ICC)* (pp. 1-6). IEEE. 10.1109/ICC.2018.8422830

Khasawneh, M., Kajman, I., Alkhudaidy, R., & Althubyani, A. (2014, March). A survey on Wi-Fi protocols: WPA and WPA2. In *International conference on security in computer networks and distributed systems* (pp. 496-511). Springer. 10.1007/978-3-642-54525-2_44

Lu, H. J., & Yu, Y. (2021). Research on WiFi penetration testing with Kali Linux. *Complexity, 2021*, 2021. doi:10.1155/2021/5570001

Nair, R. S., Babu, A., & Vijayan, V. P. (2018). A Survey on Wi-Fi Security Techniques. *International Research Journal of Engineering and Technology, 5*(4), 4705–4707.

Nam, S. Y., Jurayev, S., Kim, S. S., Choi, K., & Choi, G. S. (2012). Mitigating ARP poisoning-based man-in-the-middle attacks in wired or wireless LAN. *EURASIP Journal on Wireless Communications and Networking, 2012*(1), 1–17. doi:10.1186/1687-1499-2012-89

Pimple, N., Salunke, T., Pawar, U., & Sangoi, J. (2020, March). Wireless security—an approach towards secured wi-fi connectivity. In *2020 6th international conference on advanced computing and communication systems (ICACCS)* (pp. 872-876). IEEE.

Reddy, B. I., & Srikanth, V. (2019). Review on wireless security protocols (WEP, WPA, WPA2 & WPA3). *International Journal of Scientific Research in Computer Science, Engineering and Information Technology*, 28-35.

Restuccia, F., D'Oro, S., Al-Shawabka, A., Rendon, B. C., Chowdhury, K., Ioannidis, S., & Melodia, T. (2020). Hacking the waveform: Generalized wireless adversarial deep learning. *arXiv preprint arXiv:2005.02270*.

Sato, T., Moungnoul, P., & Fukase, M. A. (2011, May). Compatible WEP algorithm for improved cipher strength and high-speed processing. In The 8th Electrical Engineering/Electronics, Computer, Telecommunications and Information Technology (ECTI) Association of Thailand-Conference 2011 (pp. 401-404). IEEE. doi:10.1109/ECTICON.2011.5947859

Sen, S., Maity, S., & Das, D. (2020). The body is the network: To safeguard sensitive data, turn flesh and tissue into a secure wireless channel. *IEEE Spectrum*, *57*(12), 44–49. doi:10.1109/MSPEC.2020.9271808

Sheldon, F. T., Weber, J. M., Yoo, S. M., & Pan, W. D. (2012). The insecurity of wireless networks. *IEEE Security and Privacy*, *10*(4), 54–61. doi:10.1109/MSP.2012.60

Software Testing Help. (2023). *Best Wifi Packet Review.* Software Testing Help. https://www.softwaretestinghelp.com/best-wifi-packet-sniffer-review/

Tiernan, C., Comyns, T., Lyons, M., Nevill, A. M., & Warrington, G. (2022). The association between training load indices and injuries in elite soccer players. *Journal of Strength and Conditioning Research*, *36*(11), 3143–3150. doi:10.1519/JSC.0000000000003914 PMID:33298712

Compilation of References

Register. (2018). Gits Club GitHub Code Tub With Record-Breaking 1.35Tbps DDoS Drub. [Online]. *The Register.* https://www.theregister. co.uk/2018/03/01/github_ddos_biggest_ever/

Rodríguez, G. E., Benavides, D. E., Torres, J., Flores, P., & Fuertes, W. (2018, January). Cookie scout: An analytic model for prevention of cross-site scripting (XSS) using a cookie classifier. In *International Conference on Information Technology & Systems* (pp. 497-507). Springer, Cham. 10.1007/978-3-319-73450-7_47

Chaudhary, R., Singh, P., & Agarwal, A. (2012). A security solution for the transmission of confidential data and efficient file authentication based on DES, AES, DSS and RSA. *International Journal of Innovative Technology and Exploring Engineering, 1*(3), 5–11.

Agarwal, A., Bora, N., & Arora, N. (2013). Goodput enhanced digital image watermarking scheme based on DWT and SVD. *International Journal of Application or Innovation in Engineering & Management, 2*(9), 36–41.

Kaur, G., Pande, B., Bhardwaj, A., Bhagat, G., & Gupta, S. (2018, January). Defense against HTML5 XSS attack vectors: a nested context-aware sanitization technique. In *2018 8th International Conference on Cloud Computing, Data Science & Engineering (Confluence)* (pp. 442-446). IEEE. 10.1109/CONFLUENCE.2018.8442855

Hou, X. Y., Zhao, X. L., Wu, M. J., Ma, R., & Chen, Y. P. (2018, April). A dynamic detection technique for XSS vulnerabilities. In *2018 4th Annual International Conference on Network and Information Systems for Computers (ICNISC)* (pp. 34-43). IEEE. 10.1109/ICNISC.2018.00016

Zubarev, D., & Skarga-Bandurova, I. (2019, June). Cross-site scripting for graphic data: vulnerabilities and prevention. In *2019 10th International Conference on Dependable Systems, Services and Technologies (DESSERT)* (pp. 154-160). IEEE. 10.1109/DESSERT.2019.8770043

Taha, T. A., & Karabatak, M. (2018, March). A proposed approach for preventing cross-site scripting. In *2018 6th International Symposium on Digital Forensic and Security (ISDFS)* (pp. 1-4). IEEE. 10.1109/ISDFS.2018.8355356

OWASP. (2018). *Owasp top 10 application security risks - 2017, 2018.* OWASP. https://www.owasp.org/index.php/Top_10-2017_Top_10

OWASP. (2018). *Proyecto owasp api de seguridad empresarial (esapi)*. OWASP. https://www.owasp.org/index.php/Category:OWASP_Enterprise_Security_API/es

Khanzode, C. A., & Sarode, R. D. (2016). Evolution of the world wide web: from web 1.0 to 6.0. *International journal of Digital Library services, 6*(2), 1-11.

Polop, C. (n.d). *HackTricks*. HackTricks. https://book.hacktricks.xyz/welcome/readme

OWASP. (2018). *net antixss library*. OWASP. https://www.owasp.org/index.php/.NET_AntiXSS_Library

Patil, K. (2021). A Study of Web 1.0 to 3.0. AGPE THE ROYAL GONDWANA RESEARCH JOURNAL OF HISTORY, SCIENCE. *ECONOMIC, POLITICAL AND SOCIAL SCIENCE, 2*(2), 40–45.

Swissky. (n.d.). *PayloadOfAllThings*. Github. https://github.com/swisskyrepo/PayloadsAllTheThings

Caster. (2023, February 8). *C4S73R/Networknightmare: Network pentesting mindmap by caster*. GitHub. https://github.com/c4s73r/NetworkNightmare

Singh, H. A. (2020). EVOLUTION OF WEB 1.0 TO WEB 3.0.

Król, K. (2020). Evolution of online mapping: From Web 1.0 to Web 6.0. *Geomatics, Landmanagement and Landscape, 1*, 33–51. doi:10.15576/GLL/2020.1.33

Security, In. (n.d.). *Password cracking 101+1*. In Security. https://in.security/technical-training/password-cracking

Almeida, F. (2017). Concept and dimensions of web 4.0. *International Journal of Computers and Technology, 16*(7), 7040–7046. doi:10.24297/ijct.v16i7.6446

The MITRE Corporation. (2022, October 25). *MITRE ATT&CK*. MITRE. https://attack.mitre.org/versions/v12/

Praseed, A., & Thilagam, P. S. (2018). DDoS attacks at the application layer: Challenges and research perspectives for safeguarding web applications. *IEEE Communications Surveys and Tutorials, 21*(1), 661–685. doi:10.1109/COMST.2018.2870658

Rodríguez, G. E., Torres, J. G., Flores, P., & Benavides, D. E. (2020). Cross-site scripting (XSS) attacks and mitigation: A survey. *Computer Networks, 166*, 106960. doi:10.1016/j.comnet.2019.106960

Tian, Z., Luo, C., Qiu, J., Du, X., & Guizani, M. (2019). A distributed deep learning system for web attack detection on edge devices. *IEEE Transactions on Industrial Informatics, 16*(3), 1963–1971. doi:10.1109/TII.2019.2938778

Sriramya, P., Kalaiarasi, S., & Bharathi, N. (2020, December). Anomaly Based Detection of Cross Site Scripting Attack in Web Applications Using Gradient Boosting Classifier. In *International Conference on Advanced Informatics for Computing Research* (pp. 243-252). Springer, Singapore.

Abu Al-Haija, Q., & Al Badawi, A. (2022). High-performance intrusion detection system for networked UAVs via deep learning. *Neural Computing & Applications*, *34*(13), 10885–10900. doi:10.100700521-022-07015-9

Abu Al-Haija, Q., Al Badawi, A., & Bojja, G. R. (2022). Boost-Defence for resilient IoT networks: A head-to-toe approach. *Expert Systems: International Journal of Knowledge Engineering and Neural Networks*, *39*(10), e12934. doi:10.1111/exsy.12934

Abu Al-Haija, Q., & Al-Dala'ien, M. (2022). ELBA-IoT: An Ensemble Learning Model for Botnet Attack Detection in IoT Networks. *J. Sens. Actuator Netw.*, *11*(1), 18. doi:10.3390/jsan11010018

Abu Al-Haija, Q., & Alsulami, A. A. (2022). Detection of Fake Replay Attack Signals on Remote Keyless Controlled Vehicles Using Pre-Trained Deep Neural Network. *Electronics (Basel)*, *11*(20), 3376. doi:10.3390/electronics11203376

Abu Al-Haija, Q., Krichen, M., & Abu Elhaija, W. (2022). Machine-Learning-Based Darknet Traffic Detection System for IoT Applications. *Electronics (Basel)*, *11*(4), 556. doi:10.3390/electronics11040556

Abu Al-Haija, Q., Odeh, A., & Qattous, H. (2022). PDF Malware Detection Based on Optimizable Decision Trees. *Electronics (Basel)*, *11*(19), 3142. doi:10.3390/electronics11193142

Abu Al-Haija, Q., & Zein-Sabatto, S. (2020). An Efficient Deep-Learning-Based Detection and Classification System for Cyber-Attacks in IoT Communication Networks. *Electronics (Basel)*, *9*(12), 2152. doi:10.3390/electronics9122152

Adamov, A., & Carlsson, A. (2017). The state of ransomware. Trends and mitigation techniques. *2017 IEEE East-West Design & Test Symposium (EWDTS)*. IEEE. 10.1109/EWDTS.2017.8110056

Ahmad, A., AbuHour, Y., Younisse, R., Alslman, Y., Alnagi, E., & Abu Al-Haija, Q. (2022). MID-Crypt: A Cryptographic Algorithm for Advanced Medical Images Protection. *J. Sens. Actuator Netw.*, *11*(2), 24. doi:10.3390/jsan11020024

Akal, M. (2022). *Information Security Awareness in IT Companies Ufuk University*].

AL Attack Map. (2020). Distributed Denial of Service Attack. *AL Attack Map*. alattackmap.com/#anim=1&color=0&country=ALL&list=2&time=16438&view=map

Al Nabki, M. W., Fidalgo, E., Alegre, E., & De Paz, I. (2017, April). They classify illegal Tor network activities based on web textual content. In *Proceedings of the 15th Conference of the European Chapter of the Association for Computational Linguistics*, (pp. 35-43). ACL.

Albulayhi K., & Al-Haija Q.A. (2022). Security and Privacy Challenges in Blockchain Application. In *The Data-Driven Blockchain Ecosystem*. CRC Press.

Albulayhi, K., Abu Al-Haija, Q., Alsuhibany, S. A., Jillepalli, A. A., Ashrafuzzaman, M., & Sheldon, F. T. (2022). IoT Intrusion Detection Using Machine Learning with a Novel High Performing Feature Selection Method. *Applied Sciences (Basel, Switzerland)*, *12*(10), 5015. doi:10.3390/app12105015

Al-Fayoumi, M. A., Elayyan, A., Odeh, A., & Abu Al-Haija, Q. (2022). Tor Network Traffic Classification Using Machine Learning Based on Time-Related Feature. *6th Smart Cities Symposium 2022 (6SCS)*. IET. 10.1049/icp.2023.0354

Al-Haija, Q. A. (2022). Time-Series Analysis of Cryptocurrency Price: Bitcoin as a Case Study. *2022 International Conference on Electrical Engineering, Computer and Information Technology (ICEECIT)*, Jember, Indonesia.10.1109/ICEECIT55908.2022.10030536

Al-Haija, Q. A., & Alsulami, A. A. (2021). High Performance Classification Model to Identify Ransomware Payments for Heterogeneous Bitcoin Networks. *Electronics (Basel)*, *10*(17), 2113. doi:10.3390/electronics10172113

Al-Haija, Q. A., & Badawi, A. A. (2021). URL-based Phishing Websites Detection via Machine Learning. *2021 International Conference on Data Analytics for Business and Industry (ICDABI)*, Sakheer, Bahrain. 10.1109/ICDABI53623.2021.9655851

Ali, A. (2017). Ransomware: A research and a personal case study of dealing with this nasty malware. *Issues in Informing Science and Information Technology*.

Alkali, Y., Routray, I., & Whig, P. (2022a). Strategy for Reliable, Efficient and Secure IoT Using Artificial Intelligence. *IUP Journal of Computer Sciences, 16*(2).

AlkaliY.RoutrayI.WhigP. (2022b). Study of various methods for reliable, efficient and Secured IoT using Artificial Intelligence. *Available at* SSRN 4020364. doi:10.2139/ssrn.4020364

Al-Qudah, M., Ashi, Z., Alnabhan, M., & Abu Al-Haija, Q. (2023). Effective One-Class Classifier Model for Memory Dump Malware Detection. *J. Sens. Actuator Netw.*, *12*(1), 5. doi:10.3390/jsan12010005

AlSabah, M., & Goldberg, I. (2016). Performance and security improvements for Tor: A survey. [CSUR]. *ACM Computing Surveys*, *49*(2), 1–36. doi:10.1145/2946802

Altiner, I. (2021). *Evaluation of Teachers' Personal Cyber Security Awareness Levels According to Different Variables*. Ankara University.

Ambika, N. (2020). Improved Methodology to Detect Advanced Persistent Threat Attacks. In N. K. Chaubey & B. B. Prajapati (Eds.), *Quantum Cryptography and the Future of Cyber Security* (pp. 184–202). IGI Global. doi:10.4018/978-1-7998-2253-0.ch009

Ambika, N. (2022). Minimum Prediction Error at an Early Stage in Darknet Analysis. In *Dark Web Pattern Recognition and Crime Analysis Using Machine Intelligence* (pp. 18–30). IGI Global. doi:10.4018/978-1-6684-3942-5.ch002

Amin, A. M., & Mahamud, M. S. (2019). An alternative approach of mitigating arp based man-in-the-middle attack using client site bash script. *6th International Conference on Electrical and Electronics Engineering (ICEEE)* (pp. 112-115). Istanbul, Turkey: IEEE. 10.1109/ICEEE2019.2019.00029

Aminuddin, M. A. I. M., Zaaba, Z. F., Samsudin, A., Juma'at, N. B. A., & Sukardi, S. (2020, August). Analysis of the paradigm on tor attack studies. In *2020 8th International Conference on Information Technology and Multimedia (ICIMU)* (pp. 126-131). IEEE.

Anand, M., Velu, A., & Whig, P. (2022). Prediction of Loan Behaviour with Machine Learning Models for Secure Banking. [JCSE]. *Journal of Computing Science and Engineering : JCSE*, *3*(1), 1–13.

Androulaki, E., Karame, G. O., Roeschlin, M., Scherer, T., & Capkun, S. (2013). Evaluating User Privacy in Bitcoin. In A. R. Sadeghi (Ed.), Lecture Notes in Computer Science: Vol. 7859. *Financial Cryptography and Data Security, FC 2013* (pp. 34–51). Springer. doi:10.1007/978-3-642-39884-1_4

Andy, G. (2013, November). Meet the 'Assassination Market' Creator Who's Crowdfunding Murder with Bitcoins. *Forbes*, 18.

Anghel, M., & Racautanu, A. (n.d.). *A note on different types of ransomware attacks*. Computer Science Faculty, Al. I. Cuza University.

Arslan, Y. (2021). Phishing Attacks Awareness Exercise Example. *Düzce Üniversitesi Bilim ve Teknoloji Dergisi*, *9*(3), 348–358. doi:10.29130/dubited.832862

Aslay, F. (2017). Cyber Attack Methods and Current Situation Analysis of Turkey's Cyber Safety. *International Journal of Multidisciplinary Studies and Innovative Technologies*, *1*(1), 24–28.

Atmojo, Y. P., Susila, I. M. D., Hilmi, M. R., Rini, E. S., Yuningsih, L., & Hostiadi, D. P. (2021). A New Approach for Spear phishing Detection. *3rd 2021 East Indonesia Conference on Computer and Information Technology, EIConCIT 2021*, (pp. 49–54). ACM. 10.1109/EIConCIT50028.2021.9431890

Baharudin, N., Ali, F. H. M., Darus, M. Y., & Awang, N. (2015, August). Wireless intruder detection system (wids) in detecting de-authentication and disassociation attacks in ieee 802.11. In *2015 5th International Conference on IT Convergence and Security (ICITCS)* (pp. 1-5). IEEE.

Bahri, A. (2020). The Dark and Deep Web. *Medium*. https://ahedbahri.medium.com/the-dark-and-deep-web-6d629923968b

Bajpai, P. & Enbody, R. (2020). Attacking Key Management in Ransomware. *IT Professional*, *22*(2), 21-27. . doi:10.1109/MITP.2020.2977285

Bakarich, K. M., & Baranek, D. (2020). Something phish-y is going on here: A teaching case on business email compromise. *Current Issues in Auditing*, *14*(1), A1–A9. doi:10.2308/ciia-52706

Baker, F. (1995). *Requirements for IP version 4 routers*.

Bass, W. M., Gill, G. W., Jantz, R., Locard, E., Owsley, D. W., Tardieu, A. A., & Vucetich, J. (1980). Digital forensics. *History (London)*, *1*, 1990s.

BBC News. (n.d.). *Ransomware cyber-attack threat escalating*. BBC. https://www.bbc.com/news/technology-39913630

Benlloch-Caballero, P., Wang, Q., & Calero, J. M. (2023). Distributed dual-layer autonomous closed loops for self-protection of 5G/6G IoT networks from distributed denial of service attacks. *Computer Networks*, *222*, 109526. doi:10.1016/j.comnet.2022.109526

Bhardwaj, A., Avasthi, V., Sastry, H., & Subrahmanyam, G. V. (2016, April). Ransomware digital extortion: A rising new age threat. *Indian Journal of Science and Technology*, *9*(14), 1–5. doi:10.17485/ijst/2016/v9i14/82936

Bienkowski, T. (2022). 4 trends in network security to keep an eye on. *CSO INDIA*. https://www.csoonline.com/article/3649754/4-trends-in-network-security-to-keep-an-eye-on.html

Bilge, L., Han, Y., & Dell'Amico, M. (2017). Riskteller: Predicting the risk of cyber incidents. In *Proceedings of the 2017 ACM SIGSAC Conference on Computer and Communications Security*, (pp. 1299–1311). ACM. 10.1145/3133956.3134022

Bishnoi, A. & Gupta, N. (2023). Comprehensive Assessment of Reverse Social Engineering to Understand Social Engineering Attacks. Icssit. doi:10.1109/ICSSIT55814.2023.10061054

Bossler, A. M., & Holt, T. J. (2009). Online activities, guardianship, and malware infection: An examination of routine activities theory. *International Journal of Cyber Criminology*, *3*(1).

Brewer, R. (2016, September 1). Ransomware attacks: Detection, prevention and cure. *Network Security, Elsevier*, *2016*(9), 5–9. doi:10.1016/S1353-4858(16)30086-1

Butler, S. (2019). The Role of PGP Encryption on the Dark Web. *Tech Nadu*. https://www.technadu.com/pgpencryption-dark-web/57005/

Cabaj, K., & Mazurczyk, K. (2016). Using software-defined networking for ransomware mitigation: The case of cryptowall. *IEEE Network*, *30*(6), 14–20. doi:10.1109/MNET.2016.1600110NM

Caivano, D., Canfora, G., Cocomazzi, A., Pirozzi, A., & Visaggio, C. A. (2017). Ransomware at X-Rays. *2017 IEEE International Conference on Internet of Things (iThings) and IEEE Green Computing and Communications (GreenCom) and IEEE Cyber, Physical and Social Computing (CPSCom) and IEEE Smart Data (SmartData)*. IEEE. 10.1109/iThings-GreenCom-CPSCom-SmartData.2017.58

Cambiaso, E., Vaccari, I., Patti, L., & Aiello, M. (2019). Darknet security: A categorization of attacks to the TOR network. In *Italian Conference on Cyber Security*. CEUR-WS.

Cambiaso, E., Vaccari, I., Patti, L., & Aiello, M. (2019, February). Darknet Security: A Categorization of Attacks to the Tor Network. In ITASEC (pp. 1-12).

Carrier Brian, D. (2006, February). *Communications of the ACM*, *49*(2), 56–61. doi:10.1145/1113034.1113069

Carrier, B. (2002). *Open Source Digital Forensic Tools: The Legal Argument (PDF).* @stake Research Report.

Cascavilla, G., Tamburri, D. A., & Van Den Heuvel, W. J. (2021). Cybercrime threat intelligence: A systematic multi-vocal literature review. *Computers & Security, 105*, 102258. doi:10.1016/j.cose.2021.102258

Casino, F., Dasaklis, T. K., Spathoulas, G. P., Anagnostopoulos, M., Ghosal, A., Borocz, I., Solanas, A., Conti, M., & Patsakis, C. (2022). Research trends, challenges, and emerging topics in digital forensics: A review of reviews. *IEEE Access : Practical Innovations, Open Solutions, 10*, 25464–25493. doi:10.1109/ACCESS.2022.3154059

Catakoglu, O., Balduzzi, M., & Balzarotti, D. (2017, April). Attacks landscape in the dark side of the Web. In *Proceedings of the Symposium on Applied Computing* (pp. 1739-1746). ACM. 10.1145/3019612.3019796

Celiktaş, B. (2018). *The ransomware detection and prevention tool design by using signature and anomaly-based detection methods* [Master Thesis]. Istanbul Technical University.

Charlie, M. (2007). *The Legitimate Vulnerability Market: Inside the Secretive World of 0-day Exploit Sales.* Independent Security Evaluators.

Chatchalermpun, S., Wuttidittachotti, P., & Daengsi, T. (2020). *Cybersecurity Drill Test Using Phishing Attack: A Pilot Study of a Large Financial Services Firm in Thailand.* 10th Symposium on Computer Applications & Industrial Electronics (ISCAIE), Malaysia.

Check Point Software. (n.d.). *How To Prevent Ransomware Attacks.* Check Point Software. https://www.checkpoint.com/cyber-hub/threat-prevention/ransomware/how-to-prevent-ransomware/

Chen, H. (2011). *Dark Web: Exploring and data mining the dark side of the Web.* Springer Science & Business Media.

Chertoff, M. (2017). A public policy perspective of the Dark Web. *Journal of Cyber Policy, 2*(1), 26–38. doi:10.1080/23738871.2017.1298643

Choi, K.-S. (2008). Computer crime victimization and integrated theory: An empirical assessment. *International Journal of Cyber Criminology, 2*(1).

Chopra, G., & Whig, P. (2022). Energy Efficient Scheduling for Internet of Vehicles. *International Journal of Sustainable Development in Computing Science, 4*(1).

Christensen, J. B., & Beuschau, N. (2017). *Ransomware detection and mitigation tool.* Tech. Univ. Denmark, Lyngby.

Ciancaglini, V., Balduzzi, M., & Goncharov, M. (n.d.). Cibercrime and digital threats. *Trend Micro.* https://www.trendmicro.com/vinfo/pl/security/news/cybercrime-and-digital-threats/deep-web-and-cyber crime-its-not-all-about-tor.

Čisar, P., & Čisar, S. M. (2018). Ethical hacking of wireless networks in kali linux environment. *Annals of the Faculty of Engineering Hunedoara, 16*(3), 181–186.

Cisco. (2021). *Encrypted Traffic Analytics* (White Paper). Cisco Enterprise Security. https://www.cisco.com/c/en/us/solutions/collateral/enterprise-networks/enterprise-network-security/nb-09-encrytd-traf-anlytcs-wp-cte-en.html.

Cobb, C. L., Meca, A., Branscombe, N. R., Schwartz, S. J., Xie, D., Zea, M. C., Fernandez, C. A., & Sanders, G. L. (2019). Perceived discrimination and well-being among unauthorized Hispanic immigrants: The moderating role of ethnic/racial group identity centrality. *Cultural Diversity & Ethnic Minority Psychology, 25*(2), 280–287. doi:10.1037/cdp0000227 PMID:30284850

Coccaro, R. (2017). *Evaluation of Weaknesses in US Cybersecurity and Recommendations for Improvement* [Doctoral dissertation]. Utica College.

Constella. (2021). The Data Breach Era. Constella. https://info.constellaintelligence.com/white-paper-the-data-breach-era.

Council of Europe. (2013). *Electronic Evidence Guide*. Council of Europe.

CryptoDeFix. (2021). *Indian authorities will ban the purchase of goods with cryptocurrency.* CryptoDeFix. https://cryptodefix.com/articles/cryptocurrency-in-india-will-be-regulated-as-asset

Cui, A. & Stolfo, S. (2022). *A Quantitative Analysis of the Insecurity of Embedded Network Devices: Results of a Wide-Area Scan Proceedings of the 26th Annual Computer Security Applications*. NYU.

Cyber Threat Alliance. (2015). *Lucrative ransomware attacks: Analysis of the cryptowall version 3 threat.* Cyber Threat Alliance. https://www.cyberthreatalliance.org/resources/lucrative-ransomware-attacksanalysis-cryptowall-version-3-threat/

Cyber, C. I. P. (2015). Parrot Security OS for Pentesting and Computer Forensics [Online]. *EHacking.* https://www.ehacking.net/2015/06/parrot-security-os-for-pentesting-and.html

Da Costa, F. (2013). *Rethinking the Internet of Things: A scalable approach to connecting everything.*

Dan.Is. (n.d.). *TOR Nodes List*. Dan.Is. https://www.dan.me.uk/tornodes

Dangi, R., Jadhav, A., Choudhary, G., Dragoni, N., Mishra, M., & Lalwani, P. (2022). ML-Based 5G Network Slicing Security: A Comprehensive Survey. *Future Internet, 14*(4), 116. doi:10.3390/fi14040116

de Jackson, C. (2021). A comprehensive analysis of social learning theory linked to criminal and deviant behaviour. American International Journal of Contemporary Research, 11.

Del Pozo, I., Iturralde, M., & Restrepo, F. (2018). Social engineering: Application of psychology to information security. *Proceedings - 2018 IEEE 6th International Conference on Future Internet of Things and Cloud Workshops, W-FiCloud 2018*, (pp. 108–114). IEEE. 10.1109/W-FiCloud.2018.00023

Deloitte. (2016) *Ransomware holding your data*. Deloitte Threat Intelligence and Analytics. https://www2.deloitte.com/content/dam/Deloitte/us/Documents/risk/us-aers-ransomware.pdf

Demuro, P. R. (2017). Keeping internet pirates at bay: ransomware negotiation in the healthcare industry keeping internet pirates at bay: ransomware negotiation in the healthcare industry. *Nova Law Review*, *41*(3), 5.

Diagnostic analytics . (n.d.). BetterBI. https://www.betterbi.dk/diagnostic-analytics/

Drenick, A. H. (2017). *The 2017 Equifax Hack: What We Can Learn.*

Erdin, E., Zachor, C., & Gunes, M. H. (2015). How to find hidden users: A survey of attacks on anonymity networks. *IEEE Communications Surveys and Tutorials*, *17*(4), 2296–2316. doi:10.1109/COMST.2015.2453434

Erdogan, S. E. (2020). *Building an Information Security Management System, Implementation of IEC / ISO 27001 Standard in A Civil Aviation Organization*. Istanbul Kültür University.

Eroglu, C. (2023). The Size of Cyber Crimes that Businesses Are Exposed. *Journal of Security Sciences*, *12*(1), 69–96. doi:10.28956/gbd.1264593

Everett, C. (2016). Ransomware: To pay or not to pay? *Computer Fraud & Security*, *2016*(4), 8–12. doi:10.1016/S1361-3723(16)30036-7 PMID:27382895

Evers, B., Hols, J., Kula, E., Schouten, J., den Toom, M., van der Laan, R. M., & Pouwelse, J. A. (2015). *Thirteen years of tor attacks*. Github. https://github.com/Attacks-on-Tor/Attacks-on-Tor.

Evers, B., Hols, J., Kula, E., Schouten, J., Den Toom, M., van der Laan, R. M., & Pouwelse, J. A. (2016). *Thirteen Years of Tor Attacks.*

Extrahop. (2023). Extrahop System User Guide [Online]. *Extrahop*. https://docs.extrahop.com/8.9/eh-system-user-guide/

Faizan, M., Khan, R. A., & Agrawal, A. (2022). Ranking potentially harmful Tor hidden services: Illicit drugs perspective. *Applied Computing and Informatics*, *18*(3/4), 267–278. doi:10.1016/j.aci.2020.02.003

Federal Bureau of Investigation (FBI). (2016). Internet Crime Complaint Center (IC3). *IC3 Annual Report, 2016 State Reports*. FBI. https://www.ic3.gov/Home/AnnualReports?redirect=true

Federal Bureau of Investigation (FBI). (2017). Internet Crime Complaint Center (IC3). *IC3 Annual Report, 2017 State Reports*. FBI. https://www.ic3.gov/Home/AnnualReports?redirect=true

Ferrag, M., & Maglaras, L. (2019). DeliveryCoin: An IDS and Blockchain-Based Delivery Framework for Drone-Delivered Services. *Computers*, *8*(3), 58. doi:10.3390/computers8030058

Ferrag, M., Shu, L., Djallel, H., & Choo, K.-K. (2021). Deep Learning-Based Intrusion Detection for Distributed Denial of Service Attack in Agriculture 4.0. *Electronics (Basel)*, *10*(11), 1257. doi:10.3390/electronics10111257

Ferry, N. (2019). *Methodology of dark web monitoring*. IEEE. doi:10.1109/ECAI46879.2019.9042072

Filiol, E., Mercaldo, F., & Santone, A. (2021). A method for automatic penetration testing and mitigation: A red hat approach. *Knowledge-Based and Intelligent Information & Engineering Systems: Proceedings of the 25th International Conference KES2021*. 192, pp. 2039-2046. Szczecin, Poland: ELSEVIER.

FinCEN's Financial Trend Analysis. (2021). Fincen. https://www.fincen.gov/sites/default/files/2021-10/Financial%20Trend%20Analysis_Ransomware%20

Finkle, J. (2016). *Ransomware: Extortionist hackers borrow customer service tactics*. Reuters. https://www.reuters.com/article/us-usa-cyber-ransomware-idUSKCN0X917X

FinkleaK. (2015). https://digital.library.unt.edu/ark:/67531/metadc700882/

Formby D, Durbha S, & Beyah R. (2017). *Out of control: Ransomware for industrial control systems*. InRSA.

Frattini, F., Giordano, U., & Conti, V. (2019, September). Facing cyber-physical security threats by PSIM-SIEM integration. In *2019 15th European Dependable Computing Conference (EDCC)* (pp. 83-88). IEEE. 10.1109/EDCC.2019.00026

Fu, X. & Qian, K. (2008). SAFELI-SQL Injection Scanner Using Symbolic Execution. *Proceedings of the workshop on Testing*, (pp. 34 – 39). IEEE.

Gagneja, K. K. (2017). Knowing the ransomware and building defense against it - specific to healthcare institutes. *2017 Third International Conference on Mobile and Secure Services (MobiSecServ)*, Miami Beach, FL, USA. 10.1109/MOBISECSERV.2017.7886569

Gehlot, A., Singh, R., Singh, J., & Sharma, N. R. (Eds.). (2022). *Digital Forensics and Internet of Things: Impact and Challenges*. John Wiley & Sons. doi:10.1002/9781119769057

Ghania, A. S. (2016). Analyzing Master Boot Record for Forensic Investigations. *International Journal of Applied Information Systems*, *10*, 2249–0868.

Girinoto, P. D. F., Yulita, T., Zulkham, R. K., Rifqi, A., & Putri, A. (2022). OmeTV Pretexting Phishing Attacks: A Case Study of Social Engineering. *IWBIS 2022 - 7th International Workshop on Big Data and Information Security, Proceedings*, (pp. 119–124). IEEE. 10.1109/IWBIS56557.2022.9924801

Glassberg, J. (2016). Defending against the ransomware threat. *POWERGRID International*, *21*(8), 22–24.

Glynn, G. (2023). *The Ultimate OSINT Collection*. Startme page. https://start.me/p/DPYPMz/the-ultimate-osint-collection

Grigas, L. (2023). *What is end-to-end encryption and how does it work?* Nord Pass. https://nordpass.com/blog/what-is-end-to-end-encryption/

Grossman, J. (2007). *Cross-Site Scripting Worms & Viruses - The Impending Threat & thee Best Defense.* White Hat Sec. https://www.whitehatsec.com/assets/WP5CSS0607.pdf

Gulati, H., Saxena, A., Pawar, N., Tanwar, P., & Sharma, S. (2022, January). Dark Web in Modern World Theoretical Perspective: A survey. In *2022 International Conference on Computer Communication and Informatics (ICCCI)* (pp. 1-10). IEEE. 10.1109/ICCCI54379.2022.9740785

Gulhan, B. (2021). *Awareness of Information Security in Higher Education Institutions: The Case of Bahçeşehir.* University Bahçeşehir University.

Gunduzalp, C. (2021). University Employees' Awareness of Digital Data and Personal Cyber Security (A Case Study of IT Departments). *Journal of Computer and Education Research,* 9(18), 598–625. doi:10.18009/jcer.907022

Gupta, A. (2018). *The dark Web as a phenomenon: a review and research agenda* [Doctoral dissertation, University of Melbourne].

Gupta, S., & Isha, B. A., & Gupta, H. (2021). Analysis of Social Engineering Attack on Cryptographic Algorithm. *2021 9th International Conference on Reliability, Infocom Technologies and Optimization (Trends and Future Directions), ICRITO 2021,* (pp. 1–5). IEEE. 10.1109/ICRITO51393.2021.9596568

Gupta, S., Singhal, A., & Kapoor, A. (2016). *A Literature Survey on Social Engineering Attacks : Phishing Attack.* Semantic Scholar.

Halawa, H., Beznosov, K., Boshmaf, Y., Coskun, B., Ripeanu, M., & Santos-Neto, E. (2016). Harvesting the low-hanging fruits: defending against automated large-scale cyber-intrusions by focusing on the vulnerable population. In *Proceedings of the 2016 New Security Paradigms Workshop,* (pp. 11–22). ACM. 10.1145/3011883.3011885

Heater, B. (2016, May). How ransomware conquered the world. *PC Magazine Digital Edition,* 109-118.

Held, M. (2018). *Detecting Ransomware* [Thesis]. University Konstanz.

Hernandez-Castro, J., Cartwright, E., & Stepanova, A. (2017). Economic analysis of ransomware. arXiv preprint arXiv:1703.06660.

Hoelz, B. W., Ralha, C. G., & Geeverghese, R. (2009, March). Artificial intelligence applied to computer forensics. In *Proceedings of the 2009 ACM symposium on Applied Computing* (pp. 883-888). ACM. 10.1145/1529282.1529471

Hoffman, M. (2018). *Your OSINT Graphical Analyzer.* Myosint. https://yoga.myosint.training

Hwang, I., Wakefield, R., Kim, S., & Kim, T. (2021). Security Awareness: The First Step in Information Security Compliance Behavior. *Journal of Computer Information Systems*, *61*(4), 345–356. doi:10.1080/08874417.2019.1650676

IBM. (2023). *IBM Security QRadar SIEM [Online]*. IBM. https://www.ibm.com/products/qradar-siem

Ibrahim, R. F., Abu Al-Haija, Q., & Ahmad, A. (2022). DDoS Attack Prevention for Internet of Thing Devices Using Ethereum Blockchain Technology. *Sensors (Basel)*, *22*(18), 6806. doi:10.339022186806 PMID:36146163

Idoko, N. (n.d.). *How to Protect Your Data From Cyber Attacks*. Nicholas Idoko Blog. https://nicholasidoko.com/blog/2023/01/12/how-to-protect-your-data-from-cyber-attacks/

Ileri, Y. Y. (2017). Information Security Management in Organizations, Enterprise Integration Process and A Case Study. *Anadolu University Journal of Social Sciences*, *17*(4), 55–72. doi:10.18037/ausbd.417372

Ileri, Y. Y. (2018). Security in Accessing Enterprise Information Resources: A Research on Password Management of Physicians. *International Journal of Health Management and Strategies Research*, *4*(1), 15–25.

Internet Engineering Task Force. (2019). *Learning About the Domain Name System (DNS) from its Terminology*. Internet Engineering Task Force (IETF). https://www.ietf.org/blog/dns-terminology/

Iqbal, A., Saleem, R., & Suryani, M. (2020). Internet of Things (IOT): ongoing Security Challenges and Risks. *International Journal of Computer Science and Information Security*, *14*.

Jain, K., Gupta, M., & Abraham, A. (2021). A Review on Privacy and Security Assessment of Cloud Computing. *Journal of Information Assurance and Security.*, *16*(5), 161–168.

JajooA. (2021). *A study on the Morris Worm*. https://arxiv.org/abs/2112.07647

Jayasuryapal, G., Pranay, P. M., & Kaur, H., & Swati. (2021). A Survey on Network Penetration Testing. *Proceedings of 2021 2nd International Conference on Intelligent Engineering and Management, ICIEM 2021*, (pp. 373–378). IEEE. 10.1109/ICIEM51511.2021.9445321

Jeetendra, P. (2017). Introduction to Cyber Security Ahmad Rahayu(2022). A systematic literature review of routine activity theory's applicability in cybercrime. *Journal of Cyber Security and Mobility, 11*.

Jiang, P., Wu, H., Wang, C., & Xin, C. (2018, May). Virtual MAC spoofing detection through deep learning. In *2018 IEEE International Conference on Communications (ICC)* (pp. 1-6). IEEE. 10.1109/ICC.2018.8422830

Jupalle, H., Kouser, S., Bhatia, A. B., Alam, N., Nadikattu, R. R., & Whig, P. (2022). Automation of human behaviors and its prediction using machine learning. *Microsystem Technologies*, *28*(8), 1–9. doi:10.100700542-022-05326-4

Kanellis, P. (Ed.). (2006). *Digital crime and forensic science in cyberspace*. IGI Global. doi:10.4018/978-1-59140-872-7

Karresand, M., Axelsson, S., & Dyrkolbotn, G. O. (2019, July 1). Using NTFS cluster allocation behavior to find the location of user data. *Digital Investigation, 29*, S51–S60. doi:10.1016/j.diin.2019.04.018

Karunanayake, I., Ahmed, N., Malaney, R., Islam, R., & Jha, S. (2020). Anonymity with Tor: A survey on Tor attacks. *arXiv preprint arXiv:2009.13018.*

Kaspersky Lab. (2016). *Ransomware in 2014-2016 Technical report*. Kaspersky Lab. https://media.kasperskycontenthub.com/wp-content/uploads/sites/43/2018/03/07190822/KSN_Report_Ransomware_2014-2016_final_ENG.pdf

Kaur, S., & Randhawa, S. (2020). Dark web: A web of crimes. *Wireless Personal Communications, 112*(4), 2131–2158. doi:10.100711277-020-07143-2

Kenney, D. (2012). Firearm Microstamp Technology: Failing Daubert and Federal Rules of Evidence 702. *Rutgers Computer & Technology Law Journal, 38*, 199.

Kharraz, A., Arshad, S., Mulliner, C., Robertson, W. K., & Kirda, E. (2016). Unveil: A largescale,automated approach to detecting ransomware. In *USENIX Security Symposium*, (pp. 757–772). USENIX.

Kharraz, A., Robertson, W., Balzarotti, D., Bilge, L., & Kirda, E. (2015). Cutting the gordian knot: A look under the hood of ransomware attacks. In *Proceedings of the Detection of Intrusions and Malware, and Vulnerability Assessment Conference* (pp. 3-24). Springer International Publishing. 10.1007/978-3-319-20550-2_1

Khasawneh, M., Kajman, I., Alkhudaidy, R., & Althubyani, A. (2014, March). A survey on Wi-Fi protocols: WPA and WPA2. In *International conference on security in computer networks and distributed systems* (pp. 496-511). Springer. 10.1007/978-3-642-54525-2_44

Khera, Y., Whig, P., & Velu, A. (2021). efficient effective and secured electronic billing system using AI. *Vivekananda Journal of Research, 10*, 53–60.

Köksal, S., Dalveren, Y., Maiga, B., & Kara, A. (2021). Distributed denial-of-service attack mitigation in network functions virtualization-based 5G networks using management and orchestration. *International Journal of Communication Systems, 34*(9), e4825. doi:10.1002/dac.4825

Kotov, V. & Rajpal, M. S. (2014). In-Depth Analysis of the Most Popular Malware Families. *Understanding Crypto-Ransomware Report*. Bromium.

Kozlosky, B. (2021). *5 Challenging Network Security Problems & the Fixes*. TechSling Weblog. https://www.techsling.com/5-challenging-network-security-problems-the-fixes/.

Kruse, W. G. II, & Heiser, J. G. (2001). *Computer forensics: incident response essentials*. Pearson Education.

Kumar, A. (2022a, September 16). Deep dive into VPN & Proxies: How to stay safe online. *Medium.* https://medium.com/@abhijeet-secops/deep-dive-into-vpn-proxies-how-to-stay-secure-online-79731b9b654

Kumar, A. (2022b). *Hitchhiker's Guide To URL Analysis*. Github. https://github.com/wand3rlust/Hitchhikers-Guide-To-URL-Analysis

Laliberte, B. (2023). *End-to-End network visibility and management?* Enterprise Strategy Group. https://www.esg-global.com/research/topic/networking

Laszka, A., Farhang, S., & Grossklags, J. (2017). On the economics of ransomware. In *Decision and Game Theory for Security: 8th International Conference, GameSec 2017,* (pp. 397-417). Springer International Publishing.

Lee, B. (2019). *What is Ransomware? The Major Cybersecurity Threat Explained*. Spin Backup. https://spinbackup.com/blog/what-is-ransomware-the-major-cybersecurity-threat-explained/.

Lee, B., & Liles, S. (2013). *Applying the OSCAR Forensic Framework to Investigations of Cloud Processing*. CERIAS. https://www.cerias.purdue.edu/assets/symposium/2013-posters/23E-5F7.pdf,2013

Lee, J., & Lee, K. (2017, November 15). Spillover effect of ransomware: Economic analysis of web vulnerability market. *Research Briefs on Information and Communication Technology Evolution., 3*, 193–203. doi:10.56801/rebicte.v3i.59

Lévesque, F. L., Fernandez, J. M., & Somayaji, A. (2014). Risk prediction of malware victimization based on user behavior. In *Malicious and Unwanted Software: The Americas (MALWARE), 9th International Conference* (pp. 28–134). IEEE. 10.1109/MALWARE.2014.6999412

Levesque, F. L., Nsiempba, J., Fernandez, J. M., Chiasson, S., & Somayaji, A. (2013). A clinical study of risk factors related to malware infections. In *Proceedings of the ACM SIGSAC conference on Computer & communications security* (pp. 97–108). ACM. 10.1145/2508859.2516747

Li, Q., Feng, X., Wang, H., & Li, Z. L SunTowards Fine-grained Fingerprinting of Firmware in Online Embedded Devices IEEE INFOCOM 2018-IEEE Conference on Computer Communications, p. 2537 – 2545 Conference (ACSAC '10), p. 97 – 106 Posted: 2010

Liska, A., & Gallo, T. (2016). *Ransomware: Defending Against Digital Extortion*. O'Reilly Media, Inc.

Li, T., Wang, K., & Horkoff, J. (2019). Towards effective assessment for social engineering attacks. *Proceedings of the IEEE International Conference on Requirements Engineering, 2019,* (pp. 392–397). IEEE. 10.1109/RE.2019.00051

Liu, H., Crespo, R., & Martínez, O. (2020). Enhancing Privacy and Data Security across Healthcare Applications Using Blockchain and Distributed Ledger Concepts. *Health Care, 8,* 243. PMID:32751325

Liu, T., Li, Z., Long, H., & Bilal, A. (2023). NT-GNN: Network Traffic Graph for 5G Mobile IoT Android Malware Detection. *Electronics (Basel)*, *12*(4), 789. doi:10.3390/electronics12040789

Li, X., & Wang, Y. (2011). Security enhanced authentication and key agreement protocol for LTE/SAE network. *7th International Conference on Wireless Communications, Networking and Mobile Computing* (pp. 1-4). Wuhan, China: IEEE. 10.1109/wicom.2011.6040169

Lu, H. J., & Yu, Y. (2021). Research on WiFi penetration testing with Kali Linux. *Complexity*, *2021*, 2021. doi:10.1155/2021/5570001

Lu, N., Li, D., Shi, W., Vijayakumar, P., Piccialli, F., & Chang, V. (2021). An efficient combined deep neural network based malware detection framework in 5G environment. *Computer Networks*, *189*, 107932. doi:10.1016/j.comnet.2021.107932

Luo, X., & Liao, Q. (2007). Awareness education is the key to ransomware prevention. *Information Systems Security*, *16*(4), 195–202. doi:10.1080/10658980701576412

Madhu, M., & WHIG, P. (2022). A survey of machine learning and its applications. *International Journal of Machine Learning for Sustainable Development*, *4*(1), 11–20.

Maier, G., Feldmann, A., Paxson, V., Sommer, R., & Vallentin, M. (2011). An assessment of overt malicious activity manifest in residential networks. In *International Conference on Detection of Intrusions and Malware, and Vulnerability Assessment,* (pp. 144–163). Springer. 10.1007/978-3-642-22424-9_9

Maigida, A. M., Abdulhamid, S. M., Olalere, M., Alhassan, J. K., Chiroma, H., & Dada, E. G. (2019). Systematic literature review and metadata analysis of ransomware attacks and detection mechanisms. *Journal of Reliable Intelligent Environments*, *5*(2), 67–89. doi:10.100740860-019-00080-3

Mamza, E. S. (2021). Use of AIOT in Health System. *International Journal of Sustainable Development in Computing Science*, *3*(4), 21–30.

Manworren, N., Letwat, J., & Daily, O. (2016). Why you should care about the Target data breach. *Business Horizons*, *59*(3), 257–266. doi:10.1016/j.bushor.2016.01.002

Mercaldo, F., Nardone, V., & Santone, A. (2016). *Ransomware Inside Out*. 2016 11th International Conference on Availability, Reliability and Security (ARES), Salzburg, Austria. 10.1109/ARES.2016.35

Mercaldo, F., Nardone, V., Santone, A., & Visaggio, C. A. (2016). Ransomware Steals Your Phone. Formal Methods Rescue It. In E. Albert & I. Lanese (Eds.), Lecture Notes in Computer Science: Vol. 9688. *Formal Techniques for Distributed Objects, Components, and Systems. FORTE 2016*. Springer. doi:10.1007/978-3-319-39570-8_14

Milne, G. R., Labrecque, L. I., & Cromer, C. (2009). Toward an understanding of the online consumer's risky behavior and protection practices. *The Journal of Consumer Affairs*, *43*(3), 449–473. doi:10.1111/j.1745-6606.2009.01148.x

Mirea, M., Wang, V., & Jung, J. (2019). The not so dark side of the darknet: A qualitative study. *Security Journal, 32*(2), 102–118. doi:10.105741284-018-0150-5

MITRE. (2019). *Mitre att & ck*. MITRE. https://attack.mitre.org/

Modi, J. (2019). *Detecting ransomware in encrypted network traffic using machine learning* [PhD dissertation]. UVIC.

Mohammad, A. H. (2020, February). Ransomware evolution, growth, and recommendation for detection. *Modern Applied Science, 14*(3), 68. doi:10.5539/mas.v14n3p68

Morgan Office. (n.d.). *What is Ransomware?* Morgan Office. https://www.morganoffice.co.uk/help-advice/what-is-ransomware

Murphy, E. V., Murphy, M. M., & Seitzinger, M. V. (2015). Bitcoin: Questions, Answers, and Analysis of Legal Issues. *Congressional Research Service*, 1-36. https://fas.org/sgp/crs/misc/R43339.pdf

Mursul, D., & Kaya, A. (2019). A Review of Public Institutions in Terms of National Information Security Policies: Sample of Kayseri Bar Association. *ASSAM International Refereed Journal*(Special Issue), 331-343.

Naci Unal, A., & Ergen, A. (2018). Cyber Security Behaviour: A Research Conducted in Istanbul. *Manisa Celal Bayar University Journal of Social Sciences, 16*(2), 191–216. doi:10.18026/cbayarsos.439489

Nagaraj, A. (2021). Introduction to Sensors in IoT and Cloud Computing Applications. UAE: Bentham Science Publishers. doi:10.2174/97898114793591210101

Nagaraj, A. (2022). Adapting Blockchain for Energy Constrained IoT in Healthcare Environment. In K. Kaushik, S. Tayal, S. Dahiya, & A. O. Salau (Eds.), *Sustainable and Advanced Applications of Blockchain in Smart Computational Technologies* (p. 103). CRC press. doi:10.1201/9781003193425-7

Naick, B. D., & Bachalla, N. (2016). Application of Digital Forensics in Digital Libraries. [IJLIS]. *International Journal of Library and Information Science, 5*(2), 89–94.

Nair, R. S., Babu, A., & Vijayan, V. P. (2018). A Survey on Wi-Fi Security Techniques. *International Research Journal of Engineering and Technology, 5*(4), 4705–4707.

Nam, S. Y., Jurayev, S., Kim, S. S., Choi, K., & Choi, G. S. (2012). Mitigating ARP poisoning-based man-in-the-middle attacks in wired or wireless LAN. *EURASIP Journal on Wireless Communications and Networking, 2012*(1), 1–17. doi:10.1186/1687-1499-2012-89

Nance, K., & Ryan, D. J. (2011, January). Legal aspects of digital forensics: a research agenda. In *2011 44th hawaii international conference on system sciences* (pp. 1-6). IEEE. 10.1109/HICSS.2011.282

Naseem, I., Kashyap, A. K., & Mandloi, D. (2016). Exploring unknown depths of invisible web and the digi-underworld. *International Journal of Computer Applications, NCC*, (3), 21–25.

Nasr, M., Bahramali, A., & Houmansadr, A. (2018). DeepCorr: Strong flow correlation attacks on tor using deep learning. In *Proceedings of the 2018 ACM SIGSAC Conference on Computer and Communications Security* (pp. 1962–1976). ACM. 10.1145/3243734.3243824

Nayerifard, T., Amintoosi, H., Bafghi, A. G., & Dehghantanha, A. (2023). Machine Learning in Digital Forensics: A Systematic Literature Review. *arXiv preprint arXiv:2306.04965*.

Nazah, S., Huda, S., Abawajy, J., & Hassan, M. M. (2020). Evolution of dark web threat analysis and detection: A systematic approach. *IEEE Access : Practical Innovations, Open Solutions*, 8, 171796–171819. doi:10.1109/ACCESS.2020.3024198

NCCIC. (2016). *Grizzly Steppe - E – Russian Malicious Cyber Activity Summary*. NCCIC.

Nezgitli, S., & Gokcearslan, S. (2022). Review on Information Security Awareness for Public Institutions and Private Sectors. *Instructional Technology and Lifelong Learning*, 3(1), 19–44. doi:10.52911/itall.1115701

Ngo, F. T., & Paternoster, R. (2011). Cybercrime victimization: An examination of individual and situational level factors. *International Journal of Cyber Criminology*, 5(1).

Nickerson, C., Kennedy, D., Riley, C. H., Smith, E., Amit, I. I., Rabie, A., Friedli, S., Searle, J., & Knight, B. (2014). *Home*. Penetration Testing Execution Standard. http://www.pentest-standard.org/index.php/Main_Page

Nordine, J. (2022). *OSINT Framework*. OISNT. https://osintframework.com

O'Gorman, G., & McDonald, G. (2012). *Ransomware: A growing menace*. Symantec Corporation.

O'Neill, P. H. (2013). Inside the Bustling, Dicey World of Bitcoin Gambling. *The Daily Dot*.

Offensive Security. (n.d.). *Google Hacking Database*. Exploit DB. https://www.exploit-db.com/google-hacking-database

Olivier, M. S. (2009, March). On metadata context in database forensics. *Digital Investigation*, 5(3-4), 115–123. doi:10.1016/j.diin.2008.10.001

Oltsik, J. (2021). *Network Security Without Borders*. NetScot. https://www.netscout.com/whitepaper/esg-wp-network-security-without-borders, August .

Otway, H. J., & Von Winterfeldt, D. (1982). Beyond acceptable risk: On the social acceptability of technologies. *Policy Sciences*, 14(3), 247–256. doi:10.1007/BF00136399

Ovelgönne, M., Dumitra, T., Prakash, B. A., Subrahmanian, V. S., & Wang, B. (2017). Understanding the relationship between human behavior and susceptibility to cyber-attacks: A data-driven approach. *ACM Transactions on Intelligent Systems and Technology*, 8(4), 51. doi:10.1145/2890509

OWASP Foundation. (2022, December 03). *OWASP Web Security Testing Guide*. OWASP. https://owasp.org/www-project-web-security-testing-guide/v42/

Ozdemir, A. & Uluyol, C. (2021). Information Security Awareness in Public Organizations. *The Journal of Turkish Social Research, 25*(3), 649–666. doi:10.20296/tsadergisi.815635

Ozdemir, D., & Aslay, F. (2016). *Examination of The Effects of Information Security Awareness Training on Staff of Erzincan Public Health Administration*. International Erzincan Symposium, Erzincan, Turkey.

Palmer, G. (2001). *A road map for digital forensics research-report from the first Digital Forensics Research Workshop (DFRWS)*. Utica, New York.

Pal, O., Saxena, A., Saquib, Z., & Menezes, B. L. (2011). Secure Identity-Based Key Establishment Protocol. *International Conference on Advances in Communication, Network, and Computing* (pp. 618-623). Bangalore, India: Springer, Berlin, Heidelberg.

Park, Y., Mccoy, D., Shi, E., Tapia, M. G., Texas, A., Tyagi, A. K., Aghila, G., Hendriksen, H., Andriesse, D., Rossow, C., Stone-Gross, B., Plohmann, D., Bos, H., December, I., Vidros, S., Kolias, C., Kambourakis, G., Coletta, A., Van Der Veen, V., ... Brinkmann, U. (2015). Carbanak APT The Great Bank Robbery. *Computer Fraud & Security, 2015*(June), 1–5. http://www.trendmicro.com/vinfo/us/security/special-report/cybercriminal-underground-economy-series/%5Cnhttp://www.tandfonline.com/doi/full/10.1080/17440572.2016.1157480%5Cnhttp://www.sophos.com/threatreport%5Cnhttp://security-sh3ll.blogspot.com/2011/05/what-is-ze

Part, B. H., Quadir, B. Y. S., Aneez, S., Bergin, T. O. M., Layne, N., Das, K. N., & Spicer, J. (2017). THE The Bangladesh Bank Heist typo helped. *Journal of Consumer and Commercial Law*.

Patel, P. B., Thakor, H. P., & Iyer, S. (2019, March). A comparative study on cybercrime mitigation models. In *2019 6th International Conference on Computing for Sustainable Global Development (INDIACom)* (pp. 466-470). IEEE.

Pimple, N., Salunke, T., Pawar, U., & Sangoi, J. (2020, March). Wireless security—an approach towards secured wi-fi connectivity. In *2020 6th international conference on advanced computing and communication systems (ICACCS)* (pp. 872-876). IEEE.

Poolz.Finance. (n.d.). *Ransomware: - Blog*. Poolz Finance. https://blog.poolz.finance/glossary/ransomware

Poudyal, S., Subedi, K. P., & Dasgupta, D. (2018). A Framework for Analyzing Ransomware using Machine Learning. *2018 IEEE Symposium Series on Computational Intelligence (SSCI)*, Bangalore, India. 10.1109/SSCI.2018.8628743

Prashant, M. (2019). *A Textbook on cybercrimes and penalties*. Snow White India.

Qureshi, S., Tunio, S., Akhtar, F., Wajahat, A., Nazir, A., & Ullah, F. (2021). Network Forensics: A Comprehensive Review of Tools and Techniques. *International Journal of Advanced Computer Science and Applications, 12*(5), 879–887. doi:10.14569/IJACSA.2021.01205103

Rafiuddin, H. (2017). *A Dark Web story in-depth research and study conducted on the dark web based on forensic computing and security in Malaysia*. IEEE. doi:10.1109/ICPCSI.2017.8392286

Rapid7. (n.d.). *Types of Attacks*. Rapid7. https://www.rapid7.com/funda mentals/types-of-attacks/

Rawat, R., Mahor, V., Chouhan, M., Pachlasiya, K., Telang, S., & Garg, B. (2022). Systematic Literature Review (SLR) on social media and the Digital Transformation of Drug Trafficking on Darkweb. In *International Conference on Network Security and Blockchain Technology* (pp. 181-205). Springer, Singapore. 10.1007/978-981-19-3182-6_15

Reddy, B. I., & Srikanth, V. (2019). Review on wireless security protocols (WEP, WPA, WPA2 & WPA3). *International Journal of Scientific Research in Computer Science, Engineering and Information Technology*, 28-35.

Redmiles, E., Kross, S., Pradhan, A., & Mazurek, M. (2017). *How well do my results generalize? comparing security and privacy survey results from mturk and web panels to us*. Technical report.

Reedy, P. (2023). Artificial intelligence in digital forensics. In *Encyclopedia of Forensic Sciences* (3rd ed., pp. 170–192). Elsevier. doi:10.1016/B978-0-12-823677-2.00236-1

Resnick, P. (2008). *RFC 5322*. RFC. https://www.rfc-editor.org/rfc/rfc5322

Restuccia, F., D'Oro, S., Al-Shawabka, A., Rendon, B. C., Chowdhury, K., Ioannidis, S., & Melodia, T. (2020). Hacking the waveform: Generalized wireless adversarial deep learning. *arXiv preprint arXiv:2005.02270*.

Richardson, R., & North, M. M. (2017). *Ransomware: Evolution, Mitigation and Prevention*. Faculty Publications. 4276. https://digitalcommons.kennesaw.edu/facpubs/4276

Rittinghouse, J., Hancock, W. M., & CISSP, C. (2003). *Cybersecurity operations handbook*. Digital Press.

Rode, K., Patil, S., Patil, A., & Dahotre, R. (2022). Network Security and Cyber Security. *International Journal of Research in Applied Science & Engineering Technology*. https://www.ijraset.com/research-paper/network-security-and-cyber-security, June

Rudesill, D. S., Caverlee, J., & Sui, D. (2015). *The deep web and the darknet: A look inside the internet's massive black box*. Ohio State Public Law Working Paper No. 314.

Runciman, B. (2020). Cybersecurity Report 2020. *ITNOW, 62*(4), 28–29. doi:10.1093/itnow/bwaa103

Salahdine, F., & Kaabouch, N. (2019). Social Engineering Attacks: A Survey. *Future Internet, 11*(4), 89. doi:10.3390/fi11040089

Saleem, J., Islam, R., & Kabir, M. A. (2022). The Anonymity of the Dark Web: A Survey. *IEEE Access : Practical Innovations, Open Solutions, 10*, 33628–33660. doi:10.1109/ACCESS.2022.3161547

Saleh, S., Qadir, J., & Ilyas, M. U. (2018). Shedding light on the Internet's dark corners: A survey of tor research. *Journal of Network and Computer Applications*, *114*, 1–28. doi:10.1016/j.jnca.2018.04.002

Salvi, M. H. U., & Kerkar, M. R. V. (2016). Ransomware: A cyber extortion. Asian Journal of Convergence in Technology, 2(3), Solander, A. C., Forman, A. S., & Glasser, N. M. (2016). Ransomware—Give me back my files! *Employee Relations Law Journal*, *42*(2), 53–55.

Samtani, S., Chai, Y., & Chen, H. (2022). Linking exploits from the dark Web to known vulnerabilities for proactive cyber threat intelligence: An attention-based deep structured semantic model. *Management Information Systems Quarterly*, *46*(2), 911–946. doi:10.25300/MISQ/2022/15392

Sato, T., Moungnoul, P., & Fukase, M. A. (2011, May). Compatible WEP algorithm for improved cipher strength and high-speed processing. In The 8th Electrical Engineering/Electronics, Computer, Telecommunications and Information Technology (ECTI) Association of Thailand-Conference 2011 (pp. 401-404). IEEE. doi:10.1109/ECTICON.2011.5947859

Sattar, D., & Matrawy, A. (2019). Towards secure slicing: Using slice isolation to mitigate DDoS attacks on 5G core network slices. *IEEE Conference on Communications and Network Security (CNS)* (pp. 82-90). Washington, DC, USA: IEEE. 10.1109/CNS.2019.8802852

Satterfeld, J. (2016). FBI Tactic in National Child Porn Sting under Attack. *USA Today*. https://www.usatoday.com/story/news/nation-now/2016/09/05/fbi-tactic-child-pornstingunder-%0Aattack/89892954/

Scaife, N., Carter, H., Traynor, P., & Butler, K. (2016). Cryptolock (and drop it): stopping ransomware attacks on user data. In *Distributed Computing Systems (ICDCS) 36th International Conference*. IEEE.

Scanlon, M. (2009). *Enabling the remote acquisition of digital forensic evidence through secure data transmission and verification*. University College Dublin.

Scarfone, K., Souppaya, M., Cody, A., & Orebaugh, A. (2008). *SP 800-115 Technical Guide to Information Security Testing and Assessment*. National Institute of Standards and Technology (NIST). https://csrc.nist.gov/publications/detail/sp/800-115/final

Securities and Commission. (2017). Cybersecurity: Ransomware alert. *Natl Exam Progr Risk Alert*, *5*(4), 15–16.

Sen, J. (2009). A Survey on Wireless Sensor Network Security. *International Journal of Communication Networks and Information Security (IJCNIS)*, *1*(2).

Senie, D. (1998). Network ingress filtering: defeating denial of service attacks which employ IP source address spoofing. *Network*.

Sen, S., Maity, S., & Das, D. (2020). The body is the network: To safeguard sensitive data, turn flesh and tissue into a secure wireless channel. *IEEE Spectrum*, *57*(12), 44–49. doi:10.1109/MSPEC.2020.9271808

Sheldon, F. T., Weber, J. M., Yoo, S. M., & Pan, W. D. (2012). The insecurity of wireless networks. *IEEE Security and Privacy, 10*(4), 54–61. doi:10.1109/MSP.2012.60

Singh, A., & Jain, K. (2022). An Automated Lightweight Key Establishment Method for Secure Communication in WSN. *Wireless Personal Communications, 124*(4), 2831–2851. doi:10.100711277-022-09492-6

Singh, A., & Jain, K. (2022). An efficient, secure key establishment method in a cluster-based sensor network. *Telecommunication Systems, 79*(1), 1–14. doi:10.100711235-021-00844-4

Siuda, P., Nowak, J., & Gehl, R. W. (2023). Darknet imaginaries in Internet memes: the discursive malleability of the cultural status of digital technologies. *Journal of Computer-Mediated Communication, 28*(1), zmac023.

Sleuthkit. (2023). Open-Source Digital Forensics [Online]. Sleuthkit. https://www.sleuthkit.org/autopsy/

Software Testing Help. (2023). *Best Wifi Packet Review.* Software Testing Help. https://www.softwaretestinghelp.com/best-wifi-packet-sniffer-review/

Solomon, M. G., Rudolph, K., Tittel, E., Broom, N., & Barrett, D. (2011). *Computer forensics jumpstart.* John Wiley & Sons.

Srivastava, S., Tiwari, A., & Srivastava, P. K. (2022). Review on quantum safe algorithms based on Symmetric Key and Asymmetric Key Encryption methods. *2022 2nd International Conference on Advance Computing and Innovative Technologies in Engineering, ICACITE 2022,* (pp. 905–908). IEEE. 10.1109/ICACITE53722.2022.9823437

Strecher, V. J., & Rosenstock, I. M. (1997). The health belief model. Cambridge handbook of psychology, health and medicine, 113–117.

Sulaiman, M. A., & Zhioua, S. (2013, July). Attacking Tor through unpopular ports. In *2013 IEEE 33rd International Conference on Distributed Computing Systems Workshops* (pp. 33-38). IEEE. 10.1109/ICDCSW.2013.29

Symantec. (2017a). *Internet security threat report, 22.* Symantec. https://www.symantec.com.

Symantec. (2017b). *2017 internet security threat report, 22.* Symantec.

Taner, E., & Kilic, I. (2019). A Study on Determining Information Security Awareness of Security Forces. *The Journal of Security Sciences, 8*(2), 253–269. doi:10.28956/gbd.646321

Telecommunication Standardization Sector of ITU. (2009). Series X: data networks, open system communications and security. In *Overview of cybersecurity.* ITU.

Temel, M. H. (2019). Research Review: Traffic Analysis Attack Against Anonymity in TOR and Countermeasures. School of Informatics. The Eindhoven University of Technology.

The Latest Ransomware Targeting Businesses and Individuals in Nepal. (n.d.-a). LinkedIn. https://www.linkedin.com/pulse/latest-ransomware-targeting-businesses-individuals-nepal-bashyal

Thethi, N., & Keane, A. (2014, February). Digital forensics investigations in the cloud. In 2014 IEEE international advance computing conference (IACC) (pp. 1475-1480). IEEE. doi:10.1109/IAdCC.2014.6779543

Tian, K., Zhang, B., Mouftah, H., Zhao, Z., & Ma, J. (2009, June). Destination-driven on-demand multicast routing protocol for wireless ad hoc networks. In *2009 IEEE International Conference on Communications* (pp. 1-5). IEEE. 10.1109/ICC.2009.5198907

Tiernan, C., Comyns, T., Lyons, M., Nevill, A. M., & Warrington, G. (2022). The association between training load indices and injuries in elite soccer players. *Journal of Strength and Conditioning Research, 36*(11), 3143–3150. doi:10.1519/JSC.0000000000003914 PMID:33298712

Tomar, U., Chakroborty, N., Sharma, H., & Whig, P. (2021). AI based Smart Agricuture System. *Transactions on Latest Trends in Artificial Intelligence, 2*(2).

Tor Project. (n.d.a). *Relay Search.* Tor Project. https://metrics.torproject.org/rs.html

Tor Project. (n.d.b). *Types of Relays.* Tor Project. https://community.torproject.org/ relay/types-of-relays/

Tor Project. (n.d.c). *Welcome to the Tor Bulk Exit List exporting tool.* Tor Project. https://check.torproject.org/cgi-bin/TorBulkExitList.py

Tugal, I., Almaz, C., & Sevi, M. (2021). Cyber Security Issues and Awareness Training at Universities. *Journal of Information Technology, 14*(3), 229–238. doi:10.17671/gazibtd.754458

United States Department of Justice. (2015). Financial crime fraud victims. *Technical report.* The United States Attorney's Office, Western District of Washington. https://www.justice.gov/usao-wdwa/victim-witness/victim-info/financial-fraud

Upadhyaya, R., & Jain, A. (2016). Cyber ethics and cyber crime: A deep dwelved study into legality, ransomware, underground web and bitcoin wallet. *2016 International Conference on Computing, Communication and Automation (ICCCA),* Greater Noida, India. 10.1109/CCAA.2016.7813706

VPN.com. (n.d.). *Malware, Adware, Spyware, and Ransomware: What Do These Terms Mean.* VPN. https://www.vpn.com/guides/malware-adware-spyware-and-ransomware-what-do-all-these-terms-mean

Wayne, R. (2022). Best Practices for Protecting Businesses Against Ransomware Attacks. *Medium.* https://medium.com/@rick.wayne.2022/best-practices-for-protecting-businesses-against-ransomware-attacks-740221ebaf5d

Wazid, M., Katal, A., Goudar, R. H., & Rao, S. (2013, April). Hacktivism trends, digital forensic tools and challenges: A survey. In *2013 IEEE Conference on Information & Communication Technologies* (pp. 138-144). IEEE.

Weimann, G. (2016). Going dark: Terrorism on the dark Web. *Studies in Conflict and Terrorism, 39*(3), 195–206. doi:10.1080/1057610X.2015.1119546

Wen Huang, F. Y., Yu, C., Hung Hang, C., Tsai, D. T., Lee, S., & Yenkuo, A. (2022). Testing Framework for Web Application Security Assessment. *Journal of Computer Networks, 5*, 739–761.

Whig, P., & Ahmad, S. N. (2014b). Simulation of linear dynamic macro model of photo catalytic sensor in SPICE. *COMPEL: The International Journal for Computation and Mathematics in Electrical and Electronic Engineering.*

Whig, P., Nadikattu, R. R., & Velu, A. (2022). COVID-19 pandemic analysis using application of AI. *Healthcare Monitoring and Data Analysis Using IoT: Technologies and Applications,* 1.

Whig, P., Velu, A., & Sharma, P. (2022). Demystifying Federated Learning for Blockchain: A Case Study. In Demystifying Federated Learning for Blockchain and Industrial Internet of Things (pp. 143–165). IGI Global. doi:10.4018/978-1-6684-3733-9.ch008

Whig, P. (2019). Exploration of Viral Diseases mortality risk using machine learning. *International Journal of Machine Learning for Sustainable Development, 1*(1), 11–20.

Whig, P., & Ahmad, S. N. (2014a). Development of economical ASIC for PCS for water quality monitoring. *Journal of Circuits, Systems, and Computers, 23*(06), 1450079. doi:10.1142/S0218126614500790

Whig, P., Kouser, S., Velu, A., & Nadikattu, R. R. (2022). Fog-IoT-Assisted-Based Smart Agriculture Application. In *Demystifying Federated Learning for Blockchain and Industrial Internet of Things* (pp. 74–93). IGI Global. doi:10.4018/978-1-6684-3733-9.ch005

Whig, P., Velu, A., & Bhatia, A. B. (2022). Protect Nature and Reduce the Carbon Footprint With an Application of Blockchain for IIoT. In *Demystifying Federated Learning for Blockchain and Industrial Internet of Things* (pp. 123–142). IGI Global. doi:10.4018/978-1-6684-3733-9.ch007

Whig, P., Velu, A., & Naddikatu, R. R. (2022). The Economic Impact of AI-Enabled Blockchain in 6G-Based Industry. In *AI and Blockchain Technology in 6G Wireless Network* (pp. 205–224). Springer. doi:10.1007/978-981-19-2868-0_10

Whig, P., Velu, A., & Nadikattu, R. R. (2022). Blockchain Platform to Resolve Security Issues in IoT and Smart Networks. In *AI-Enabled Agile Internet of Things for Sustainable FinTech Ecosystems* (pp. 46–65). IGI Global. doi:10.4018/978-1-6684-4176-3.ch003

Whig, P., Velu, A., & Ready, R. (2022). Demystifying Federated Learning in Artificial Intelligence With Human-Computer Interaction. In *Demystifying Federated Learning for Blockchain and Industrial Internet of Things* (pp. 94–122). IGI Global. doi:10.4018/978-1-6684-3733-9.ch006

Why go for TISAX if you have already reached ISO27001 ? (n.d.). LinkedIn. https://www.linkedin.com/pulse/why-go-tisax-you-have-already-reached-iso27001-guido-b%C3%BCcker

Witman, P. D., & Mackelprang, S. (2022). The 2020 Twitter Hack – So Many Lessons to Be Learned. *Journal of Cybersecurity Education, Research and Practice, 2*(2), 1–13. https://digitalcommons.kennesaw.edu/jcerpAvailableat:https://digitalcommons.kennesaw.edu/jcerp/vol2021/iss2/2

Wyke, J., & Ajjan, A. (2015). *The current state of ransomware.* Sophos Labs.

Yang, L. Y. (2019). Dark Web forum correlation analysis research. ITAIC 2019, Chongqing, China. doi:10.1109/ITAIC.2019.8785760

Yang, Y., Wu, L., Yin, G., Lifie, L., & Hongbin, Z. (2017). A Survey on Security and Privacy Issues. Internet-of-Things IEEE Internet of Things Journal.

Yang, M., Luo, J., Ling, Z., Fu, X., & Yu, W. (2015). Deanonymizing and countermeasures in anonymous communication networks. *IEEE Communications Magazine*, *53*(4), 60–66. doi:10.1109/MCOM.2015.7081076

Yildirim, E. Y. (2018). *Cyber Attacks Directed Information Systems (IS) and Maintenance of Cyber Security.* 2nd International Vocational Science Symposium (IVSS), Antalya, Turkey.

Yorke, C. (2010). CYBERSECURITY AND SOCIETY: Bigsociety.com. *The World Today*, *66*(12), 19–21. https://www.jstor.org/stable/41963033

Zane, P. (2014). How to Navigate the Deep Web. *ISSUU*, (3).

Zavarsky, P., & Lindskog, D. (2016). Experimental analysis of ransomware on windows and android platforms: Evolution and characterization. *Procedia Computer Science*, *94*, 465–472. doi:10.1016/j.procs.2016.08.072

Zhang, N. (2020). A generative adversarial learning framework for breaking text-based CAPTCHA in the Dark Web. National Science Foundation.

Zillman, M. P. (2015). *Deep Web Research and Discovery Resources 2015*. LLRX. https://www.llrx.com/2019/01/deep-web-research-and-discovery-resources-2019/

Zitar, R. A., & Mohammad, A. H. (2011). Spam detection using genetic assisted artificial immune system. *International Journal of Pattern Recognition and Artificial Intelligence*, *25*(8), 1275–1295. doi:10.1142/S0218001411009123

About the Contributors

Keshav Kaushik is an experienced educator with over eight years of teaching and research experience in Cybersecurity, Digital Forensics, and the Internet of Things. He has published 50+ research papers in International Journals and has presented at reputed International Conferences. He is a Certified Ethical Hacker (CEH) v11, CQI and IRCA Certified ISO/IEC 27001:2013 Lead Auditor, Quick Heal Academy certified Cyber Security Professional (QCSP), and IBM Cybersecurity Analyst. He was a keynote speaker and delivered 50+ professional talks on various national and international platforms. He has edited over ten books with reputed international publishers like Springer, Taylor and Francis, IGI Global, Bentham Science, etc. He has chaired various special sessions at international conferences and served as a reviewer in peer-reviewed journals and conferences.

Akashdeep Bhardwaj achieved his PhD from University of Petroleum & Energy Studies (UPES), Post Graduate Diploma in Management (PGDM), Engineering graduate in Computer Science. He has worked as Head of Cyber Security Operations and currently is a Professor in a leading university in India. He has over 24 year experience working as an Enterprise Risk and Resilience and Information Security and Technology professional for various global multinationals.

***

Kritka is a Master of Technology(M.Tech) in Computer Science and Engineering and is awarded with Young Researcher Award 2023 by Institute of Scholars. The author is currently working with the Government of India and has obtained several certifications in the field of cyber security. The author is a Certified Cyber Hygiene Practitioner issued by Ministry of Electronics and Information Technology and has obtained international certifications from prestigious institutions, namely, ISC2 and EC-Council along with research papers published in the field of digital forensics and e-governance.

Qasem Abu Al-Haija received his Ph.D. from Tennessee State University (TSU), the USA, in 2020. He is currently an Assistant Professor at the Department of Computer Science/Cybersecurity, School of Computing Sciences, Princess Sumaya University for Technology (PSUT), Amman, Jordan. He is the author of more than 100 scientific research papers and book chapters. His research interests include Cybersecurity and Cryptography, Threat Intelligence, Artificial Intelligence (AI), the Internet of Things (IoT), Cyber-Physical Systems (CPS), Time Series Analysis (TSA), and Computer Arithmetic.

Aaeen Alchi is a Teaching Associate, Department of Biochemistry and Forensic Science, Gujarat University, Ahmedabad, Gujarat, India.

Muhammed Aslan has been working in the IT sector for 12 years. He graduated from Çankaya University Electronics and Communication Engineering Department in 2010. He completed his MBA education in 2014. He completed his master's thesis in the information technologies program in 2022. He continues his PhD in Business Administration.

Sandeep Dalal is an Associate Professor at Maharshi Dayanand University, Rohtak.

Kiranbhai Dodiya is a Research Scholar in the Department of Biochemistry and Forensic Science, Gujarat University, Ahmedabad, Gujarat, India.

Piyush Gupta is an Assistant Professor Department of Computer Science and Engineering School of Engineering Sciences and Technology Jamia Hamdard New Delhi-110062.

Noor Jebril has a Mater's of computer engineering.

Manoj Kumar obtained his Ph.D. Computer Science from The Northcap University, Gurugram. He did his B. Tech in computer science from Kurukshetra University. He obtained M. Sc. (Information Security and Forensics) degree from ITB, Dublin in and M. Tech from ITM University. Mr. Kumar has 9.5+ years of experience in research and academics. He published over 29 publications in reputed journals and conferences. He published 2 books and 5 patents with his team. Presently, Mr. Kumar is working on the post of assistant professor (SG), (SoCS) in university of petroleum and energy studies, Dehradun. He is a member of various professional bodies and reviewed for many reputed journals. He is Editorial Board Member in The International Arab Journal of Information Technology (IAJIT), Jordon and Journal of Computer Science Research, Singapore. He is recognized as Quarterly Franklin

Member (QFM) by London Journal Press from March 2019 onwards and Bentham Ambassador (India) on behalf of Bentham Science Publisher. He delivered various key speech and talks in national and international forums. He got best researcher award 2020 from ScenceFather research community.

Shilpa Mahajan has more than 14 years of teaching experience at postgraduate and undergraduate levels. She is a committed researcher in the field of sensor networks and has done her Ph.D. in the area of Wireless Sensor Network at Guru Nanak Dev University, Amritsar. She completed her post-graduation with distinction from Punjab Engineering College, Chandigarh. She specializes in Cyber Security, Computer Networks, Data Structures, Operating Systems, and Mobile Computing. She has introduced and designed various courses like Network Security and Cyber Security. Presently two doctoral scholars are pursuing their Ph.D. under his supervision. She has guided various M. Tech and B. Tech Projects. She has published many research papers in peer-reviewed reputed international journals and conferences. She has been a resource person in various FDP's, workshops, guest lectures, and seminars. She is a CEH and CCNA certified instructor and has also done certifications in Data Scientist Tools, Exploratory Data Analysis, and Getting and Cleaning Data from Johns Hopkins University. She is the a Lifetime member of ISTE. She is a CISCO -certified training instructor for CCNA modules- 1, 2, 3, and 4. She has been awarded an Advanced Level Instructor this year. She received an appreciation from Cisco Networking Academy for 5 years' active participation. She also set up a CISCO Networking Academy and developed a CISCO lab at NCU, Gurgaon in January 2014.

Ambika N. is a MCA, MPhil, Ph.D. in computer science. She completed her Ph.D. from Bharathiar university in the year 2015. She has 16 years of teaching experience and presently working for St.Francis College, Bangalore. She has guided BCA, MCA and M.Tech students in their projects. Her expertise includes wireless sensor network, Internet of things, cybersecurity. She gives guest lectures in her expertise. She is a reviewer of books, conferences (national/international), encyclopaedia and journals. She is advisory committee member of some conferences. She has many publications in National & international conferences, international books, national and international journals and encyclopaedias. She has some patent publications (National) in computer science division.

Tolga Pusatli is an associate professor and vice director of Graduate School of Natural and Applied Sciences at Cankaya University. He obtained his Ph.D. in information systems from University of Newcastle (Australia) in 2009. Cyber security, e-Government, online commerce, geographic information systems and electronic health records are amongst his fields of interest.

Pavika Sharma is currently working as Assistant Professor in the ECE department at Bhagwan Parshuram Institute of Technology, GGSIPU, New Delhi. She has worked with Amity School of Engineering & Technology, Amity University, Noida at the Department of ECE. With more than 12 years of teaching and research experience, she has 3 patents and published more than 20 research papers in reputed journals and conferences. She has served as a reviewer of many reputed journals and conferences including IEEE Transactions on Intelligent Transportation Systems, Future Generation Computer Systems, Elsevier, Annals of Operations Research, Parallel Processing Letters (PPL), Transactions on Asian and Low-Resource Language Information Processing and Physical Communication, Elsevier, etc. She has also served as session chair at Springer International conferences and was invited as a distinguished speaker at the National Science and Technology Entrepreneurship Board, Department of Science and Technology, Govt. of India. Her area of interest includes Wireless Communication, Physical Layer Design & Security, 6G, Beyond 6G, Internet of things, Smart Cities, FPGAs, and ASIC Design.

Hitesh Kumar Sharma 1. Ph.D. in computer science 2. 03 patent published 3. 15 Scopus indexed publication

Kamna Solanki is an Associate Professor, M.D.U., Rohtak.

Aviral Srivastava is an accomplished cybersecurity researcher and expert, specializing in server-side attacks, machine learning, data analytics, network security, and cryptography. With a strong academic background and an unyielding passion for the field, Aviral has dedicated his career to understanding the ever-evolving landscape of cybersecurity threats and developing cutting-edge solutions to combat them. His research has spanned multiple aspects of cybersecurity, resulting in numerous published articles in international conferences and journals, as well as several patents filed with the Indian Patent Office. Aviral's work has not only garnered recognition in the form of awards, such as two times Young Researcher Award and International Best Researcher award, but has also left an indelible impact on the cybersecurity community. In this book chapter on server-side attacks, Aviral brings forth his vast knowledge and experience to provide an in-depth exploration of the topic, shedding light on the underlying mechanisms, potential vulnerabilities, and mitigation strategies for these types of attacks. Drawing from real-world examples and case studies, he expertly dissects the complexities of server-side attacks and offers valuable insights for professionals, researchers, and students alike. Outside of his research work, Aviral is an active member of the cybersecurity community, participating in international conferences, collaborating on research projects, and contributing to book chapters on various subjects related to cybersecurity. His dedication to the field is not only evident in his impressive academic and research accomplishments but also in his constant pursuit of knowledge and his commitment to sharing that knowledge with others. As a respected figure in the world of cybersecurity, Aviral Srivastava continues to push the boundaries of what is possible, exploring new frontiers in the field and working tirelessly to make the digital world a safer place for all.

Index

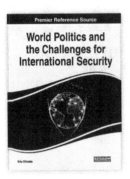

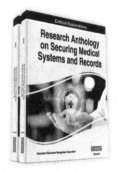

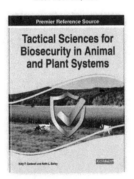

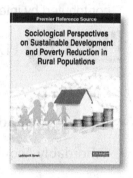

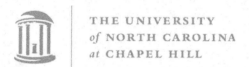

Printed in the United States
by Baker & Taylor Publisher Services